Insect Pest Management and Ecological Research

Insect Pest Management and Ecological Research explores the ecological research required for the development of strategies to manage pest insects, with particular emphasis on the scientific principles involved in the design and conduct of pest-related research. Although the connection between integrated pest management (IPM) and ecology has long been appreciated, their specific relationship to one another has until now been vague. Here, Gimme Walter develops the first general model of the entomological research requirements of IPM. He shows how to navigate through the diversity of options presented by current ecological theory, emphasising pest situations. Besides theory and principle, the book also includes practical advice on understanding and investigating species, examines the ecological problems associated with polyphagous pests and beneficial species, and scrutinises the different ways of improving insect biological control. As such, it will be an important resource for both graduate students and researchers, in IPM, insect pest management, entomology, ecology and crop protection.

G. H. WALTER is Reader in Entomology at The University of Queensland in Australia. He has lectured in entomology, evolution and ecology for 25 years in South Africa and Australia, and his personal research has focused on the host relationships and evolutionary ecology of parasitic wasps. His interest in insect pest management stems from his research on understanding the ecology of pest and beneficial species in agricultural situations.

Insect Pest Management and Ecological Research

G. H. WALTER
The University of Queensland

CAMBRIDGE
UNIVERSITY PRESS

CAMBRIDGE UNIVERSITY PRESS
Cambridge, New York, Melbourne, Madrid, Cape Town, Singapore, São Paulo

Cambridge University Press
The Edinburgh Building, Cambridge CB2 2RU, UK

Published in the United States of America by Cambridge University Press, New York

www.cambridge.org
Information on this title: www.cambridge.org/9780521800624

First published 2003
This digitally printed first paperback version 2005

A catalogue record for this publication is available from the British Library

Library of Congress Cataloguing in Publication data
Walter, G. H., (Gimme Hugh) 1954–
   Insect pest management and ecological research / G. H. Walter.
      p. cm.
   Includes bibliographical references (p.).
   ISBN 0 521 80062 5
   1. Insect pests – Integrated control. 2. Insect pests – Ecology. I. Title.
SB931 .W25 2003
632′.7–dc21 2002073611

ISBN-13 978-0-521-80062-4 hardback
ISBN-10 0-521-80062-5 hardback

ISBN-13 978-0-521-01866-1 paperback
ISBN-10 0-521-01866-8 paperback

The publisher has used its best endeavours to ensure that the URLs for the external
websites referred to in this book are correct and active at the time of going to press.
However, the publisher has no responsibility for the websites and can make no
guarantee that the sites will remain live or that the content is or will remain
appropriate.

Affectionately dedicated to my sons Greg and Scott

Contents

Preface

In agriculture, integrated pest management (IPM) is increasingly implemented for dealing with pestiferous insects, primarily to reduce our reliance on toxic chemicals. Before the introduction of the synthetic organic pesticides, pest management was linked to ecological understanding as a basis for developing appropriate control methods or combinations of methods. However, the nature of the linkage between IPM and ecological theory was more implicit than explicit, and the main purpose of this book is to re-investigate this link. Any such investigation must accommodate the crucial role of socioeconomic considerations now seen in IPM (e.g. Norton & Mumford, 1993; Stoner *et al.*, 1986). The stress from this source is on the need for timely consultation with those for whom IPM packages are intended, and even for producers to be involved in research.

The developments outlined above have generated some tensions within IPM, especially in relation to the 'reductionist science' element of insect ecology research. But what is 'reductionist science'? Why is the term so frequently used pejoratively in the IPM context? Does 'reductionist science' have a role in IPM? If so, what role? Tackling these questions provides important ancillary aims for this book.

Insect ecology research for IPM purposes is represented by a rather grey area; the linkage between theory and practice is still not explicit. To a large extent, insect ecology is portrayed in IPM texts only in the form of brief summaries, usually of one particular subject area (such as population, community or ecosystem ecology). Only in some cases does one find suggestions about what sort of data to gather. A further difficulty is that different texts frequently make quite different suggestions in this regard, and detailed or analytical discussion is rarely available. One is left to ponder such basic issues as (i) what type of data are needed if a pest species is to be adequately understood for IPM purposes, (ii) how are such data related to the overall aim of IPM, (iii) whether IPM theory or ecological theory should provide the direction for the ecological research that is intended to underpin IPM practice, and (iv) why so many divergent ecological research options are still suggested in the IPM literature. Besides the complication to understanding already mentioned, an inspection of the recent ecological literature reveals changing perceptions of the way in which ecological systems are thought to 'run themselves'. Inevitably, perceptions of the linkage between IPM and ecology will need also to be changed.

The general approach I have taken is to start at a very basic level and to examine such aspects as scientific method and the relationship between 'applied science' and 'pure science', because perceptions at this level influence attitudes, research and interpretation. This broad approach is narrowed with successive chapters, to deal with issues that get closer to ecological topics relevant to IPM. In this way, by the end of Part 1, a perception of the place of insect ecology research for IPM is defined and the nature of the research can be specified. In Part 2, specific ecological topics are covered in terms of the principles developed in the earlier chapters. The range of topics tackled is intentionally small and is designed to be illustrative rather than comprehensive. Sufficient detail can thus be given to the subtleties that so frequently attend ecological research and the various ways in which it can be approached. The book closes with a synoptic chapter that considers the suggestions made about ecological research and IPM, particularly in relation to social attitudes to science, which have changed so dramatically in the past century and continue to do so.

This book is intended for all who have an interest in IPM. It is not so much a statement of 'what is', as being a question about where we are headed and whether that is the best direction. It is hoped that it will stimulate introspection and discussion about an area of applied ecology that has yet to develop a strong theoretical underpinning, in spite of some noteworthy practical successes. It is a call to IPM students, in the broadest sense of the term 'student', to slow down sufficiently to contemplate deeply the structure of IPM and the multitude of activities related directly and even indirectly to it, and to specify those relationships explicitly. Only through such adjustment can we secure the very basis of the discipline and help to ensure the increased rate of practical successes to which we are so frequently exhorted.

Gimme Walter
Brisbane

Acknowledgements

Numerous individuals have offered me advice in relation to the views expressed in this book. Some suggestions have been unsolicited and some not very flattering. All have, however, been mentally 'gnawed', turned over and re-examined before modification, implementation or rejection, although this may not seem evident to all from the finished product. To all involved: I have appreciated the stimulating opportunity to participate in the whirlpool of research, contemplation and discussion that makes up science, and I hope this book is taken in this spirit and seen as a serious contribution to scientific development.

For specific comments on various parts of the manuscript and extensive discussion at various stages in its development I thank Dav Abeeluck, Jenny Beard, Flor Ceballo, Anuchit Chinajariyawong, Tony Clarke, Dhileepan, Priyanthie Fernando, Bernie Franzman, Mike Furlong, Rob Hengeveld, Andrew Loch, David Logan, Jennifer Marohasy, David Merritt, Melina Miles, John and Manthana Milne, Chris Moore, Nurindah, Dan Papacek, Hugh and Shirley Paterson, Chris Pavey, Greg Quimio, Raghu, Michael Ross, Jo Rungrojwanich, Michael Ryan, Katherine Samuelowicz, Perly Sayaboc, Stefan Schmidt, Rolli Tiongson, Rieks van Klinken, Irene Vänninen, Rey Velasco, Vijay Vijaysegaran, Lewis Wilson and Myron Zalucki. My postgraduate students from various parts of the world have all contributed substantially to my understanding of insect ecology and IPM needs, for which I am most grateful.

I am appreciative, too, of the faith in my teaching held by my various heads of department, for that has fuelled this manuscript substantially. They include Cliff Moran, Robin Crewe, Hugh Paterson and Gordon Gordh. Anthony O'Toole, Ros Schumacher, Caroline Meacham, Liz Snow and Michelle Larsen all assisted substantially and cheerfully with bibliographic, editorial and illustrative work. I am impressed with the patience, perception and willingness of the editorial team at Cambridge University Press and thank Tracey Sanderson and Erica Schwarz, in particular, for their input. Finally, the help and tolerance of Jenny Beard is appreciated far more than might be obvious. Thank you.

Introduction

Insect pest management is an apparently complex subject which is often too difficult to comprehend fully. Even separate components of the subject present diverse and complicated interactions...[and]... unless the system is formalized in some way, it is very difficult to maintain a balanced and holistic perspective. This can lead to biases in emphasis, a narrowing of approach and adopted options, poor decision making and communication.

D. DENT (2000, p. 330)

Identifying the problems

Integrated pest management (IPM) is today a common approach to dealing with insect pests of agriculture, although reliance on synthetic organic pesticides remains high. As an applied science, IPM has a structure that incorporates knowledge and information from several subdisciplines and technologies, a central one of which is insect ecology. At least, insect ecology is widely acknowledged to have such a role, and most of the entomological inputs into IPM are comprehensively covered in a range of textbooks (e.g. Dent, 1997, 2000; Kennedy & Sutton, 2000; Norton & Mumford, 1993; Pedigo, 1999; Pimentel, 1997; Ruberson, 1999; Speight *et al.*, 1999), at least one of which is undergoing progressive development on the World Wide Web (Radcliffe & Hutchison, 2002). However, the actual relationship of insect ecology to IPM remains somewhat abstract for it has never really been specified in concrete terms. Even those ecological aspects that relate directly to IPM are not given good coverage in IPM texts. Neither do insect ecology texts spell out the way in which the ecological principles they cover are significant to IPM, or at least their claims in this regard are not particularly convincing (Walter, 1995a). How one sets out to acquire an appropriate understanding of the biology and ecology of pest and beneficial organisms for IPM purposes is usually neglected in ecology texts. This occurs despite claims by Perkins (1982, pp. 81, 97, 261) and others that IPM provides an intellectual framework not only for practice, but also for research. The story of IPM is therefore incomplete.

Working from a partially developed framework is undesirable; it reduces the chances of successful outcomes and that is in turn economically unsatisfactory. The relationship that IPM shares with ecological considerations is perhaps omitted because the term 'applied' has two distinct meanings in relation to IPM. The first deals with the 'application of understanding'. The development of each control method now available was contingent on at least a certain amount of understanding of the pest's life cycle, physiology, behaviour or ecology, and this knowledge was then applied to develop a practical solution to a problem. Textbooks virtually never deal explicitly with this meaning, but tend to focus on the second meaning, which deals with the 'application of techniques'. Control techniques (or management practices) that are already available, and which were perhaps developed independently of the problem situation at hand, are recommended to growers and applied by them to suppress insect numbers. Textbooks tend to provide instruction on application of the second type perhaps because the information is available and the methods have usually worked. The original problem and the difficulties encountered and lessons learned in solving that problem (the development of understanding) are rarely considered. This leaves an educational gap that is eventually likely to carry over into attempts to solve new IPM problems, with probable negative effects.

Although the two complementary aspects covered by the term 'applied' have been long and widely recognised by applied entomologists (e.g. DeLong, 1934; Smith *et al.*, 1976, p. 2), the emphasis in the general pest management literature has ensured that the first one, the 'application of understanding', has never been fully or meaningfully crafted into the tradition of IPM. Almost surely, practice has been disadvantaged as a consequence, for mistakes are inevitable when people are left to cope as best they can with an inadequate framework. The persistence of such a deficiency is surprising, for almost all who are scientists at heart would presumably agree with Louis Pasteur's statement (Dubos, 1951, p. 67) that the development of technology, or the application of science, requires at least a certain degree of theoretical understanding to ensure its most efficient and effective development and use.

The textbook picture of IPM may also explain why we commonly see a 'follow-the-leader' approach to dealing with problems in applied entomology. Projects that are considered successful are used as templates (or exemplars) for generating solutions to other pest problems. Research may therefore be by-passed or simply repeated, despite the differences in systems, localities and species involved. Successful cases are also frequently used, in review, in efforts to extract generalisations to enhance success in applied entomology (e.g. Mills, 1994a). Outwardly, the 'mimetic' approach to practice may appear to make good intuitive sense and it is easy to 'sell' to funding agencies, cheap to initiate and has worked (at least in some cases). Also, it will work again. The other side of the 'mimetic' story is that the approach is likely to prove expensive if failure is ongoing, a situation that is not uncommon (see Chapters 6–9). Mimetic methods provide a distinct contrast to the approach that seeks sound principles for generating understanding about the specific systems that are to be manipulated.

Another problem for IPM is that even the research that is conducted to generate understanding about pests for improvement of IPM has problems. Texts on IPM provide little research direction, with some simply stating 'Know the ecology of the pest' or 'Know the limiting ecological factors' (Flint & van den Bosch, 1981), so research for IPM purposes is widely perceived to be somewhat routine in comparison with the ecological research conducted for more theoretical purposes. Applied scientists frequently harbour reciprocal views of 'pure science' being a speculative pursuit and thus something of a luxury and unwarranted expense. More starkly, 'applied science' is frequently seen by IPM practitioners as factual, 'pure science' as esoteric and unrealistic.

In summary, IPM seems to be far too reliant on mimicry and outdated perceptions of the way in which a sound knowledge base may be acquired, which leaves scientists involved in pest management to cope as best they can. Although some may do so admirably, for successful IPM programmes are regularly developed, the overall performance could undoubtedly be improved. Indeed, performance must be improved if IPM is to gain in credibility as well as in uptake in the field.

Understanding is the most appropriate basis for the development of effective technology or control programmes for dealing with insects in the field. Not all applied scientists accept this principle, or they do not apply it, which amounts to the same thing in practice. One need only contemplate the immense difficulties encountered in developing pheromone-based technologies for monitoring or controlling pests (see Booth, 1988). The development and application of pheromone-related technology was retarded by the premature claims of entomologists that the manipulation of pheromones would virtually eliminate insecticide use. Scientific understanding of the pest species and their sexual communication systems was clearly not available and the development of useful pheromone-based approaches was delayed considerably. Such influences are insidious, for they work against science by negating the importance of scientific principle and they have negative influences on public perceptions of the abilities of science, a point raised again much later (Chapter 10).

The main thrust of this book is first to argue the benefits of a 'problem-identification and problem-solving' approach to the generation of understanding for IPM purposes. Then, an approach to identifying and building a sound theoretical framework for IPM-related research on insects of economic significance is developed. This approach is illustrated with reference to ecological aspects that relate to host interactions of pest and beneficial insects, and to the use of beneficial insects in the biological control component of IPM. The implications of these developments for other areas of IPM are left for others to explore.

Structure of this book

The issues covered in this book have been selected to support a particular argument, that IPM practice is likely to be improved substantially if we move away from reliance on research approaches that are inadequate. For this, three things are needed. First, the formal structure of IPM needs to be strengthened, to ensure that the different aspects covered by the single term 'application' are assigned their appropriate place. Second, we need to deal realistically with the relationship between IPM research and scientific method. Third, the role of insect ecology research in the practice of IPM needs to be clarified. The rest of this introductory chapter outlines the way in which these three issues are addressed. Subsequent chapters extend and justify the arguments outlined briefly above, and also offer solutions to the associated problems.

An initial background sketch (Chapter 2) explains the way in which scientific information is acquired. Practising scientists, whether they regard themselves as 'pure' scientists or 'applied' scientists, sometimes consider they have a sufficiently good 'feel' for scientific method, perhaps even that it would be unproductive to spend time on the matter. But the issue of method is raised because applied entomologists tend to pay relatively little attention to the theoretical and methodological aspects of biology or to the developments in these areas. Of underlying importance in this regard is that in scientific terms the 'mimetic' and 'fact-gathering' approaches outlined in the previous subsection are flawed, simply because they are inconsistent with the ways in which understanding tends to be achieved in science. Although new techniques and solutions are still being developed and applied in pest management, their effectiveness in the field is in danger of being compromised by an undermining of the very basis of applied science – its reliance upon scientific method for identifying and solving problems. In addition the place of problem identification in applied entomology, as an essential step to understanding a problematic system, is all too frequently overlooked or dismissed (e.g. Newman, 1993, p. 2). Should one wish to invoke economic efficiency in defence of avoiding underlying principles, bear in mind that the sorts of shortcuts and 'fact-gathering' methodologies outlined above (and criticised later) are likely to prove much more costly than an initial investment in understanding. Sufficient examples exist to justify such a claim (e.g. see Chapter 6). To gain understanding of a problem situation one first needs to investigate it scientifically, which implies there is no escape from science in the strict sense, nor from its associated methodology.

Scientific method is crucial to developing interpretations of particular phenomena, but scientific method and scientific curiosity do not drive any scientific agenda on their own. This point is illustrated, in Chapter 3, with reference to the various influences that have governed the theory and application of pest management. An understanding of the external factors that have influenced our current attitudes and approaches to pests and pest management is useful in the identification of weaknesses in current views and also for recognising the limitations under which science and application operate in relation to IPM. This is not a negative exercise, for the identification of weakness is the first step to correcting problems. Chapter 3 also describes the emergence of IPM, which is an approach or philosophy that guides us in dealing with pest and beneficial organisms and thus influences the overall direction of the research we conduct on them. Clearly, the central role of society's influence on research direction is critical in this regard. The question of how those involved in pest management can influence such general perceptions in a positive way is important, but is discussed mainly in the final chapter.

A detailed inspection of what information should constitute the basis of IPM programmes reveals (in Chapter 4) that we are dealing with complex systems of knowledge. Simultaneously, we deal with agricultural production systems that themselves are not simple. The design of an effective IPM programme therefore demands an extremely broad knowledge base. This, in turn, implies a considerable amount of integration of information from disparate areas, leading to the 'recipe' of how management techniques will be integrated. Effective integration requires a wariness as to the quality of the information that is integrated and an open-minded inquisitiveness as to what further scientific tests are required. For example, what does previous research on the pest species' ecology reveal? Is the information reliable, or is it open to alternative interpretation, and what further tests are required? How should spray programmes be slotted into biological control activities so that the latter are not unnecessarily disrupted? What suite of recommendations will be considered reasonable by growers? Are these measures likely to work in areas other than that region for which the IPM package was originally defined? And so on.

The issue of integration needs to be considered from another angle – and one that is more important for the general aims of this book – namely, the development of an understanding of the ecology of the species that are central to the production system under consideration. What integration, if any, is required when it comes to the entomological research that underpins IPM? What advice does one get in this regard by consulting IPM books? The general conclusion (still in Chapter 4) is that IPM theory does not provide biologists, usually entomologists, with research direction. In many ways this is not a critical omission on the part of IPM theory, because other theoretical constructs or bodies of theory serve this purpose. Perhaps the principal aim for the biological side of IPM is the location of the appropriate biological theories to assist in deriving high quality research results and interpretations that are of practical value. This particular issue is tackled in depth in Chapter 5, specifically in relation to ecological theory.

What aspects of ecological theory are relevant to the study of pest and beneficial organisms? Chapter 5 illustrates that this is a deceptively difficult question. First, ecological theory has several major components, including population ecology (or population dynamics), autecology (the ecology of species), community ecology (or synecology) and ecosystem ecology. The validity of each and their relationships to one another remain unresolved. Second, disagreement among ecologists is rife, even among those who work solely within any one of these component areas. Third, alternative theoretical constructs that have been offered within any of the components listed above are frequently ignored or portrayed incorrectly. Consequently, students who wish to undertake serious and constructive investigation of pests or beneficial insects are faced with a dual problem, for orientation through ecological theory is as complex as tackling any ecological problem in the field. Chapter 5 has therefore been designed to provide a relatively brief 'map' to assist students of IPM in their dealings with ecology. The approach that is supported here leads directly to autecology, which is based upon an understanding of individual organisms and their species-specific characteristics. Significantly, this is an approach that has been advocated and developed to a considerable extent by some entomologists with an interest in pest management (e.g. Andrewartha, 1984; Andrewartha & Birch, 1954, 1984; Wellington, 1977). Although autecology has been eclipsed by the promise of the other strands of ecology, the growing appreciation of the idiosyncrasy of species is ensuring that autecology is now more widely appreciated.

Chapter 5 closes by drawing in detail the relationship between ecological theory and evolutionary theory. Understanding this relationship is crucial because the way in which organisms behave in nature relates to their adaptations or properties (which are combinations of biochemical, physiological and morphological mechanisms). All individuals carry numerous such adaptations and each one is complex, meaning that the successful completion of any ecologically important act by organisms, be it achieving fertilisation or acquiring, manipulating and digesting food, needs several important steps to be performed in an appropriate sequence and in the appropriate way. In other words, the ecology of individuals is dictated by their genetic constitution and their interaction with local habitat conditions. This, essentially, is autecology. Populations, communities and their dynamics are thus consequences of such interactions, rather than being the primary drivers of local ecology through their demographic properties. The genetically coded traits that an individual carries are determined, primarily, by the gene pool from which it arose. Such gene pools match what we commonly call species. So, what are species? How should they be defined? And how should we investigate their ecology? Consideration of these questions brings the first part of the book to a close.

Part 1 asserts, therefore, that the understanding of the species status of pest and beneficial organisms receives too little attention as a basis for understanding their ecology. This is not simply another call for better taxonomy and more taxonomy. That is an undoubted requisite, but not the entire solution. Here, IPM itself does not provide research direction and we need to look elsewhere. Species theory is therefore covered in some depth (Chapter 6). An understanding of species is most realistically and accurately based on an appreciation of the sexual behaviour of the constituent individuals in their usual habitat and the consequent population genetics of the system. Sexual species are given prominence at this point because most species relevant to IPM are sexual and because an understanding here should direct the way in which the ecology of pest and beneficial organisms is researched and interpreted. Dealing with asexually reproducing organisms presents less difficulty, so they are covered separately and in less detail.

As an illustration of why species need attention, consider the frequent encouragement given to biological control workers to introduce different 'biotypes' to enhance control (e.g. Ruberson et al., 1989). Recent developments in species theory, however, warn against over-reliance on 'biotypes', because 'biotype' is a subjective designation that almost invariably is inappropriate (Clarke & Walter, 1995). Consequently, 'biotypes' should be tested for possible genetic species status before release. If the results demonstrate that we are dealing with only one species, we take a particular course of action. A quite different course of action would be recommended if the results demonstrated that we actually had several morphologically similar species erroneously combined under one name.

Tests of species status are surprisingly difficult to design appropriately and interpret accurately, mainly because various outcomes are possible from each test and only some of the possible outcomes yield unambiguous interpretations. This point is not widely appreciated, which implies that many interpretations in the literature are premature, if not erroneous. Chapter 6 therefore ends on a somewhat extended methodological note. But it is not simply method oriented; the emphasis is on using techniques and theory interactively. How should tests of species status be designed, what techniques are appropriate, which results are unambiguous, how does one interpret the results, and how should the study be extended if the outcome is ambiguous?

The principles that relate to species and the associated interpretation of adaptation, developed in Chapter 6, are applied in Chapters 7–9, which represent other problems of interpretation faced in understanding pest and natural enemy ecology for IPM purposes. Chapter 7 addresses a difficulty presented to IPM by polyphagous pests, parasitoids and predators. Many severe pests worldwide are polyphagous, including several bollworm species (*Heliothis* and *Helicoverpa*) (Matthews, 1999), various tephritid fruit flies (White & Elson-Harris, 1994), planthoppers (Denno & Perfect, 1994) and heteropteran bugs (Schaefer & Panizzi, 2000). Interpreting the ecology of such polyphagous pests is outwardly simple: they reproduce on whatever alternative host species are available, although they do have 'preferences' for certain hosts at particular times. This line of reasoning is, however, unproductive because all observations are so readily accommodated that the theory is not tested. The concept of polyphagy generates virtually no predictions that are accurate or useful, and is also inconsistent with the underlying principles of species theory developed in Chapter 6. Similar difficulties hold for polyphagous parasitoids and predators. An alternative approach is needed and one is derived in Chapter 7 with reference to recent studies on a diversity of 'polyphagous' species.

Chapters 8 and 9 both deal with biological control of insect pests, with each of the chapters tackling the discipline from a different perspective. Chapter 8 details the current, demographically driven approach to the biological control of insect pests. It outlines the directions in which theory is currently advancing, the expectations of that theory, and the research and practical recommendations developed. These developments are contrary to the principles developed in relation to autecology (Chapter 5) and species (Chapter 6). To illustrate the application of these principles in a specific context, a detailed criticism of the demographic approach to biocontrol (which includes aspects of the 'mimetic' methodology outlined in Chapters 1 and 2) is developed in Chapter 8. The intention here is not simply to be negative, but rather to illustrate how to penetrate a system of thought in science when stronger principles are available and the system needs improvement. Chapter 9 balances these criticisms by proposing an alternative integration of theory and practice in insect biocontrol. It draws on the principles of Chapters 5–7 to demonstrate that improvements to the practice of biocontrol can be achieved through the application of suitable theory. The details presented demonstrate again that theory and practice should be developed and applied interactively.

The synoptic final chapter (Chapter 10) restates the arguments developed within the various chapters and links them to one another. The stress remains on the significance of understanding the fundamental theory relevant to the problem at hand and the critical importance of phrasing and researching the most appropriate questions about the organisms under consideration. In this way emphasis is shifted from methods that are 'mimetic' or dedicated to 'gathering the facts' to a broader approach that gives prominence to the identification and resolution of specific problems through the use of the appropriate fundamentals. By concentrating on unsolved problems in IPM through Chapters 6–9, the relevance of appreciating the appropriate level of investigation in IPM, as defined in Chapter 4, is emphasised, as well as the necessity of using appropriate ecological and evolutionary theory for the solution of such problems. In short, a much stronger theoretical framework is possible for IPM and, if that is coupled with the ecological theory that deals directly with organisms (autecology), a better basis is already available for understanding insect pests and beneficials as a basis for their more effective management.

The introspective view of IPM-related research offered in this book raises questions also about the role of science in modern society. For a long time science, used almost synonymously with technology, was seen as humanity's saviour. In many ways the goods have been delivered: think of our understanding of infectious diseases, the gains of the green revolution, nuclear power, increased human life span and so on. But almost invariably such progress has been accompanied by ills, social or otherwise, and one of the social consequences seems to be an increasing disillusionment with science. This should not be the case, and a general understanding of what science can do will clarify its potential contribution to society and thus its place among the many activities of humans that make up that society.

The application of scientifically derived technology is more an ethical issue than a strictly scientific one. Science delivers options and information about the application of those options. Predictions of the consequences of application are often in the form only of probabilities. So science does not decide which option to take, although scientists may well deliver timely warnings and may sit on advisory panels. Ultimately, decisions to deploy technology are usually the responsibility of government. An aspect for which a subset of scientists has been responsible has sometimes been the promise of near-miraculous results, usually to the media or to funding agencies, and this has frequently been based on hope or greed rather than on understanding. The practice is an unfortunate one as it has the potential to undermine valid but much slower progress through generating an unwarranted wariness of particular research avenues; IPM has seen its share of these consequences. Finally, the cultural role of science and technology in understanding the place of our species in space and time should also not be underestimated (Harris, 1981).

Part 1

The place and nature of insect ecology research for IPM

Pest management as an applied science: the place of fact, theory and application

Though the theory is worthless without the well-observed facts, the facts are useless without the frame of the theory to receive them.

NORA BARLOW ON CHARLES DARWIN'S (1958, p. 158) VIEW OF SCIENCE

Science tells us what we can know, but what we can know is little, and if we forget how much we cannot know we become insensitive to many things of very great importance.

BERTRAND RUSSELL (1961, p. 14)

Introduction: Factual information and generalisation

Close examination of an example that stands squarely in the realms of pest management will help to introduce the methodological problems faced in IPM-oriented ecological research on pest or beneficial species. The development of ecological understanding of this nature invariably involves a range of theoretical intricacies.

Table 2.1 is a reproduction of the first page, of five, listing the names of all host plants of *Helicoverpa armigera* and *H. punctigera* in Australia (Zalucki *et al.*, 1986). Consider, with regard to host plant use by *H. punctigera*, what actually represents a fact. Almost invariably respondents regard, as a fact, the observation that *H. punctigera* may feed on any one of numerous host species. By slight extension, it becomes a fact that *H. punctigera* is polyphagous. But these are not facts; rather, they are generalisations. Although such generalisations in biology have different facets, they are seldom specified in their entirety. In the example above, they specify pattern (*H. punctigera* larvae feed on many host plant species) as well as process. Process in biology has two equally relevant aspects, 'mechanical' aspects such as the behaviour and physiological mechanisms that 'connect' the

Table 2.1. *Host records of* Helicoverpa armigera *and* H. punctigera
in Australia

Botanical name	Common name	Host type[a]	Species recorded
Aizoaceae			
Trianthema pilosa	Pigweed	W, e	*punctigera*
T. portulacastrum	Black pigweed	W, n	*armigera + punctigera*
Zaleya galericulata	Hogweed	W, n	*punctigera*
Amaranthaceae			
Amaranthus interruptus	Amaranth	W, n	*punctigera*
A. viridus	Green amaranth	W, e	*Heliothis* spp.
Gomphrena globosa	Globe amaranth	W, e	*punctigera*
Asteraceae			
Arctotheca calendula	Capeweed	W, e	*punctigera*
Bidens pilosa	Cobbler's pegs	W, n	*Heliothis* spp.
Calendula sp.	Marigold	C, e	*punctigera*
Callistephus chinensis	Aster	C, e	*Heliothis* spp.
Calotis lappulaceae	Yellow daisy burr	W, n	*Heliothis* spp.
Carthamus lanatus	Saffron thistle	W, e	*armigera*
C. tinctorius	Safflower	C, e	*punctigera*
Conyza canadensis	Canadian fleabane	W, e	*Heliothis* spp.
Conyza sp.		W, n	*Heliothis* spp.
Dahlia pinnata	Aztec dahlia	C, e	*armigera*
Eupatorium adenophorum	Hemp agrimony	W, e	*Heliothis* spp.
Gerbera jamesonii	Gerbera	C, e	*armigera*
Gnaphalium japonicum	Cudweed	W, e	*Heliothis* spp.
Gnaphalium sp.		W, e	*Heliothis* spp.
Guizotia abyssinica	Niger seed	C, e	*punctigera*
Helianthus annuus	Sunflower	C, e	*armigera*
Helichrysum spp.	Everlastings	W, n	*punctigera*
Lactuca sativa	Lettuce	C, e	*armigera + punctigera*
L. serriola	Prickly lettuce	W, n	*Heliothis* spp.
Sonchus oleraceus	Common sowthistle	W, e	*punctigera*
Xanthium pinnata	Noogoora burr	W, e	*armigera + punctigera*
X. spinosum	Bathurst burr	W, e	*Heliothis* spp.
Zinnia elegans	Common zinnia	C, e	*punctigera*
Balsaminaceae			
Impatiens balsamina	Balsam	C, e	*punctigera*
Bignoniaceae			
Tecomaria capensis	Cape honeysuckle	C, e	*armigera*
Boraginaceae			
Echium plantagineum	Paterson's curse	W, e	*armigera + punctigera*
Brassicaceae			
Brassica campestris dichotoma	Brown sarson	C, e	*punctigera*
B. c. sarson	Yellow sarson	C, e	*punctigera*
B. c. toria	Toria	C, e	*punctigera*
B. juncea	Indian mustard	W, e	*punctigera*

Table 2.1. *(cont.)*

Botanical name	Common name	Host type[a]	Species recorded
B. napus	Rape	C, e	*armigera* + *punctigera*
B. nigra	Black mustard	C, e	*armigera* + *punctigera*
B. oleracea var. *botrytis*	Cauliflower	C, e	*punctigera*
B. o. var. *capitata*	Cabbage	C, e	*armigera* + *punctigera*
B. o. var. *italica*	Broccoli	C, e	*armigera*

[a] W, wild (uncultivated); C, cultivated (field, garden or horticultural crop); e, exotic; n, native.
Source: Zalucki *et al.* (1986), where the original authorities can be traced.

insects to the host plants (and which are called proximate mechanisms) and the evolutionary processes (ultimate mechanisms) (Mayr, 1961) believed to have moulded those mechanisms. The evolutionary processes most commonly invoked at present include random mutation and natural selection, and these are tied to the selective background imposed by the surroundings to explain how the proximate mechanism evolved.

All scientists trade in generalisations, for that tells them what is important when they come to deal with specific instances and what is not so important. Generalisations in science are represented by assumptions, theories and hypotheses, and we cannot do without theory of this nature. Scientists construct their investigations around such generalisations and they use them to aid communication. The only issue about generalising in this way is whether we generalise (or theorise) well or badly (Emmet, 1968, p. 18).

Returning to the example, what transpires if one probes the information represented in the table and asks questions about the nature and quality of this information? Consider the first host plant listed for *H. punctigera*, *Trianthema pilosa*. How did it come to be placed in the table? The table legend indicates that Richards (1968) recorded it as a host. To establish Richards' reasons for specifying it a host, one must first locate a copy of his thesis and then check the data. In Table 27 of Richards' thesis *T. pilosa* is listed as a host for *H. punctigera* on the basis that 'the lepidopteron was obviously breeding (and feeding) on it, as indicated by the presence of eggs and/or early-instar larvae and by the ability of the larvae to complete development on the plant' (p. 136). No record is presented of how many times *H. punctigera* had been found on that host, nor any indication given of what numbers of individuals may be expected on that species. Also, no *H. punctigera* survival data on *T. pilosa* have been presented.

In other words, the observational data (the 'facts') required to establish whether *T. pilosa* is host to *H. punctigera* are simply not available for scrutiny. We also do not know how frequently eggs or larvae must be recorded on a particular plant species, and in what numbers, before that plant is regarded as a host (Walter & Benfield, 1994). Such frequencies make sense mostly in relation to the occurrence of *H. punctigera* on other plant species. The interpretation of the respondents, that '*H. punctigera* is polyphagous', is thus clearly influenced by considerations that are not so much factual as interpretive or theoretical in nature.

To specify the interplay between fact and theory in a more general way, research for pest management needs to be examined in relation to scientific method in general. The critical importance of generalisation will become clear and so, too, will the point that alternative generalisations or interpretations are almost invariably available for any specific information of an ecological or pest management nature. Even some of the more realistic alternative generalisations may not be stated at the outset. The role of the scientist, including applied scientists, is to judge among alternatives and select the most realistic and appropriate for further use.

Applied science and basic science

Because pest management is an applied science, communication takes place among scientists of different disciplinary backgrounds, experience and aims. Some conduct research, or extend knowledge; others apply knowledge (see later). Although conflicting standpoints tend to develop as a consequence, instructive benefits can be gleaned from the background to some of these differences of opinion.

Applied science frequently sidesteps scientific enquiry: 'We ... need an answer now ... and the challenge in applied ecology is often to reach the best decisions possible on the basis of present information' (Newman, 1993, p. 2). An immediate need for action is a reality, but such calls should not be used as a refuge if challenges about method and understanding are raised. Any remedial action developed immediately on contemporary understanding is likely to need considerable adjustment, if not major change, if a more efficient solution or a more acceptable long-term one is to be developed. Research is required for the development of the most favourable solutions, and these are the research programmes to which attention is directed in this book. This is the research conducted or advocated by applied ecologists themselves in search of the most acceptable solutions.

'Applied science' is still widely portrayed as different from 'basic science' or 'pure science'. The former is seen to deliver material gain; the latter is perceived to exist only for self interest (Medawar, 1984). An influential implication, still commonly discernible in pest management writing, is that the methods of the two kinds of science differ from one another and even that 'applied science' has a life independent of 'pure science'. This perceived dichotomy persists despite the differences between science and technology having been, for some time, virtually imperceptible (Keller, 1985). The point, in any case, is that both should be equally scientific in terms of their reliance on common elements of method (Drew, 1994; Medawar, 1984, p. 34; Romesburg, 1981). Applied science, as the application of understanding (meaning 1 in Chapter 1, p. 2), is underpinned by scientifically derived interpretation, and each specific interpretation is based upon the generalisations or theories mentioned above. This means that all applied scientists must generalise the scientific interpretations they use. Specific examples of how this is currently done for IPM purposes are spelled out in the chapters (6–9) that make up Part 2 of this book.

The warning of Pasteur is again relevant: 'There are not two different kinds of science; there is science and there are the applications of science' (Dubos, 1988, p. 31). In other words, no strict or fundamental division exists in science, including the biological sciences, between 'pure science' and 'applied science'. What we should worry about more is the distinction between good science and science that does not stand scrutiny for quality. Unfortunately, the appellation 'applied science' appears sometimes to be used as a blind for work that does not represent good science. The implicit justification is that it is useful or urgently required and good for that reason alone. And that begs the question of just how useful it could be (Drew, 1994).

Scientific method

Scientific and technological progress are not delivered by following a generally accepted 'recipe'. Nevertheless, certain aspects of their advance do require careful attention to methodological strictures (Medawar, 1984).

Philosophers and many practising scientists talk about general method in science, or the method of gaining knowledge. Epistemology, the technical term for general method, has long been discussed (see, e.g., Chalmers, 1999; Medawar, 1984; Quine & Ullian, 1978). Surprisingly perhaps, advances are still made in our understanding of this area. Another significant aspect, which is unfortunately little appreciated even by some scientists, is that previously accepted ways of doing science have been discarded by most epistemologists and practising scientists who are aware of those advances (see Lovtrup, 1984; Medawar, 1984). I elaborate on these aspects later in this chapter.

How do these considerations of science in general terms relate to 'applied science'? In dealings with those who conduct research related to pest management one encounters, not infrequently, a disinterest in theory and a dedication to the 'facts of the matter' that make up the 'real world'. From such statements two strong implications emerge:

1   The negative connotations for those who concern themselves with theory.
2   That theory can be ignored and one can legitimately concentrate simply and effectively on facts.

The spirit embodied in this way of thinking is portrayed in the following passage from Carr's (1987) insightful analysis of method in history. In reading the extract below, 'scientists' and 'science' can legitimately be substituted for 'historians' and 'history', respectively.

> The nineteenth century was a great age for facts. 'What I want,' said Mr Gradgrind in *Hard Times* [by Charles Dickens], 'is Facts . . . Facts alone are wanted in life.' Nineteenth-century historians on the whole agreed with him . . . First ascertain the facts . . . then draw your conclusions from them . . . Facts, like sense-impressions, impinge on the observer from outside and are independent of his consciousness. The process of reception is passive: having received the data, he then acts on them . . . This is what may be called the commonsense view of history. History consists of a corpus of ascertained facts.
>
> CARR (1987, p. 9)

An identical commonsense view prevails in science, but it is as inappropriate and misleading as demonstrated by Carr (1987) for history. The rest of the current chapter is dedicated to considering general methods in science. Note that scientific techniques, including the use of statistical tests and mathematical models, do not represent 'methods' in the sense intended here.

Claims from scientists as to the pre-eminence of facts and the consequent irrelevance of theory and philosophy do not reflect much about scientific method. They may say much more about the claimants' understanding and quality of science (Feyerabend, 1970). A better understanding of the nature of facts and the place of facts in scientific advance would undoubtedly induce more reticence with regard to such claims. The assessment of the validity and quality of factual evidence would also be more rigorous. Such an appreciation of scientific method would undoubtedly affect not only interpretation, but also practice. For example, it would help to free more scientists from reliance on those types of mythology that derive from unwarranted speculation, from persuasion by data gathered unscientifically, and from recognisably inappropriate preconception. Such myths are more common in science than is generally appreciated, and good examples of refutations of biological myths that are today still retold in textbooks can be found, fully resolved, in papers by Cox & Knox (1988), Mitchell (1991) and Witmer & Cheke (1991). These examples involve the relatively concrete phenomena of pollination and seed distribution. Myths involving more abstract constructs, or interpretations, infiltrate the broader scientific community much more readily and have far wider and deeper ramifications for practice.

How do patently wayward myths become accepted and perpetuated, especially when so many scientists insist so forcefully on the primacy of facts? Should we be surprised? Remember that:

> we scientists may have more than an eye on the main chance and, especially nowadays, the threat to our financial support . . . Our motives are no more pure than those of business people, politicians, bureaucrats or the military whom we often blame for the misuse of science.

> REES (1993, p. 204)

This view is readily illustrated. The public has recently been privy to a 'monumental dispute of international proportions' over who could actually claim credit for the discovery of the viral cause of AIDS in humans (Schoub, 1994, p. 9). Here, financial considerations, camouflaged in terms of national prestige, intruded in sinister fashion (Grmek, 1990). The facts did not speak for themselves, or, at the very least, were not allowed to do so. Ronald Reagan, who was then the American president, and

Jacques Chirac, soon to be the French president, felt obliged to arrange a formal agreement between the disputants. The deal required, in part, that the research leader on each side collaborate in a joint account of the chronology of AIDS research (Gallo & Montagnier, 1987). As one might expect from such a reconciliatory act, 'All that a historian finds litigious or somber in this affair was carefully glossed over in silence' (Grmek, 1990). Note that this issue is of immediate practical concern to people and involves material entities (viruses) and events (the first recognition of the viruses) that cannot be dismissed simply as theoretical notions.

The nature of knowledge: How should we generalise?

Beginnings
During the Dark Ages knowledge was conserved by the Scholastics, or Schoolmen. With a few admirable exceptions (Russell, 1961), they relied entirely on knowledge that had been generated much earlier by the Ancient Greeks. Their attitude is often illustrated by the following story, apocryphal as it may be. If any of the Scholastics wished to know how many teeth horses have, they are said to have consulted the writings of Aristotle rather than look in a horse's mouth. The word of authority was sacrosanct. The Renaissance yielded the first perceptible developments of modern science.

The modernisation of science gained impetus at the beginning of the seventeenth century and was profoundly influenced by the thinking of Francis Bacon. He sought to formalise scientific procedures and was the first of many scientifically minded philosophers to emphasise the importance of induction (which infers interpretive generalisations from specific observations), as opposed to deduction (which reasons from the generalisation to the specific, and can thus test the general), as the primary principle in scientific method (Medawar, 1984, p. 33; Russell, 1961, p. 527). Although both these processes are critical to scientific advance (see later), the process of inductive interpretation has proved utterly impossible to formalise. The deductive process is much more amenable in this respect and is dealt with later in this chapter.

Bacon's writings initiated a view of science that is still a major influence in biological science and especially in applied biological fields, as detailed by Romesburg (1981). Surprisingly, this influence persists, despite being dismissed as misleading well over a century ago by many influential thinkers. The 'Baconian view' of how science progresses, entitled empiricism, can be summarised as follows (Hamilton & Chiswell, 1987).

1   Scientists carry out experiments or make observations. That is, they make carefully quantified measurements of aspects of the world.
2   Scientists as a group accumulate a bank or network of carefully measured data. The data are considered to be reliable because they are obvious to the senses and scientists are objective.
3   The data are shared, usually through publication in learned journals, and gradually a large number of related facts begins to build.
4   As data accumulate generalisations become clear. That is, they 'emerge'. Scientists are now able to recognise and formulate general hypotheses.
5   Scientists try to verify hypotheses so that law-like statements can be developed.
6   A body of scientific knowledge slowly grows in this way. It is added to by the accumulation of small increments of data or scientifically established facts. As one increment of knowledge is added, so the amount of ignorance is decreased by a proportional amount. Such a view assumes that knowledge is finite.

The paradox, discussed in the following subsection, is that despite many scientists' accepting the above, even if only implicitly, they practise science rather differently. The following advice suggests how to develop a research programme in entomology: '*Understand one species well*, based on observations and experiments. Get the facts, so the critical issues in ecology are known' (Price, 1996). This begs the question, though, of how one decides which 'facts' to collect. The dislocation between belief in how science is done and the actual practice of science ensures that method and scientific advance both suffer.

How scientists actually go to work

Ironically, the empiricist prescription does not explain the way in which most scientists actually work, and neither does it reflect accurately the subtlety of Bacon's thinking (e.g. Hampshire, 1956; Medawar, 1984; Platt, 1964; Russell, 1961). Indeed, the outline summarised above 'would be an intellectually disabling belief if anyone actually believed it, and it is one for which John Stuart Mill's methodology of science must take most of the blame' (Medawar, 1984, p. 33). Its principal weakness is its 'failure to distinguish between the acts of mind involved in *discovery* and *proof*' (Medawar, 1984, p. 33). In other words, a vast gulf separates the acts of developing interpretations (through induction) and testing ideas for their validity (through deduction).

All practising scientists actually direct their enquiry in particular ways; they collect only certain observations and leave others. Some form of deductive reasoning is used to identify which observations they will collect. Scientists cannot simply rely on sensory data to be picked up objectively, for 'we *learn* to perceive' (Medawar, 1984, p. 116). As discussed later, the range of observations that are open to collection in any particular situation is not only immense, but shifts as we alter the fundamental principles from which we work. This has at least two significant consequences. First, from the outset of their research, the scientist already perceives the problem in a particular way, or from a pre-existing slant. Second, empiricist beliefs tend to lock scientists very tightly into a single system of thought. The real problem, then, is that the original perception, or slant, is almost certainly not the only one, or even the best one, for developing the most robust understanding. The following simple examples are commonly reproduced for amusement, but they do illustrate the ease with which we miss options in visual perception (see Fig. 2.1). A more realistic and telling example is found in Gombrich's book *Art and Illusion* (Gombrich, 1977, p. 74). Two paintings of the same scene are reproduced (Fig. 2.2), one by a European artist and the other by an Oriental one. The different social backgrounds of the two have obviously provided them with perceptions of the scene that differ remarkably from one another. Alternatively, what each has been taught to see, perhaps not even directly, constrains them in what they do see. They have learnt to perceive. If we can be influenced so easily in developing or interpreting a diagrammatic representation of what we can actually observe, we are presumably more easily 'deluded' by more complex phenomena in the world around us.

Examination of biological examples reveals that fundamental differences in perception are not confined to visual representation. For example, do the sterile individuals in insect societies (Fig. 2.3) actually 'work' for a queen to their own genetic disadvantage, or are they 'selfishly' manipulating the queen for their own advantage? Each of these alternative perceptions is current in the literature. They are based upon alternative fundamental principles and they generate different research programmes from one another. Since they mutually exclude one another, such different principles cannot be reconciled with one another, although fairly elaborate contortions are sometimes made to unify them (e.g. Holldobler & Wilson, 1990, pp. 182ff.). The issue lies in how does one decide between

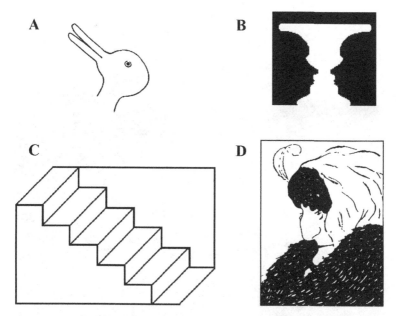

Figure 2.1. Each of these simple diagrams can be interpreted in at least two ways. **A.** Rabbit or duck. **B.** White vase or dark silhouette of two faces. **C.** Stairway, seen either from above or from below. **D.** Upper third of a young woman's body, or head and shoulders of an old lady.

them? This example illustrates a problem that ramifies through all science (e.g. Lewontin, 1991). We need to extract the most robust fundamental principles upon which to build and research more specific aspects. In doing so, we need to remember that 'ways of seeing' are important (Hughes & Lambert, 1984), as embodied in Einstein's succinct summary: 'It is the theory which decides what can be observed' (Carr, 1987, p. 164).

The problems with induction and verification

The description, above, of how scientists actually set about their work presents a strong case against the Baconian approach to science, but is not the sole reason for rejecting it. Epistemologists long ago realised that empiricism is built on shaky ground. It is based on the logic of induction, which is demonstrably too unreliable for developing workable generalisations. Because induction sees many similar observations collapsed into a general statement, repeat observations are used to identify and verify the general point, the one purported to be most 'truthful'. But if observations are made in the same way each time, alternative interpretations

Figure 2.2. Paintings of similar Derwentwater scenes by two artists of vastly different background. The upper one was painted by the Chinese artist Chiang Ye in 1936, the lower one by an anonymous painter, undoubtedly European, from the Romantic period (1826). From Gombrich (1977).

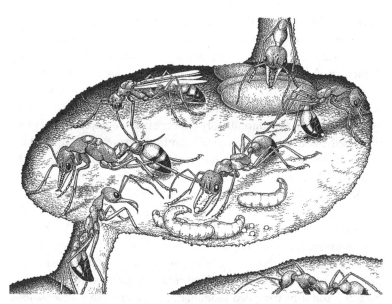

Figure 2.3. Social life of a primitive Australian myrmicine ant. Do the sterile defender and worker morphs of social insects benefit selfishly from their actions, or does their queen benefit? The mother queen, which deposits the eggs that will be permitted to hatch and develop, is the large wingless individual on the left. Behind her is a winged male. The other adult ants are workers. The one on the right is laying a trophic egg, which will be fed to the larvae in the foreground. From Wilson (1971).

are unlikely to surface because all we have is an association between observables (Romesburg, 1981). That science has been led into error so many times demonstrates that the procedure of letting facts speak for themselves does not lead inevitably to truth (Medawar, 1984, p. 91). Consider how such pursuit led to the single-minded portrayal and wide acceptance of honeybee dance language as fact. Alternatives were simply not considered in the early development of the story, although contrary evidence and obvious alternatives were available (Vadas, 1994).

A further difficulty for induction is that no matter how many times we verify something, we can never be absolutely certain that the same course of events will follow once again. A favoured example of philosophers has us observing the sun rise daily for our entire lives, then pointing out that we nevertheless cannot be certain that it will rise tomorrow. On the surface this may seem absurd to an applied biologist. That it is not (see Rouvray, 1997) should be more widely appreciated. No matter how many times we verify an interpretation by induction, we cannot be certain that we have the correct interpretation. At some future stage an observation may be made that conflicts with what we now consider factual. Because scientists are so familiar with their subject matter many of them, perhaps most, feel confident that such a destabilising possibility is unlikely in their area of expertise. In reality, a good proportion of scientists has had to face the contradiction when it has surfaced. Frequently, it seems, the contradictory evidence is downplayed or otherwise dismissed. Because of the nature of scientific interpretation, covered below, such contradictory information continues to surface. In other words, this type of destabilisation of scientific interpretation cannot be dismissed simply as historical curiosity, something that used to happen in the more primitive past. Contemporary scientists are regularly confronted with the situation.

The reservations outlined above can be applied to the bollworm example that started the chapter. Once *Helicoverpa punctigera* had been recognised as separate from the more widespread pest *H. armigera* (in the 1950s), host lists could be compiled (Common, 1953) and further observations made about the species. The inductive approach has simply returned more host records, which seems to be a somewhat standard procedure for polyphagous species (see Chapter 7). The problem with such 'fact-gathering' comes if the observations are seen as verification of the species' polyphagous nature and alternative interpretations or bases for generalisation are not considered. This point is expanded later in the present chapter, and the issue of polyphagous pest and beneficial species is analysed in Chapter 7.

Various other problems that have been associated with induction as the primary approach to scientific investigation are summarised elsewhere (e.g. Chalmers, 1999; Medawar, 1984, p. 97). A replacement model, called the hypothetico-deductive method, provides a more accurate description of how scientists actually go about scientific investigation, often unconsciously so. This overcomes the problems of induction to a considerable extent and it helps scientists to develop more appropriate tests of the generality or validity of interpretation. It does not, however, insure against reliance on inappropriate fundamental principles (see later). Induction does, however, play a creative role in science (also expanded below). Whewell, a contemporary critic of Mill, recognised this when he said induction and deduction are not mutually exclusive alternatives, but are instead mutual adjuncts in the development of understanding (Ruse, 1976).

Empiricism still influences practice

Despite the problems with empiricist methods, one can see widespread evidence of scientists' attempting to apply them (as explained with respect to applied ecology by Romesburg, 1981, for example). They do so as a basis for gathering data, or as a justification for conducting a particular piece of research. Even the most modern molecular technologies are often applied with the hope that 'something' will emerge from the banks of accumulated data. The widely acclaimed Human Genome Project seems also to have been construed in this way. The project is widely supported because the sequence data are expected to yield genetic therapies, but how such cures can be developed from sequence data is still not clear (Commoner, 2002; Kmiec, 1999; Lewontin, 1991, p. 66). In practice, therefore, the inductive approach encourages work designed to verify hypotheses and, in some ways, discourages the scrutiny of underlying assumptions, perhaps because those premises are believed to have been recognised through the collection of good, objectively derived information.

A practical consequence of the approach just outlined is that science 'completed' is frequently considered sacrosanct, to the point that contributions in a particular area of endeavour may even be considered to have finalised that research. Industries that fund applied entomological and ecological research frequently treat science in this way. The approach also has detrimental effects in education. Students, who are far too frequently allowed to pick up the Baconian approach by default, mainly from textbooks, may gain little insight into judging what information is important. More significantly, they may fail to discern which questions are crucial to progress. The critical and creative scientific skill of identifying significant questions to address cannot emerge from such an education. Influence from the Baconian approach is readily detectable in those published reviews and students' written assignments in which equal effort and space is given to all observations. For example, long descriptions of the eggs, larvae and so on are frequently presented without any justification for the inclusion of such information. Ultimately, students are thus easily led to seek research projects that comprise work, however mundane, that has 'not been done', in areas about which 'nothing is known', or to 'fill gaps', without realising that this attitude, without qualification, could be used with equal validity to justify even the study of something that is clearly trivial relative to their ultimate ambitions.

Consider why empiricism is still influential: '... inductivism, even if logically unsound, is still the accepted method for explaining the progress of science in the minds of most working scientists. Because scientists cut their teeth on textbooks, which still imply that steady accretion of facts yields laws which allow further predictions of how to collect useful facts, one can hardly be surprised that this is still the case' (Hamilton & Chiswell, 1987, p. 9). But what about the empiricism of working scientists? The theories that most scientists have accepted apparently become so entrenched and so widely believed that they come to hold a factual status. The theoretical nature of interpretation is forgotten and belief is thus strengthened. The economic necessity and social dynamics related to winning research grants tend not to disrupt such a view, and neither do other external influences (e.g. Wieland, 1985).

The warning above should generate a constant awareness that all interpretations are based, in one way or another, on assumptions. And we should never forget that, in the past, the beliefs that large numbers of scientists held very comfortably have been entirely discarded or have had their pre-eminence revised. Such changes have occurred in physics (Newtonian to Einstein's relativity) and in biology (natural theology to Darwinian evolution and Galen's interpretation of blood flow to that of Harvey (Mowry, 1985; O'Neill, 1969; Schamroth, 1978)). Kuhn (1962) has described how such shifts are initiated and instigated. Further changes of this magnitude will take place, and major shifts have been foreshadowed in several areas of biology (e.g. Rose, 1997; Smocovitis, 1996; Walter & Hengeveld, 2000). The shifting theoretical ground that is relevant to understanding the ecology of pest and beneficial organisms is dealt with in later chapters. The structure of theory in ecology (Chapter 5) and concepts of species (Chapter 6) are particularly relevant.

Sadly for science, as well as for society, the situation is even less straightforward than that sketched above. The dominant research direction, the one that sets the agenda of what is actually done by scientists, is all too frequently set up with significant influence from external forces, inevitably represented by dominant elements of society and their interests. They may even derive from the will and financial interests of one or relatively few individuals, as happened with the push to use hybrid corn varieties, instead of selected lines, for financial reasons (Lewontin, 1991). Molecular biology was under similar financial pressure to steer a course through the Human Genome Project that is more likely to reward financially and selectively than scientifically and generally (Lewontin, 1991). Consider also the origin of the gene-centred view that dominates current interpretations to the point that it has generated the ideological power even to interpret and pronounce on the human condition and to offer remedies for our social ills (Rose, 1997, p. 273). Rose reminds us that it was the financially powerful Rockefeller Foundation that shaped this agenda as long ago as the 1930s, and 'ensured that alternative understandings of biology withered'. The way in which scientists see the world 'is not the result of simply holding a true reflecting mirror up to nature: it is shaped by the history of our subject, by dominant social expectations and by the patterns of research funding' (Rose, 1997, pp. 273–274). The metaphor of the all-controlling gene that has grown from these beginnings (e.g. Dawkins, 1986) still holds us hypnotised (Rose, 1997, pp. 120–121).

In summary

The belief that scientists approach their task dispassionately and objectively is widespread. They are seen to gather facts uninfluenced by others, until a general picture or law or theory emerges. The implication is that whoever makes a scientific investigation of a particular aspect (e.g. ecology, applied ecology) will come up with the same interpretation as any other researcher if they are objective and dispassionate. But scientists, even those who believe the interpretation just sketched, behave quite differently. Even if they claim they are interested only in facts, and that they gather facts objectively, they are not correct. All scientists are directed to gather observations from the perspective of a particular *theoretical* framework. Carr (1987, p. 163) develops the point that it is the interaction between hypothesis and observation that results in discovery. Furthermore, imagination plays a critical role in the generation of alternative hypotheses, a facility that is entirely beyond recipe, and this is where we see great variation among individual scientists. Readable accounts of how science has progressed in this way are available in O'Neill's (1969) work *Fact and Theory*. Finally, an important aspect to remember is that certain observations or experimental results may clash with accepted dogma, an issue that is central to the following section.

Acquisition of new theories, and the consequences

The ways in which new theories or 'ways of seeing' arise are far from clear. Such 'ways of seeing' are now widely called 'paradigms' to indicate they represent a general intellectual framework on which observations or facts can be arranged conveniently for interpretation, memory and so on. Unfortunately, the concept of paradigm is not straightforward and is even

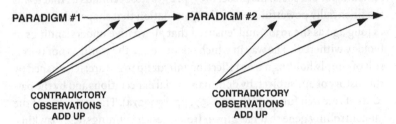

Figure 2.4. Diagrammatic representation of the behaviour of scientists in relation to changes in the dominant theoretical constructs, or paradigms, of the day.

invoked for rather trivial differences in perspective or even for alternative hypotheses that are rather specific. It is reserved here for that set of theories that accord with a common set of fundamental assumptions. An ecological analysis that follows this stricture is presented by Walter & Hengeveld (2000).

Kuhn (1962) introduced the concept of scientific paradigm and outlined the way in which he believed scientists work. Once a paradigm is in place, scientists tend to conduct science in relation to that framework, although alternatives may persist for decades without any clear resolution (see Hengeveld & Walter, 1999, for such a situation in ecology). Their questions are directed by the paradigm, as are their interpretations of the data gathered. However, attention is drawn periodically to results that contradict interpretations derived from that paradigm. In general, the value of such results tends not to be recognised. But when the contradictions become more frequent, and the deficiencies more widely appreciated, an alternative paradigm is sought, as illustrated in Fig. 2.4. Any replacement paradigm is almost invariably accompanied by dispute, for scientists have vested interests and, as mentioned earlier, are not as objective as is widely believed (Wieland, 1985). Frequently enough, interest-groups develop and these can dictate the passage of science for considerable periods. Hull (1988) provides a compelling example, again from biology. Even without ulterior motives, the tendency is for maintenance of the status quo for, as Margolis (1993) argues, scientists are taught 'habits of mind' in which a mental process of pattern recognition is formed and that is what appears correct to the individual. Such pattern recognition imposes a habit of mind and thus a barrier to change. Novel intellectual development demands that barriers be broken, which is not easily achieved even when it is patently warranted. For example, even something that we see as relatively concrete, the organisation and workings of the human circulatory system, was the subject of strong disputation when Harvey outlined the interpretation we accept today. Twenty years lapsed before the supporters of Galen's older view finally went quiet, despite the strength of Harvey's evidence and the rational and clear structure of his argument (Mowry, 1985).

How is a new paradigm developed? Unfortunately no recipe is available for guidance. Perhaps the best succinct summary of what is required was made by a Nobel Laureate in Physiology and Medicine, Albert Szent-Györgyi: 'To see what everyone has seen, to think what no one has thought' (Kaminer, 1988). Here we see the reason there can be no recipe. Development at this point is purely inductive, drawing from personal observation and experience, an eye for synthesis and a feel for the extension of knowledge. Creative skills will always be independent of recipes, and different scientists are liable to return different results and interpretations should they tackle the same problem independently of one another.

An important lesson from the above is that paradigms are bodies of theory. An obvious point about theoretical constructs that is too frequently forgotten, in our talk of natural laws for example, is that all theory represents the perspective of people and is developed by people. Theory does not emerge inevitably from concrete fact, so scientific theory should not be perceived as immutable. No penalty accrues to changing any theory if it has shortcomings. The history of science demonstrates the point comprehensively and compellingly (Brush, 1974). We should remove any mental block to resisting change in this direction. The complex task for scientists, whether they wish to consider themselves 'pure' scientists or 'applied' ones, is to balance an open mind with not being gullible. Individuals have to learn to assess the quality of evidence and must learn to question intelligently, to ensure the highest possible quality of interpretations and their overall personal knowledge.

Whenever we make a theoretical statement, which is necessarily a generalisation, we encapsulate what we feel is characteristic of the phenomenon. Simultaneously, and inevitably, we also exclude certain other aspects related to the phenomenon. This is a crucial feature of theorising and has been described metaphorically by Bronowski (1978, p. 69) as making a 'cut'. The 'Bronowski cut' is important. It clears our mind of triviality, but we achieve this through exclusion. Shift theory to incorporate something new and we push something else into the background. As the focus of our interpretation shifts, what was previously 'focal knowledge' shifts into the background and becomes 'tacit knowledge' (Petrie, 1976, after Michael Polanyi). Petrie illustrates the point with reference to such ambiguous diagrams as in Fig. 2.1. Focus on the old lady in Fig. 2.1D and the line of her mouth is focal. Her far eye is part of the background, and can be considered 'tacit'. The eye becomes focal in discerning the young lady, whereas the 'mouth' (now a choker) becomes subsidiary. This view justifies why (i) we cannot build a comprehensive interpretation, (ii) we are restricted to views that include a fraction of the external world, and (iii) knowledge can never be finite. We have no other option. The insidious influence of seeing knowledge as finite information, and thus a commodity, is discussed in Chapter 10, where the ideas developed about pest management and insect ecology research in the rest of the book are integrated and extended.

Paradigm identification and quality assessment

Paradigmatic structures

The series of interpretations that represent a paradigm is not always easy to recognise, and this is especially true of biological theory. The most fundamental assumptions in biology, those representing the paradigmatic core, seem generally to remain unstated and unquestioned. Perhaps they seem so obvious that they do not bear repetition or perhaps it is a means of shielding them, for it seems to be mainly critics who uncover them (e.g. Brady, 1982). These hidden assumptions are the most cherished ones. They remain unquestioned, even taken as self-evident or factual and beyond denial (see Ruse, 1976, p. 231). They are the ones defended most vigorously against attack and rejected most reluctantly (Popper, 1983). More superficial interpretations and assumptions tend to be far more readily rejected or altered by scientists. Relatively superficial criticism is commonplace and openly debated; deeper criticism remains hidden.

Another problem in defining a paradigm is that each one is a 'nested set' of subsidiary concepts or theoretical structures that surround the central core of assumptions or premises. The logical consequences of one theory thus represent the starting point of a more specific theory, in providing the underlying assumptions or hypotheses for the latter (Medawar, 1984, p. 103). One therefore has first to penetrate the outer aspects of the paradigm and thus penetrate ever deeper to the core (as exemplified in the following subsection with reference again to the bollworm example). This reveals the difficulty in assessing the scientific qualities of a paradigm.

The central core of a paradigmatic structure is not readily tested directly (see Brady, 1982, p. 379) because all predictive statements contain a tacit clause of all things being equal (*ceteris paribus*). In addition to this qualification, all predictions (besides the extremely simple) are supported by other theories. Three possible explanations are therefore available if a prediction fails a test: the theory, the *ceteris paribus* assumption or the background theory could be at fault. Ad hoc additions are readily made to the first two components, an expedient that has always been basic to science (all from Brady, 1982). Progress will suffer, though, if such additions are used to prop up exhausted theory, as described by Paterson (1985) with regard to species concepts in biology. For core theories such as natural selection, the *ceteris paribus* clause is indeterminate because organisms are multifaceted and present so many possibilities (Brady, 1982). This implies that we cannot simply design an experiment that will yield singular evidence that is sufficient to decide the validity of a paradigm. A range of evidence must be taken into account, and assessment is aided if viable alternatives are available (see Brady, 1982, for example). We again see why knowledge cannot be finite.

Perhaps the most useful practical advice about recognising a paradigm is to seek out the most fundamental concept that is driving research in a particular area of science. In certain biological disciplines, including ecology, this is more easily said than done, because of the difficulties in ecology outlined in the previous chapter and detailed in Chapter 5. Ecology is therefore said to fall short of being a 'mature' science (Aarssen, 1997; Loehle, 1987). By contrast, physicists apparently agree on which are the major research questions for the advance of their discipline, although cracks in quantum physics, for example, are apparently becoming evident. The task is perhaps even more difficult in pest management theory than in ecological theory because pest management has the added complexity of socioeconomic influences on practice to take into account, an issue expanded on in Chapter 3 and formalised in Chapter 4.

In setting out to test the validity of paradigmatic statements, whether they be called theories, hypotheses, or whatever, several aspects of the paradigm can be profitably investigated.

1   Are the assumptions realistic? One can design tests (experiments and/or observations) to assess whether the assumptions reflect nature in a realistic way.

2   One can test predictions that derive from the hypothesis, theory or interpretation.

3   The historical development of the theory can be probed to establish whether its development was soundly based empirically and whether lines of reasoning were sound. Whereas this third approach does not necessarily identify faults with the theory, it may indicate weaknesses worthy of further thought, investigation and testing. (For an ecological example, see Simberloff, 1984, and Walter, 1988b.) Sole reliance on modern electronic databases is short sighted, as history and breadth are eliminated simultaneously.

In the area of testing interpretations or assumptions, the philosopher Karl Popper made an important contribution (Magee, 1973; Medawar, 1984). Remember back to induction; no number of verifying observations gives us certainty. For this reason Popper suggested a change in emphasis, because he realised that only one falsifying observation is sufficient to yield certainty. Results of acceptable experiments or observations that contradict theory demonstrate conclusively that the theory is not generally applicable. However, such situations sometimes devolve into disputes as to what constitutes an acceptable test. Many years may pass before certain bodies of evidence are accepted and their consequences for general interpretation become widely appreciated (see McIntosh, 1975, Mowry, 1985, and Allen *et al.*, 1986, for case histories from ecology, human anatomy and geology, respectively).

Constructing tests

As stated earlier, theory tends to come in nested sets. With reference to the host relationships of *Helicoverpa punctigera*, the situation could be teased apart as follows. Beginning with the interpretive or theoretical statement of least depth, we state: '*H. punctigera* is polyphagous'. Somewhat less superficial than this are the few statements that have been made about polyphagous species in general. For example, polyphagous species are considered to be generalists so that they can readily maximise fitness on alternative hosts that are available (e.g. Fitt, 1989). Alternatively, herbivorous insects may be polyphagous only for 'insurance' purposes. Should their major breeding hosts, usually only one or a few species, be unavailable, then survival may be maintained on alternative hosts (e.g. Milne & Walter, 2000; Velasco & Walter, 1993; Wint, 1983). Moving deeper yet, we encounter general theories on insect–host plant relationships (Chapter 7). These, in turn, are based on the more general theories of evolution, including theories of adaptation and natural selection and concepts of species (Chapter 6). They, in turn, are based on further assumptions.

Such a breakdown of the relationships between theories illustrates two important points. First, the more fundamental one's knowledge base, the greater will be one's grasp of more specific theories. Alternative explanations will be easier to see and tests will be more representative and realistic. Further, a grasp of fundamentals helps to set priorities among possible alternatives, an aspect that is dealt with more comprehensively in Chapter 6, when species concepts are covered.

Second, if one wanted to know more about the ecology of *H. punctigera*, as a means of designing ecologically desirable management plans for example, one could test the *H. punctigera* polyphagy hypothesis. Such an approach is justified by the need to know where populations of this species are at different times, as well as what they are doing in terms of maintaining themselves (e.g. in diapause) and in terms of reproduction. The alternative option, favoured by many, would be acceptance of the polyphagy interpretation and to extend research from that point. A different research direction is thus generated and it may encompass anything from population modelling to investigating host choice. Research suggested by this approach is clearly different from that conducted under the approach that directly tests the polyphagy hypothesis, which, to date, has not been comprehensively done.

A test of the *H. punctigera* polyphagy hypothesis should be built around an initial consideration of the set of observations that suggested the hypothesis in the first place. Insects classified in the taxon *H. punctigera* are regularly found, as eggs or larvae, on a great number of plant species in the field. How could such a situation possibly be explained? Here induction (or retroduction) is used to list all possible alternative explanations (Chamberlin, 1897; Platt, 1964; Romesburg, 1981), as illustrated in Table 2.2. In the environmentally related sciences, there is a tendency to stop at the first retroductively derived hypothesis (Romesburg, 1981). The negative consequences, often long term, are illustrated by Vadas's analysis of honeybee foraging behaviour (Vadas, 1994). This approach is common, too, in insect pest management, although it is weak for developing reliable knowledge (Romesburg, 1981). For each alternative explanation an experiment or a set of observations should be designed to test the proposition in

Table 2.2. *Analysis of observed pattern and the alternative explanations that could account for the observation*

Observation:

Helicoverpa punctigera eggs and larvae have been collected from many plant species in a diversity of families.

Possible explanations:

1 The taxon *H. punctigera* comprises more than one cryptic species, each of which is more host specific than the group of species taken as a whole.

2 Although the taxon *H. punctigera* comprises only one true species, it uses a great number of host species to maximise reproductive output at all times.

3 The single species, *H. punctigera*, uses a diverse range of host species, but most of them are used only under conditions when adult *H. punctigera* females are stressed physiologically.

4 Although the single species *H. punctigera* may be collected from a diversity of host species, mostly it is confined to a small subset of primary host species. Other records came about for incidental reasons.

---

as strong a way as possible. Popper thinks of this as putting the explanation or theory at risk. He saw little value in putting a theory to a test that is not risky, for relatively little is likely to be learnt (Magee, 1973; Medawar, 1984).

To set up a test, a prediction of the following type should be made: 'If this explanation is correct, then I predict x if I do experiment y'. The next step is to design an experiment with an appropriate control, a series of controlled observations, or an appropriately designed sampling programme (see Romesburg, 1981). If the prediction is not met, that explanation has been falsified and can be eliminated. Through this approach all but one of the explanations may be eliminated, and the more replications carried out with the same outcome, the stronger the support (Romesburg, 1981). That explanation may then be accepted provisionally because we have eliminated, by falsification, all of the others and have failed to falsify the successful one in a risky test.

In reality the suite of tests and answers is seldom tackled as clinically as described above, and perhaps could never be conducted in such a way. Each is complex in its own right and each has its own difficulties and requirements in terms of background knowledge. Hypotheses may need modification as one proceeds with the research and acquires more information. Also, mistakes may be made. For example, the design of the experiment may be inappropriate, or the prediction made may not represent the theory accurately. Such situations, should they lead to claims of falsification, have been termed 'naive falsification'. Clearly, resolution of the issue is a painstaking, extended and expensive process. Industry and society must acknowledge this if we are to achieve the quality of pest management that will inevitably be demanded by society (see Chapter 10).

The lessons for pest management

The ideas covered so far in this chapter have some far-reaching implications for the application of knowledge for insect pest management purposes. The three principal ones can be summarised as follows.

1   Interpretations and actions are both influenced by theory

The term 'observation' provides a useful replacement for 'fact'. The latter seems to deaden scientific discussion and it promotes the popular notion that 'common sense' is capable of appropriately directing research and application. Recourse to common sense is also potentially misleading, as shown by many widely held interpretations that have since been demonstrated to be false (Strong, 1983). An appreciation of caution in this regard has touched all spheres of human endeavour. Here is Vincent van Gogh: '... for even if one knows ever so much *by instinct*, that is just the reason to try ever so hard to pass from *instinct* to *reason*' (van Gogh writing in 1884; italics in original).

Research questions are identified and given priority by theory, even in those research programmes aimed directly at applied outcomes. Such 'applied research' may well be at particular risk of failure because direct paths are taken to specify what research should be done. Subtleties are by-passed, perhaps inadvertently, and the selected research direction is frequently bolstered by such phrases as 'practical research' or 'relevant research'. But such utilitarian bases for judging science can have dire consequences, although they are still commonly used by applied biologists. Yet, even in economic terms it is folly not to investigate a possibility that could undermine investment in a particular approach to solving a practical problem (see Sinclair & Solemdal, 1988, for a fisheries example and Thorne, 1986, for one in crop production). Only scientific considerations can indicate relevance in this connection.

The dependence of actions on theory is readily illustrated with specific reference to insect pest management. Several authors have suggested that in pest management we should deal with pests of lesser importance

first, because any human-induced reduction in the numbers of the primary pest species will allow the secondary pests to increase in number
since they would previously have been kept in check by the main pest
(Barker, 1983; Wallace, 1985). Such a view is encouraged by the preconception that competition between species is a dominant ecological influence
(Simberloff, 1984; Walter, 1988b). As Wallace (1985) comments: 'The paradox arises ... when seeking support for an eradication program whose
early efforts would be spent not on "real" pests but, rather, on the elimination of insect species none of which is causing appreciable harm'. Because
so many biologists seem to have a common sense notion of the importance
of interspecific competition (Simberloff, 1984; Strong, 1983), extension
of the idea to pest management warrants serious scrutiny. Fortunately,
this approach has not been taken seriously. Nevertheless, such ideas, or
closely related ones, are revived regularly (e.g. Ehler, 1994; Liss *et al.*,
1986; Mills, 1994b). This aspect of ecology is considered in more detail in
Chapter 5.

2   Application of knowledge vs. development (or extension) of knowledge

All applied science comprises two aspects, the application of knowledge
and the development (or extension) of knowledge. The distinction requires qualification if it is not to remain simplistic and misdirect attention. It is best appreciated with reference to an activity that all scientists
practise, whether they consider themselves 'applied' or 'pure' scientists.
All engage in problem solving (Roman, 1993) and here the cotton bollworm example is again used in illustration. A scientist who wished to
contribute to the pest management effort against *Helicoverpa punctigera*
could tackle it in one of two ways with respect to any particular control tactic that might be employed. As an example, behavioural manipulation is considered here. A practical programme could be developed to manipulate the behaviour of the moths in the field, based on
what is already known of the pheromones of various *Helicoverpa* and
Heliothis species. This would require the solving of certain problems, including dispenser design, optimal release rates and so on. Such problems are accepted to be best solved by following scientific method. Our
understanding of H. *punctigera* in the field is thus likely to be extended.
What is often overlooked, although undoubtedly appreciated, is that
in this approach particular interpretations about the behaviour and

ecology of *H. punctigera* have already been accepted, whether consciously or unconsciously, as a basis for solving the problem that has been tackled. We can think of the acceptance and use of such information in this way as the *application of knowledge*. Earlier sections of this chapter make clear that the quality of the accepted information about *H. punctigera* is uncertain.

The second approach, hinted at previously, may proceed as follows. Although scientists may feel that a pheromonally based programme may be the best option for bollworm management, they may have an element of doubt about the available knowledge on *H. punctigera*. Such doubts may be generated by insights of one or more origins. They may, for example, derive from an understanding of theoretical developments in biology, or be based on experience with other pheromone systems. Scrutiny of current interpretations may follow. Ideally, knowledge would be extended or developed through solving particular problems identified with regard to our current understanding of *H. punctigera*, our current perceptions of sex pheromone systems, or both. Scrutiny of the evidence upon which interpretations are based may yield weaknesses in the purported knowledge base, or attempts to repeat reported observations may be similarly revealing. In turn, fruitful new questions and approaches may well be perceived. Examples are detailed in later chapters.

In this second approach the knowledge about *H. punctigera* is not being applied, it is being *extended*. Such development of knowledge about a particular species does not imply that the investigation is immune from theoretical considerations. In extending knowledge about the life cycle and ecology of a particular species, for example, the questions asked are inevitably underpinned, in turn, by theoretical interpretation and its attendant assumptions. The theoretical aspects may also be revised if they are tested, directly or indirectly, by the observations.

Leeway therefore exists for the 'applied' researcher to hold an underlying interpretation of their specific subject area that may be weaker, or stronger, than others available. This revelation returns to the original point of this chapter. Different biologists setting out to understand a particular species or situation in the field may well generate different interpretations of that situation. The various interpretations may be similar to one another, and overlapping to some extent, or even mutually exclusive. This illustrates that the facts do not simply emerge inevitably from the research to provide the correct interpretation.

Finally, scientists may test interpretations that are more general (i.e. more fundamental) or more specific. Changes in understanding at more fundamental levels are sure to influence interpretation in more specific areas as well, so such fundamental investigation is likely, in turn, to impact significantly on practice at some point.

3   Application of knowledge vs. application of technology

Ideally, scientifically derived understanding is used as a basis for developing technology or techniques. For example, a good understanding of the pheromonal communication system of *H. punctigera* would be needed before an effective trap could be developed. In other words, knowledge is applied to develop the technology. Only when we have a suitable trap are we in a position to use (or apply) the technology. Unravelling the differences between the application of knowledge and the application of technology indicates that the contraction 'applied science' is somewhat misleading because it hides two processes, each quite different from the other, under a single phrase. When the contraction is used without clarification of which stage in the process is involved, the potential for misunderstanding is considerable.

Before one can apply understanding, knowledge has to be developed or extended in some way. Usually this involves research of one kind or another. The extension of knowledge through research has important parallels in education, but this is mentioned here only in passing, to indicate the depth to which the ideas covered in this chapter ramify into other areas. An educator may conserve knowledge through emphasis on the factual side. This approach tends to be easy to cope with, but it is not very satisfying to good students. Refuge may even be taken in such an approach. Many students are comfortable in this zone because description is emphasised at the cost of analysis. By contrast, one may extend knowledge. In extending knowledge, evaluation and interpretation are of utmost importance. Although the extension of knowledge is more difficult than the conservation of knowledge, it is a much more satisfying approach to learning (see Clanchy & Ballard, 1991, pp. 13, 45). And of course, exercise in the extension of knowledge is good preparation for participating in empirical research. Although calls for all environmentally oriented sciences to include education in scientific method have been made, to provide a means for the reliable extension of knowledge (Romesburg, 1981), this is not always practised.

The next chapter

The interpretations developed above, about extending knowledge and applying it and about the extraneous influences on scientific interpretation, provide a basis for considering IPM in a broad context in the following chapter. A historical approach is taken because it demonstrates that perceptions of pests have changed with time, as have the ways in which society has dealt with pests. Current perceptions of pest-related matters are thus exposed and so, in turn, are potential weaknesses in certain of our ways.

3
_____

Historical trends in pest management: paradigms and lessons

Why did those entomologists most responsible for moving insect control toward less reliance on chemicals have such vastly different opinions about the research needs for their science?

J. H. PERKINS (1982, p. v)

Introduction

Pest management poses many problems and they are diverse. Whereas some are scientific or technical in nature, others relate primarily to the development of policy or to the socioeconomic influences that impact on growers' needs and abilities. The relative mix of these various aspects has changed with time. Their combined end product, the general approach to dealing with pests, has consequently also changed. These changes are considered in this chapter, but only with reference to insects. Although no definitive history of pest management and its major influences has yet been written, many historical events have been documented, from as long as several thousand years ago (e.g. Ordish, 1976; Smith *et al.*, 1973). The few critical analyses of important transitions in pest management that are available yield insights crucial to the ongoing development of theory and practice (e.g. Kogan, 1998; Perkins, 1982; Whorton, 1974).

Analytical historical reviews are beneficial. They help to identify 'general guides for future action' (Carr, 1987) and they alert us to how early interpretations influence our perception of reality (Sinclair & Solemdal, 1988). Through such investigation, for instance, we might identify a contemporary set of circumstances that had previously led to an undesirable outcome. Another such outcome could therefore be anticipated, which suggests pre-emptive measures be taken. Thus, a knowledge of past developments in one's discipline should help to prevent the futile resurrection of theories or interpretations that have already, for good reason, been discarded. One can thus learn from the mistakes and insights of others. Furthermore, accurate and realistic planning in pest management, and the identification of crucial directions and questions for research, may be facilitated by an understanding of both historical and current trends (Perkins, 1982, p. 285). For instance, from an analysis of how extension entomology has changed, Allen & Rajotte (1990, p. 392) concluded: 'The forces that will shape the roles of extension entomologists in the future are likely to be more extensively rooted in the nonagricultural sector'. Changing public opinion had not been recognised before as playing such a prominent or direct role in the extension side of pest management.

The previous chapter covered in a general way the concept of 'paradigms of knowledge'. Inevitably the concept was imported into the pest management literature, by Perkins (1982). Parts of this chapter précis aspects of Perkins' work, but his insights are also extended, primarily in relation to research on pest and beneficial insects. The primary sources of information consulted by historians are frequently obscure documents and inaccessible publications. Here the records mentioned by historians are referred to only by page number in the historical collation or analysis, to provide indirect access to a primary source.

Early dealings with pest insects

Many mechanisms that were developed to reduce the impact of insect pests have a long history, perhaps not much shorter than the development of plant cultivation and the storage of agricultural produce (Jones, 1973). Even insecticidal chemicals have been in use as long as any other control technique, and they were used in some surprisingly subtle ways (Panagiotakopulu et al., 1995). The Sumerians apparently applied sulphur for pest control as early as 4500 years ago (Jones, 1973; Pedigo, 1999), the Egyptians used various mineral and organic chemicals (Panagiotakopulu et al., 1995), and at least 3000 years ago the Chinese treated seeds with toxins extracted from plants (Klemm, 1959), presumably for storage purposes. About this time the Chinese also burnt toxic plants to fumigate against insects, and applied arsenical compounds, the inorganic by-products of copper smelting, to the roots of rice plants during replanting to control pests that arrived subsequently (Konishi & Ito, 1973).

Whether treatments recorded from antiquity were effective or magical cannot, however, be easily discerned (Panagiotakopulu *et al.*, 1995).

Biological control was first used in China by the third century AD. People bought colonies of tailor ants (*Oecophylla smaragdina*) to place in their citrus trees, so the ants would prey on herbivorous insects and reduce damage (Huang & Yang, 1987; Klemm, 1959; Konishi & Ito, 1973). Cultural practices were also well developed by this stage, and in some places different methods were subtly integrated. Balinese rice cultivation entailed a blend of rotational irrigation and cooperative fallow and planting schedules, all of which were centrally planned and controlled by priests over entire watershed areas (Stevens, 1994). Early policy was also evident in China, where laws for control of locusts were promulgated by the twelfth century (Klemm, 1959). In summary, poisoning, biocontrol, cultural control and probably mechanical control were all available to reduce pest attack, at least in parts of the Middle and Far East. In some cases methods were integrated in complex ways, and some actions were even legally imposed.

By comparison the beginnings of Western pest control look somewhat ridiculous. Some Greek and Roman remedies were clearly based on understanding, and were presumably effective, for example the siting and construction of granaries (Beavis, 1988, p. 179). Belief in spontaneous generation of plant-feeding pests supported at least some suggestions that could not have worked, including the enticement of herbivores to sheep guts filled with dung (Beavis, 1988, p. 243). During biblical times of the Old Testament insects were seen, almost exclusively, in religious terms and without the pragmatic aspects apparent in the Far East. They were regularly regarded as vehicles of divine punishment for sinners (Harpaz, 1973). Perhaps we could take this as an illustration, albeit somewhat extreme, of how one's action is dictated by one's underlying philosophy. Faced with pest outbreaks or invasions, all that producers could bring themselves to do was pray or make offerings, depending upon their religion. Although some 'chemical methods', often derived from classical writings, were also used during this period, they were strongly associated with mysticism (Harpaz, 1973; Ordish, 1976). In early Christian times and the Dark Ages, and even well into the Renaissance, special saints were implored and there were prayers, sacrifices and religious processions around the affected fields.

Failure was inevitable, and this may explain why insects and other pests were taken to court (Beier, 1973; Ordish, 1976, pp. 43ff.). Presumably the pests were tried *in absentia*, but they were defended by an officer of the court who would argue that demands to have them burned were illegal because procedure had not been followed. The accused should first have been requested to leave the country within a specified period. If they had not done so by then, their defence argued, the proper sentence was ex-communication. Tuchman (1978, p. 236), in her revelation of daily life in fourteenth-century Europe, explains why such an unusual approach was tolerated. The vision of lay people was 'clouded by the metaphysics of tran-substantiation', which even encouraged belief in the sacramental wafer having magical powers: 'Placed on cabbage leaves in the garden, it kept off chewing insects, and placed in a beehive to control a swarm, it induced the pious bees, in one case, to build around it a complete chapel of wax with windows, arches, bell tower, and an altar on which the bees placed the sacred fragment'. The last such ecclesiastical judgement handed down to pest insects evidently took place in Europe in 1733 (Beier, 1973), although Ordish (1976, p. 45) says it may have been as recently as 1830.

Such activities are bizarre by modern standards, but they do hold a con-temporary message. Our current beliefs and actions, perhaps even cen-trally important ones, may well be similarly scorned in future. We will be even more open to reproach because we have enough background infor-mation and advantages of historical hindsight to know better. But we still, for example, fall easy prey to what Orr (1994) has called 'technological fun-damentalism'. And notions about a 'balance of nature' continue to resur-face and influence both theoretical and applied ecology (see Chapter 5), despite the obvious idealistic origins of a 'natural balance' in early theo-logical writings (e.g. Worster, 1977).

The beginnings of recent Western pest management

With the Enlightenment, dealings with pests became more practical and methods became more akin to the pragmatic technology of the Chinese. New advantages available in the West included the development of per-sonal observation and experimental method as means of investigating natural phenomena, the growing willingness to question authority, and the development of technological aids such as the microscope (Ordish, 1976, pp. 65–73). So, with increasing frequency, the causes of all sorts of observed phenomena were sought, and this often proved advantageous for practical considerations (Ordish, 1976, p. 79).

Progress with insecticides was initiated by the rediscovery or introduc-tion of several plant-derived substances (Jones, 1973; Ordish, 1976, p. 97;

Whorton, 1974). For example, tobacco leaf infusions were used in Europe in the seventeenth century against bugs on fruit trees. *Pyrethrum*, a daisy plant known as flea-grass, was used at the beginning of the nineteenth century to prepare an insecticidal powder, and was portrayed as 'the insecticide of the future' (Whorton, 1974). Rotenone, hellebore and derris dust were also used at that time. Other control measures advocated early in the nineteenth century included mechanical control, biological control and the use of resistant plant varieties (Ordish, 1976, pp. 121, 146). Inorganic poisons such as sulphur, arsenic and mercury were also introduced in the nineteenth century. The best known of the insecticidal arsenicals was Paris green (copper aceto-arsenite), marketed originally as a paint in the USA but found serendipitously to kill herbivorous insect pests. By the 1870s Paris green was widely recommended against the rapidly spreading Colorado potato beetle (*Leptinotarsa decemlineata*) (Casagrande, 1987). London purple became more popular for a while, but it was the introduction of lead arsenate (1892) against the gypsy moth that was to have the more significant long-term impact on pest management (Whorton, 1974). Initially somewhat reluctant to spray such a toxic substance on food, growers were encouraged to do so by government entomologists and sometimes even forced to do so, through legislation. Simultaneously, medical debate was drawing attention to chronic arsenic poisoning rather than acute toxicity, but initially insecticides were not the focus of the debate. The increase in residues on produce after 1900, when adhesives were included in sprays, and increasing public and governmental awareness of the dangers of chronic toxicity helped to fuel a somewhat acrimonious interchange between growers and government officials. Ultimately economic considerations triumphed; legislation against residues remained weak and the legislature failed (until 1954) to regulate the public release of insecticides through mandatory safety tests (Whorton, 1974).

Although insecticidal chemicals became available during the nineteenth century and were used then, they were not as efficient, cheap or easy to use as the imminent synthetic organics. Chemical technology was slow to develop, and originally comprised a largely disorganised and ad hoc application of treatments. Only lead and calcium arsenate became fairly widely used, but mainly in the USA, and they seem to have been restricted to a particular subset of products (Whorton, 1974). Nevertheless, changing attitudes to the use of chemicals could be detected by early in the twentieth century. The introduction of Paris green, for example, had a profound influence on the attitude of economic entomologists: 'There is no great loss without some small gain and we may be grateful that the invasion of this beetle also brought about the use of Paris green' (Saunderson, 1921, p. 260, cited by Casagrande, 1987). The potato beetle problem had ostensibly been solved and, typically, the perceived need for additional research on this pest abated. Between 1908 and 1949 only five papers and five notes that dealt explicitly with the potato beetle were published in the *Journal of Economic Entomology*, where before there had been many (Casagrande, 1987). This enthusiasm was carried through to the introduction of lead and calcium arsenate, as described above.

The limited efficacy of the insecticides of the time, and the relatively small-scale and diverse approach to agriculture generally practised, ensured other means of control would thrive well into the twentieth century. For example, cultural and varietal control had a critical role in potato cultivation in the USA, and rotation of crops and sanitation helped to stop the build-up of pests (Casagrande, 1987). Even biocontrol had a considerable impact by the turn of the century thanks to it saving the Californian citrus industry in 1888 from the ravages of cottony cushion scale insect. This was one of the first classical biological control projects attempted. Doutt's (1958) account of it is enlightening; it reveals that the idea and cost of foreign exploration (in Australia) were considered antagonistically.

Before the introduction of the synthetic organics, management methods were, therefore, generally specific to pest species, and most often a range of techniques was suggested for each major pest species. Nevertheless, the framework for the uncritical exploitation of synthetic organics had been well laid (Dunlap, 1981, p. 17; Whorton, 1974). The synthetic organics comprise a diverse array of chemicals that developed alongside the growing discipline of organic chemistry. Unlike the arsenates they have contact toxicity and do not have to be ingested to exert their lethal effects.

Emergence of the 'chemical paradigm'

Paradigm changes do not occur in a clean or smooth way. Influences may be subtle and long-established or they may be more direct, and they may even be driven politically. Change is not inevitably underpinned by good science (see Chapter 10), and even if soundly based scientific results support a fundamental shift or an altered practice, adoption of that change may still be extremely slow or may even not take place at all.

Thoughts that seem to presage the chemical paradigm can be traced back into the nineteenth century, especially with reference to Paris green, lead arsenate and calcium arsenate, which were produced on a relatively large scale for the time. The firm establishment of the chemical era was contingent on circumstances introduced during the world wars fought in the periods 1914–1918 and 1939–1945 (Perkins, 1982, p. 4). During the First World War certain ingredients became scarce and Paris green was replaced by lead arsenate and calcium arsenate. The capacity of industry to produce arsenical insecticides increased dramatically after the war because of the sudden decrease in demand for arsenic-based wartime products such as lead shot and signal flares. International conflict thus ensured production could and would be high at the end of hostilities, especially in the USA, whose immense population and leading role in the world economy through the twentieth century ensured a high intensity and large scale of agricultural development. Interpretive histories of these changes to agriculture tend to concentrate on the USA, although similar influences and changes were evident in other countries.

The increased scale of chemical production inevitably changed the attitudes and expectations of society towards agricultural products, and thus towards the scale at which poisoning insects could take place. Ultimately, though, the properties of a single compound, DDT (p,p'-dichlorodiphenyltrichloroethane), swayed public perceptions the most. That these became widely appreciated at the time of the Second World War certainly speeded the adoption of this 'miracle' compound. The high contact killing power of DDT, its perceived safety in relation to humans and its incredibly high persistence, even under field conditions, ensured that the war would force its rapid adoption. Furthermore, DDT was effective in killing an extraordinarily broad range of insect species. Predictable results, low cost, cheap application and easy storage all added to its allure. In Switzerland, DDT was used to protect crops and to kill human lice on refugees and thus eliminate the risk of typhus (Fig. 3.1). In about 1941, the USA was looking for ways to protect soldiers from lice and fly-borne diseases in the Pacific and also to protect food crops at home (see Perkins, 1982). The shortage of other pesticides ensured the war would set the stage for the reception of DDT (Fig. 3.1).

For the first time in human history a wide range of pests, from lice and mosquitoes to scale insects, cotton boll weevil and codling moth, could be dealt with using a single control technique. That this was a vast change can be judged by comparison with the diversity of species-specific methods (such as biological, cultural and varietal control) used before. The outline above, taken primarily from Perkins' (1982) analysis, is intended to give

Figure 3.1. Illustrations of the way in which DDT was used during the Second World War. Top: Workers dumped DDT down the necks of their anti-contamination suits to protect themselves against lice as they removed the sick from the barracks at Bergen-Belsen concentration camp in 1945. Picture courtesy of AFS Intercultural Programs, Inc. Bottom: Dusting a child with DDT for typhus control (from Pedigo, 1999; reprinted courtesy of WHO, Geneva).

an impression of the difficulties of the time, the desires and expectations of the populace, and the developing potential of industry. Furthermore, the experience with lead arsenate in the USA had already prepared the psyche of growers and consumers for the ready acceptance of DDT for public use (Whorton, 1974). The years of wrestling with arsenic and lead residues had made scientists and health officials sensitive to the possible danger of small quantities of a regularly ingested chemical and to the necessity of detailed and extended laboratory studies of the chronic pathological effects of such chemicals. Samples from the earliest shipments of DDT to the USA were therefore tested, and scientists detected liver damage as a result of prolonged consumption of DDT. However, the legal power to prohibit the sale and use of insecticides until their safety could be determined became available in the USA only in 1954 (Whorton, 1974).

DDT was released for civilian use in 1945, and '[t]he United States, flush with the victory of wars on opposite sides of the globe, confidently turned to the synthesis and manufacture of complex organic chemicals to win the war against pests' (Moore, 1987). The success of DDT stimulated a search for other synthetic insecticides, as summarised by Perkins (1982, p. 11). Between 1945 and 1953 some 25 new pesticides were introduced (e.g. benzene hexachloride (BHC), parathion and dieldrin), with the development of the organophosphates having been pioneered in Germany (Jones, 1973). The success of all these chemicals and their widespread adoption changed insect control practices and their associated activities, which is hardly surprising when one considers it in relation to the 'changing capital structure of agriculture' (Perkins, 1982, p. 265). 'During this era farm productivity rose more than 60% while the number of people fed by a single farm rose from 15 to 45 (Hallberg, 1988)', and 'A substantial portion of this progress can be attributed to the use of pesticides and other agricultural chemicals (Fite, 1964)' (Allen & Rajotte, 1990, p. 386; see also Szmedra, 1991). Increased productivity was vital for feeding an expanding population and a labour force that was shifting from rural to urban industries. The deployment of surplus military aircraft for rapidly and effectively 'dusting' large areas, with an associated massive saving in labour, provided an added technological bonus, especially after the Second World War (Downs & Lemmer, 1965).

The general attitudes of people towards pests and production were changed substantially (Perkins, 1982, p. 11). Manufacture of the old insecticidal chemicals declined forthwith (see Table 2 in Jones, 1973), because the new compounds were so cheap and effective. They were adopted even for use against insects that were under effective non-chemical control, and forestry and public health agencies sprayed wilderness and suburbs, which previously had been insecticide free (Dunlap, 1981, p. 76). Ecological research on pests, beneficials and alternative control methods dropped in priority and questions about pesticide application dominated the research agenda. A general complacency developed, based on the view that pesticides could deal with all pest problems. Plant breeders, for example, felt liberated to concentrate on yield increase and other qualities, although their disregard for natural resistance to pests was to have serious consequences (Ferro, 1987, p. 196). A concomitant change in attitudes and expectations of society in general also took place. Consumers, who had previously shown a fair degree of tolerance towards pests (e.g. Perkins, 1982, pp. 15, 18), became fussy, even in relation to the cosmetics of the produce. On a more sinister note, chemicals were seen also, by many, to offer the possibility of permanent control by eradication (Lyle, 1947; Perkins, 1982, pp. 12, 184). Dissenting voices were raised (e.g. Cottam & Higgins, 1946; Dunlap, 1981; Strickland, 1945), but attracted little attention.

The chemical industry underwent enormous growth (Perkins, 1982, p. 13). For example, in 1944 about five million kilograms of DDT were produced in the USA whereas in 1951 50 million were produced. Huge loans were negotiated to establish chemical manufacturing plants, and aggressive sales techniques were initiated to service debts (Perkins, 1982, p. 14). Advertising is now more persuasive or humorous (see, e.g., Figs. 3.2 and 3.3). Farmers who did not turn entirely to chemicals risked being outcompeted and forced out of business. For instance, 'High competition in the apple industry plus the spectre of renewed poverty defined the arena into which the new insecticides were introduced. Whether they would be adopted or not was to be answered in the context of uncertainty and fear' (Perkins, 1982, p. 18, and see also pp. 13, 271–273). Control with synthetic organic chemicals was much cheaper than other methods, much less complex than reliance on a suite of control measures, and also required much less labour – labour requirements continued to decrease well into the 1980s, with a drop of 73% from 1974 to 1986 (Szmedra, 1991). In general, countries with developed economies followed this same route (e.g. Briejer, 1968). Elsewhere, chemical control protected the new high yielding 'green revolution' varieties from which virtually all resistance to pests had been bred (Shiva, 1991). Developing countries still use pesticides, imported at high cost, that are banned elsewhere, and moves to IPM have been thwarted by demands

THE MIRACLES OF MODERN INSECT CONTROL

From fields in which it was impossible to grow corn ten years ago, farmers now get 100 bushels per acre without rotation. Cotton fields that weren't worth picking now produce two bales per acre. Homes that would have rotted away from termite damage stand as sound as the day they were built. Lawns and gardens flourish, free of insect pests. These are some of the "miracles" of modern insecticides, and of the dedicated scientists who have developed safe and effective ways to use them. New pesticide chemicals have helped man increase his standard of living, his security, and his peace of mind. Those produced by Velsicol Chemical Corporation have been exceptionally useful. Velsicol is proud of them, and both grateful and indebted to the entomologists and other technicians who have made these insecticides so universally beneficial.

Figure 3.2. A persuasive advertisement published in the early 1960s in *Annals of the Entomological Society of America*. Reprinted with permission from Velsicol Chemical Corporation.

Figure 3.3. Modern examples of the way in which humour and military appeal are used to sell pesticides. The advertisement with the fighter jet was not used, apparently because its release would have coincided with the Gulf War. Top picture reprinted with permission from Incitec Ltd. The products pictured here are not the currently registered container for sale in Australia. Middle picture reprinted with permission from Applied Biotechnologies Pty Ltd. Bottom picture reprinted with permission.

from financiers for a full-season preventive control schedule as a loan condition. Governments have thus been encouraged to subsidise pesticides (Bottrell, 1996).

Reliance on chemicals, and its consequences

Two general problems became evident almost immediately after reliance on synthetic organic insecticides had been established, although this does not imply that the proponents of the technology (e.g. Brown, 1961) acknowledged or even discussed the issue. One was the hazard for humans and the environment posed by the strong toxins. Besides their impact on wildlife and waterways, the insecticides also poisoned beneficial organisms that parasitised or preyed upon the target pests. The other was failure of the technique as a result of insect populations developing resistance to the chemicals; this effect was compounded by the mortality of the natural enemies. Looking back, we naturally have a clearer view of the dangers. Yet some of these dangers persist, particularly in certain developing nations. Such 'local' environmental threats have global implications, but this is not widely enough appreciated (e.g. Shiva, 1991).

Human safety and conservation factors came under public scrutiny because of the persistence in the environment of many of the chemicals and their high toxicity to a wide range of non-target species. Initially the debate centred around the hazards to people. In the early 1950s the US government decreed that human consumers had to live with certain levels of pesticides by setting tolerance levels, or residue levels, for pesticides in food and fibre (Moore, 1987; Perkins, 1982, p. 30). This lead was followed worldwide, presumably by default in many cases. Hazards to wildlife became widely appreciated through such vehicles as Rachel Carson's (1962) book *Silent Spring* (Briggs, 1987), and the demonstration of the serious consequences of the accumulation of lipid-soluble toxins in the fat bodies of organisms higher up the food chain (see Dunlap, 1981). Carson, in particular, exposed the scale of the environmental destruction that was taking place. Others performed similar functions elsewhere, for example Briejer (1968) in the Netherlands.

The environmentalist movement, like Carson herself (Briggs, 1987), did not necessarily reject entirely the use of chemicals. Rather they objected to the indiscriminate use of the persistent broad spectrum toxins, as indeed did a small proportion of pest controllers (see Dunlap, 1981). Carson, herself a scientist, was well aware that a much higher level of scientific thinking about pest problems was readily achievable (Briggs, 1987), and she strongly favoured more research and greater reliance on non-chemical means of control. Carson wrote a great deal about the potential benefits of biocontrol. Recently the public has begun turning more strongly to 'organic produce', which parallels an increasing concern for the environment; these trends are expected to spread generally (Dent, 2000; Rural Industries Research and Development Corporation, 1990), which will undoubtedly enhance the turn away from toxins. Other influences are presently also working in society. In the late 1980s the issue of the dieldren content of Australian export beef attracted the ire of beef producers, mainly directed against banana and sugarcane growers, who were then still permitted to use dieldren against insect species that were otherwise difficult to control (Corrigan & Seneviratna, 1990; National Health and Medical Research Council, 1993). This represents a new slant; primary producers rallying against the use of a pesticide, which was subsequently banned even for special purposes (National Health and Medical Research Council, 1993).

Some target insects developed insecticidal resistance surprisingly rapidly after the introduction of the synthetic organics. Human health concerns re-emerged through disease vectors commonly developing resistance. Housefly (*Musca domestica*) populations became resistant to DDT very soon after its initial use against the species (by 1946–1948; King & Gahan, 1949). The mosquito vectors of malaria re-emerged as a health risk after a period of control (Greece, in 1951) because they acquired resistance, as did the head lice vectors of typhus (Korea, also in 1951) (Brown, 1961). Resistance also became widespread among crop pests, starting soon after the commercial adoption of the synthetic organics. For example, the Colorado potato beetle first showed signs of resistance to DDT within a maximum of 14 generations (Cutkomp *et al.*, 1958), spider mites became resistant to organophosphates by the early 1950s, and codling moth, cabbage looper and tomato hornworm soon became resistant to DDT (Brown, 1961).

Despite resistance being recognised as a problem, the need for alternative non-chemical control methods was not apprehended. Research on alternatives was considered low priority, except by some individuals, most of whom were connected to the Californian biological control schools (Kogan, 1998). H. S. Smith, for example, foresaw the problem posed by resistance as early as 1940, when he declared that chemical control is temporary (see also Smith & Allen, 1954). He even argued, because of his view of the potential of natural selection (Perkins, 1982, pp. 35, 52), that no single method of control for a species would necessarily be permanent (Smith, 1941). This message had serious implications for pest control, but was not widely heeded. A few others were also not taken in by the 'chemical fashion'. They had evidence to support their interpretation that the resurgence of pests and secondary pest outbreaks were caused by insecticides through a reduction in the impact of natural enemies (Clausen, 1936; Smith & DeBach, 1942). DeBach (1951) saw very early that only ecological understanding could protect the farmer from problems inherent in use of chemicals. The popular scientific literature also carried the message (e.g. Smith & Allen, 1954).

Nevertheless, the scientific establishment as a whole persisted in considering resistance only as an inconvenience and did not mobilise strong support for research on alternative control methods (Perkins, 1982, p. 37). The general response of scientists to resistance was simply to change to another chemical (Brown, 1961). Serious actions were taken to alleviate the resistance situation only when the problem threatened commercial agricultural production comprehensively. Ultimately, the regional economic crisis precipitated by the comprehensive and diverse resistance across the entire cotton pest complex in the Lower Rio Grande Valley of Texas sparked substantial change (Adkisson, 1973; Perkins, 1982, pp. 40ff.).

In the early 1960s boll weevils, a major pest in the southern USA since the previous century, developed resistance to chlorinated hydrocarbons. There was then a shift back to calcium arsenate in some cases, and to organophosphates, carbamates and various mixtures (Perkins, 1982, p. 41). In the meantime, secondary pests had become more common because of insecticidal depletion of their natural enemies, and their status as pests had increased. For example, *Heliothis zea* and *H. virescens*, which had never been serious pests of cotton before, became 'secondary' pests and had to be controlled with DDT, BHC, sulphur and parathion. By 1967, these bollworms were resistant to organophosphates as well as to chlorinated hydrocarbons and carbamates; 20 sprays per year could not control them, compared with 8–12 sprays in high input systems in the Southwest in the 1990s (Luttrell, 1994). Either a solution had to be found or the industry would shut down, with all the negative consequences for the surrounding communities (Harris, 2001). These events threatened the credibility of the entomological profession (Perkins, 1982, p. 43), which promptly developed a set of guidelines designed to keep boll weevil populations low without the extensive application of insecticides (see Perkins, 1982, pp. 116ff.). These techniques, including trap crops, pheromone traps and the reduction of diapausing populations by spraying at the end of the cotton season, did much less harm to the natural enemies of the bollworms. The bollworms again came under sustained biocontrol (e.g. Matthews & Hislop, 1993).

Such integrated pest management was beginning to emerge on several fronts in various parts of the world, including for Nova Scotian apples (Croft & Hoyt, 1983), citrus in California, South Africa and Israel (DeBach & Rosen, 1991), and American cotton, pecan and sorghum (Adkisson, 1973; Harris, 2001). Insecticides were not rejected in this approach, and the emphasis has now generally shifted to target-specific insecticides, technologies that deliver smaller quantities of the active ingredient more accurately, and compounds that do not persist in the environment (Dent, 2000; Matthews, 2000; van Emden & Peakall, 1996).

Development of integrated pest management

The general approach to the integrated pest management of today, and the philosophy behind it, evidently developed in the USA through resources provided by the 1972 Federal IPM Thrust (Kogan, 1998). Funding covered research, development, training and field scouting. Most authors portray IPM as having developed logically and gradually from approaches current before the advent of the pesticide era (e.g. Kogan, 1998). Perkins' view (1982) differs considerably, however; he detected a lack of clear direction brought about by the multifaceted failure of a once omnipotent technique of pest suppression. At the time of the economic crises that resulted from this failure (e.g. Adkisson, 1973; Harris, 2001), different schools of thought undoubtedly perceived solutions in different ways. According to Perkins (1982, pp. 58, 97, 150–152) the general confusion saw two competing thrusts: the original IPM, and total population management. The latter is covered in the next section.

Integrated pest management was rather vaguely defined and has remained so, as demonstrated by Kogan's (1998) disentangling of its roots and associated terminology. Indeed, development of the concept took place largely through a sequence of compromises from various quarters. Consequently 'IPM' carries quite different meanings for different groups of people (see Ehler & Bottrell, 2000). Even the ecological principles claimed to have been influential, such as density-dependent population regulation (e.g. Flint & van den Bosch, 1981; Hagen & Franz, 1973;

Perkins, 1982, p. 67), are now increasingly questioned, as discussed in Chapters 5–9. What would help IPM is the consensus on definition called for by Kogan (1998, p. 248), to aid in choice of performance criteria for IPM implementation targets.

The IPM movement had originally envisaged the shift away from chemicals should be built primarily around biological control. This influence derived mainly from C. B. Huffaker, who led through the massive federally funded (see above) 'Huffaker IPM Project' of the 1970s. His emphasis on biological control had developed from his earlier notable successes in the use of predators and parasitoids for pest control (Perkins, 1982, pp. 62ff.). However, Huffaker's project was compelled to include the systems approach to modelling and analysing 'agroecosystems', through the active participation of the International Biological Program (IBP) (see McIntosh, 1985), under the auspices of the National Science Foundation (NSF) (Croft, 1983). The thrust of the IBP and NSF was fundamental science or pure science, not applied science, so their participation hinged on their importing systems analysis and mathematical modelling, apparently as a means to develop general theories in biological control (Perkins, 1982, pp. 71, 144–149). The IPM end product was also to include economic considerations as a means of identifying optimum management techniques that could be used against pests, but this only ever reached unsophisticated levels (Perkins, 1982, pp. 142–145).

Ironically, the economic component has continued to 'drive' the use of insecticides, since no other pest control method can be instituted with as little preparation and as quickly as can chemicals nor in sufficient time to prevent pest populations overshooting threshold densities. The 'economic backbone' of IPM, which is seen as integral by so many (e.g. Kogan, 1998; Speight et al., 1999), has thus helped maintain the insecticidal (or curative) emphasis, and has even drawn attention away from the goal of working against the establishment of pests in fields and their subsequent population increase in a more ecologically constructive way (preventive IPM) (Clarke, 1995; Pedigo, 1995). Even more extreme is the recent recommendation that the default position in IPM should assume that insecticide resistance will develop, and that monitoring programmes should be run to provide baseline data for early detection of that resistance (Lemon, 1994).

In contrast to the economic emphasis outlined above, threshold values have been said to be irrelevant, even detrimental, to rice IPM (Matteson, 2000), and developments in rice represent one of the great recent IPM triumphs. In general terms, Ehler & Bottrell (2000) state that economic thresholds have proved to be largely of academic interest as they can be intractable in practice.

Such dichotomous viewpoints suggest we should not see pest management as a smooth continuum ranging from reliance on chemicals to fully developed IPM, as so frequently portrayed (e.g. Benbrook *et al.*, 1996; Kogan, 1998; Speight *et al.*, 1999). Adoption rates for IPM seem to have been high only in situations in which some sort of crisis, economic driver or political imperative has forced the issue (as in most of the examples mentioned by Harris, 2000, and others) even though authors have portrayed the shift as generally being gradual. Viewing IPM as a gradual continuum may even bode ill for timely conversion to the sort of IPM detailed by Ehler & Bottrell (2000). Nevertheless, an important early aspect of IPM entailed the involvement, at least in theory, of professional supervision. This meant that an appropriately educated entomologist (see Harris, 2000) would advise the necessary decisions.

Inclusion of a professional decision-maker to prevent the indiscriminate use of chemicals was not an innovation of IPM. The idea had been implemented early in the twentieth century, most notably at the hands of C. W. Woodworth in California and D. Isley in Arkansas, who developed the practice of supervised insect control (Ehler & Bottrell, 2000; Kogan, 1998; Perkins, 1982, pp. 74, 83). Originally the decisions were intended to prevent wastage of chemicals, but the approach was conducive to preserving natural enemies whilst lowering pest densities and the strong Californian biocontrol contingent pushed the procedure in this direction, prior to the advent of DDT (e.g. Michelbacher, 1945). Decisions were taken on the basis of what is frequently described as an understanding of the pest species' ecology (Smith & Allen, 1954). For example, Michelbacher & Smith (1943) found that the alfalfa butterfly was usually kept under control by indigenous parasitic wasps, so insecticide use was recommended only when sampling indicated that pest populations had actually escaped natural control. These workers were highly cynical of insurance or prophylactic sprays; they considered them a 'failure of expert knowledge' (Perkins, 1982, p. 76). The decision-making role was to be abandoned in favour of DDT, but insecticide resistance ensured that by the 1950s the early ideas relevant to IPM would be resurrected and promulgated by those previously involved with them (Perkins, 1982, pp. 77–78, 83) and that concepts like economic injury levels would be launched (Stern *et al.*, 1959).

Development of an IPM programme entailed the following set of steps, as outlined in the Huffaker Project Research Grant Proposal (Perkins, 1982, pp. 85–86).

1   Identify real pests, as opposed to induced pests.
2   Establish economic injury levels.
3   Identify the most important factors, called key factors, affecting populations of the real pest.
4   Identify the most important factors affecting induced pest populations.
5   Identify alternative controls for real and induced pests.
6   Use systems analysis and modelling to guide the planning and execution of control tactics and to refine the research programme.

By 1977 the achievements of the Huffaker Project were deemed 'excellent' by the Council on Environmental Quality. Implementation of the knowledge base it created was said to have allowed insecticide use in the USA to be reduced by 50% or more (Perkins, 1982, p. 89). The reality is somewhat different and is discussed later.

Development of total population management

In the USA, total population management (TPM) emerged alongside IPM as an alternative 'concept of how to organize research and practice involving different control techniques' (Perkins, 1982, p. 97). The ultimate goal and geographical scale of TPM distinguished it from IPM. The former sought the eradication of a key pest from large geographical areas, in contrast to the aim in IPM of holding populations of all relevant pests at or below predetermined or acceptable densities (Perkins, 1982, p. 189).

The basic philosophy of TPM originated with E. F. Knipling in 1937 (Perkins, 1982, pp. 105–106, 123), although earlier eradication projects (Myers et al., 1998) must have been influential. Later, as Director of Entomology in the USDA, Knipling helped implement the technique. He had been involved in two major earlier successes in pest control and these played an important role in his thinking (Perkins, 1982, pp. 101ff.). Knipling had had an instrumental role in the adoption of DDT and the rapid pest control achieved impressed him, as did the fact that people could be freed of pests like mosquitoes, lice, bedbugs and flies for the first time ever. Knipling was not naive about the problem of reliance on chemicals (resistance, resurgence, etc.) (Perkins, 1982, pp. 31, 102) and to him the need for their recurrent use meant they could only ever provide temporary relief. They would not provide the long-term solution that had once seemed such a real possibility. Knipling also knew that some pests, notably the screwworm fly, were not readily amenable to insecticidal control. The idea of permanent eradication as a cornerstone of TPM came from the success that Knipling's group achieved in releasing sterile male flies to reduce screwworm populations on Sanibel Island near the Florida coast and to eradicate them from Curaçao, a much larger island in the Caribbean. Subsequently, much larger mainland areas were treated successfully (Perkins, 1982, pp. 107–108; Spradbery, 1994), although doubts about the efficacy claimed for this technique have been raised (e.g. Carey, 1991, 1996; Readshaw, 1989).

A massive experiment, the Pilot Boll Weevil Eradication Experiment (PBWEE), was set up to test the feasibility of using TPM against crop pests. First, techniques for sterilising male boll weevils with chemicals had to be refined, as radiation affected them too severely. Because pesticides are most efficient at high densities and sterile male releases at low densities, the two techniques were set to operate in tandem. 'Reproduction-diapause' (r–d) control was a novel and somewhat unusual insecticidal practice (Perkins, 1982, pp. 116, 125). Application was timed to kill weevils on cotton at the end of the season, when the farmer had normally ceased control. The end of season weevils produce the population that diapauses through winter and these then produce the generation that infests the following year's crop. This innovation controls the pest when the cotton is no longer susceptible to weevil damage, but the weevils feed heavily on cotton at that time to accumulate fat reserves in preparation for diapause. Post-harvest defoliation and stalk destruction eradicates food and diapause sites and reduces diapause populations. In 1965 r–d achieved a 98% reduction in overwintering weevils, which encouraged the idea that complete eradication was feasible.

Eventually, in 1971, the immense PBWEE, which comprised a carefully coordinated set of techniques to be implemented over three years, was started (Perkins, 1982, pp. 119–121). The administrative superstructure was massive, as was the budget. All growers within the designated area had to cooperate if successful timing of operations was to be achieved (Perkins, 1982, pp. 120, 127, 129). The exercise even required legislation and policing to achieve maximal compliance, an unusual intrusion in a scientific experiment at that time (Perkins, 1982, p. 130). The ultimate aim was to answer the outwardly simple question: Is the technology available to eradicate boll weevil from the USA (Perkins, 1982, p. 130)? The PBWEE was not suitably designed to answer this question; indeed, it may be impossible to design an experiment to do so, and value judgements entered the interpretation (Perkins, 1982, pp. 38, 130–131). The results and their significance were subject to radically different interpretations (Perkins, 1982, pp. 127, 131–133). But what was certain from the project was that boll weevil numbers were dramatically lowered by the techniques used and that the sterile male technique needed improvement to fulfil its stated aim (Perkins, 1982, pp. 132, 135, 137).

Eradication is expected to play a more important role in agricultural entomology, depending mainly on how society's aims and expectations develop (Calkins *et al.*, 1994; Lindquist *et al.*, 1990), and on how the associated costs and benefits are assessed (Myers *et al.*, 1998). The incorporation of various other control techniques into eradication programmes (e.g. pheromone monitoring and cultural control practices to reduce overwintering populations, which were used even in the boll weevil experiment) ensured that TPM now resembled IPM, which had a very wide definition in any case. Indeed, eradication now tends to be seen as a component technique of IPM and is now usually referred to as the sterile insect technique (SIT). It is from this amalgamation that today's IPM has developed, with a further contribution from southeast Asia of grower education through participation in decision-making and certain aspects of research (Matteson, 2000; Matteson *et al.*, 1994). This educational approach was developed to help disseminate the information required for effective IPM, an aspect not before seen necessary in conjunction with the chemicals used previously for control (Dent, 2000, p. 328).

An 'area-wide' focus on a key pest species in certain IPM projects has been another important contribution of TPM (Lindquist *et al.*, 1990), although sporadic efforts in this direction had been made since the turn of the nineteenth century (Myers *et al.*, 1998; Perkins, 1982, p. 122). Despite the effective development of an anti-locust campaign over the entire distribution of this widespread and highly mobile pest (Rainey, 1989), species-wide monitoring for sources of pest migrants into agricultural systems is not yet routine. However, pests are now more frequently considered from the perspective of their ecology over larger parts of their geographical range (e.g. Kisimoto & Rosenberg, 1994; Kisimoto & Sogawa, 1995; Kogan, 1996; Zalucki *et al.*, 1994). The aim is reduction of the initial population size of the key species in the crop to a very low level so that the development of damaging population densities is delayed, at which point population suppression is initiated. In part, the development of insecticide resistance has ensured a broader spatial perspective be adopted, mainly through the imperative of not spraying the entire pest population (van Emden & Peakall, 1996). This practice has encouraged an understanding of when the pest is outside the crop, where it is and what processes are associated with its movement, and this has been carried through to the introduction of transgenic plants (Carriere *et al.*, 2001; Gould, 1998). The scale of the exercise is regional (Kogan, 1996) and this must presumably be dictated by the geographical distribution of the key pest and its movement patterns as influenced by topography and habitat availability. Kogan (1996) has formalised the amalgamation of IPM with area-wide pest management considerations.

Performance of IPM and underlying problems

The performance of IPM is not easily judged, for no criteria of a successful programme have been specified (Kogan, 1998). Consequently, success rates are readily exaggerated. Part of the problem lies in IPM having exceedingly broad limits (Bajwa & Kogan, 1996), to the extent that even exclusively insecticidal management is misleadingly included (Bottrell, 1996; Kogan, 1998). Another part is related to the problems encountered in developing a widely accepted, but sufficiently strict, definition of success itself, in a system that should be incorporating tolerance of pests to some extent (see Ehler & Bottrell, 2000). So judging performance presently seems confined to a relative measure of levels of insecticide use before and after instituting IPM, as well as a relative measure, before and after, of the effectiveness of control levels. Modern IPM has undoubtedly performed very well in some systems, including cotton in the USA, which incorporates SIT against boll weevil (Bacheler, 1995), rice in Indonesia (Matteson, 2000; Matteson *et al.*, 1994; Whitten, 1992), citrus on several continents (DeBach & Rosen, 1991; Papacek & Smith, 1998), and glasshouse cultivation of some crops in Europe (Grant, 1997; van Lenteren & Woets, 1988). The proportion of a crop under IPM, even in these systems, may be relatively low. For example, only 30% of glasshouse production in Europe is even under biological control, with the equivalent worldwide figure being estimated at 5% (Hokkanen, 1997), and a mere 1% of Asian rice farmers have graduated from Farmer Field School IPM programmes (Matteson, 2000). On the other hand, acceptable IPM methods have clearly not been achieved in other systems, at any scale, despite considerable effort. But such a conclusion has to be inferred because IPM failures are not frequently described as such in the literature. Integrated pest management in cotton in Australia has yet to be realised, for example, despite the urgency imposed by loss of effective pesticides and environmental concerns.

The last point, above, suggests that scrutiny of the ecological understanding said to underpin IPM is warranted. The problem seems to be that the instruction is straightforward – understand the interactions, population dynamics and mortality factors (e.g. Kogan, 1998; Speight *et al.*, 1999; Tait, 1987) – but that translating that into results relevant to IPM is evidently far less straightforward. When IPM successes are considered, it is not always altogether clear where the research inspiration has come from, what theoretical underpinning has driven it, or why the treatment has actually succeeded. Biocontrol, for example, has been motivated by ideas of restoring balance or imposing stability, although some of the significant successes have involved some form of climate matching. Other non-insecticidal successes have relied on disrupting a particular phase in the life cycle, whether that be, for example, mating (pheromones), oviposition (trap crops and host plant resistance), initiation of larval feeding (host resistance), or diapausing stage (altering the environment). The connection between theory and application is not obvious, even in those publications that spell out quite comprehensively the general research pathways considered to be needed (Fig. 9.5 of Dent, 2000; Dent, 1997). The ecological claims from pest management all too frequently give the impression that finding direction in ecology is relatively straightforward. Specifically, they imply that derivation of the necessary ecological understanding is easy to achieve, that it derives from relatively simple factual observation, and that the interpretation will be accessible to all who may seek it. However, ecology is not that yielding to scrutiny, as detailed in Chapter 5, and a vast number of pest situations have not been so readily solved. These views imply no more than that ecological research and understanding are not accurately portrayed in schemes or generalisations relating to IPM, or to the theory usually associated with it. An additional point is that perhaps we should not expect ecological theory to tell us what to do, only how to do the research in any chosen area. These points are justified, developed and expanded in the remaining chapters.

Several indirect measures may be used to demonstrate that IPM has not fulfilled its potential.

1   Failure in the development, deployment and adoption of particular technologies. Here, biocontrol and pheromone technologies come to mind. The former is not as successful and the latter is not as widely used as desired by protagonists, and neither is financed well enough (e.g. Greathead, 1994; Jones, 1994).

2   The persistent status of so many species as primary pests that still require insecticidal treatment for their suppression, including *Helicoverpa armigera* over most of its broad distribution and *H. punctigera* in Australia (de Souza *et al.*, 1995; Zalucki *et al.*, 1994).

3   The failure to reduce global rates of insecticide production and application (Pimentel & Lehman, 1993). Even when reduction has been claimed for a region, as in the USA after the Huffaker Project, statistics indicate no reduction in pesticide use during the period in question. Even when application rates on major crops in the USA did subsequently drop (Osteen & Szmedra, 1989), it was mainly because compounds were switched (Szmedra, 1991).

4   The claims of IPM protagonists that rates of IPM adoption are too low (Benbrook *et al.*, 1996; Dent, 1991, p. xv; Ehler & Bottrell, 2000; Grant, 1997; Matteson, 2000). 'There is now a growing awareness that IPM as envisioned by its initial proponents is not being practiced to any significant extent in U.S. agriculture ... true IPM is probably being practiced on only 4 to 8 percent of the U.S. crop acreage. Globally, the percentage is even lower ...' (Ehler & Bottrell, 2000).

The above provides only a rough indication; a thorough, quantified analysis would be preferable. In general, though, pesticide use is apparently far more common and intense than necessary, and some pest species still reduce yield over vast areas and across several crops. For these reasons, and because new pests arrive in cropping systems and changed growing practices alter the pest status of insects already present, decisions are needed on where to place emphasis (or how to identify priorities) if we wish to improve the contribution of IPM.

At least four underlying causes may hinder the deployment of IPM: (i) failure to implement practices believed to work, (ii) regression or decay of implemented practices that work (Stoner *et al.*, 1986), (iii) failure to develop an adequate ecological understanding, and (iv) reliance on new 'state of the science' techniques to shore up failing approaches (Dent, 2000, p. 10). With regard to the final point, Kogan (1998, p. 263) adds the warning: 'The excitement about genetic engineering . . . dominates the futurist literature in IPM. If there is a lesson to learn from the past 35 years, it is that a silver bullet is unlikely to come out of any of the new technologies, and nothing would have been learned from the past if genetic engineering is emphasized over all other technologies that are also blossoming'.

Naturally, different reasons for failure demand different remedies. The literature reveals, however, strong claims for the priority of particular aspects of IPM. For instance, some involved in the socioeconomic sphere, which relates to IPM system design and implementation, have accused 'specialist' researchers (also referred to, rather pejoratively, as 'reductionist scientists') of having commandeered funding opportunities to the exclusion of implementation. Although this claim may be accurate in some situations or to some degree generally, it is a hard argument to sustain, not least because evidence is simply not available. The argument is, at best, speculative, for alternative explanations are equally valid in explaining why IPM is not more widespread and successful. Add to this the fact that different explanations may well be valid for different programmes and we have a tricky situation to interpret in general terms. Other possible explanations include a lack of alternative options to chemical application (in some cases a suitable alternative may be impossible), lack of understanding of relevant aspects of the ecology of the pest or beneficial species, and lack of external pressure to impose adoption of IPM (Burn *et al.*, 1987; Ehler & Bottrell, 2000). Clarifying which possible explanation is valid for any particular situation is a complex task, at least on current information.

Consider, for example, the relatively simple question of whether an alternative control technique is available. How are we to assess this issue for a particular situation, unless of course such a technique has already been developed and has been shown to work on a large scale? How is one to know whether a suitable control tactic is likely, especially when we have no prescription from anywhere within the IPM literature of how to select the most appropriate control method? Expecting to identify a general cause of low rates of IPM adoption is premature, or even unrealistic. However, a lack of political will may well be a rather common factor (Ehler & Bottrell, 2000; Ramirez & Mumford, 1995; Whitten, 1992). More significant about the whole dispute, perhaps, is that the sort of divisions described above among participants, within an enterprise as complex as IPM, are not helpful. That the dissent represents argument for obtaining research funds at the cost of an 'opposition' is even somewhat destructive, and IPM is likely to suffer.

Conclusions

The historical information covered in this chapter leads to at least five significant conclusions about pest management. A sixth point is developed with reference to the conclusions drawn about scientific method in Chapter 2, to provide an overview of the challenges that face the research-related aspects of IPM.

1   The goals of the food and fibre production systems of society will
     largely determine which pest management practices will be adopted
     voluntarily by agriculturalists (Perkins, 1982, p. vi). The definition of
     such goals is influenced, in turn, by non-agricultural interests in
     society and by policy-makers (Ehler & Bottrell, 2000; Ramirez &
     Mumford, 1995; Whitten, 1992).

2   Production goals have shifted irreversibly with the evolution of human
     society, the invention of technology, and as economic considerations
     have forced stricter demands on the efficiency of farmers (Jones, 1973;
     Perkins, 1982, pp. 209ff.). Different goals are also evident
     geographically. Consequently, research requirements shift temporally
     and spatially. No simple reversion to past practices will suffice in
     overcoming the problems of today, as the situation has changed
     irrevocably and we also have different technologies and different
     concepts in biology, all of which influence our behaviour. In addition,
     the make-up of general approaches to pest management seem to be
     influenced considerably by the political structure of the agencies or
     institutions in which they are developed (Perkins, 1982, pp. 274–275,
     280). Some avenues, such as TPM, could never have been pursued
     unless a sufficiently strong bureaucracy had been in place to steer or
     permit developments financially. The history of TPM may well
     demonstrate also that if sufficient funds are available to support a
     specified goal, the chances of success are increased substantially. In the
     meantime, pest management methods have been readily accepted only
     when scientific and technological discoveries coincided with the needs
     of society and society's perceptions of how agricultural production
     should be organised. The demands consequent upon increasing
     human population size have also been significant, and are likely to
     continue to prove so in the coming decades and beyond. However,
     scientists have also influenced developments in pest management, as
     evidenced by the impact on society generally of the book *Silent Spring*.
     The current debates about biodiversity and genetically modified
     organisms are also likely to prove influential in pest management, the
     former probably not only because of its influence on biocontrol

practice (e.g. Simberloff & Stiling, 1996). So it is possible for agricultural scientists to take more of a leading role in shaping society's expectations and perceptions and thereby directions in pest management.

3   Certain problem species and difficult situations remain intractable even after years of trial and error adjustment to practice and a great deal of research effort. Such perennially serious pests as tephritid fruit flies, heteropteran bugs and various heliothine caterpillars still pose major management difficulties in many areas. Persistently intractable situations like these undermine the outwardly simple imperative 'Understand the ecology of the species in question'.

4   No fundamental scientific principles have ever been successfully developed to specify accurately the most appropriate control technique, or mix of control techniques, for a particular situation (Pedigo, 1995). Whether a basis exists for such a development needs serious introspection.

5   The relationship between research (to extend knowledge) and system design (to apply knowledge) has not been well enough clarified. Perkins maintains that the current IPM paradigm in insect pest management provides research direction. In a general way, he is correct. But the direction thus provided is too vague to provide the more immediate insights required by entomologists to solve the particular problems they face. Their education and experience should be sufficient for such a purpose, but the world of science changes as well, and 'applied' entomologists need to consider changes in epistemology, evolutionary biology and ecology, as well as the way in which these changes impact on perceptions of problem solving in general and in IPM in particular.

6   A major concern for IPM-related research is the use of scientific method, an issue only infrequently examined in the literature on insect pests. An approach that focuses on the empiricist gathering of what is considered 'factual' information seems to be considered satisfactory in a good proportion of IPM-related publications. That research has been influenced by Baconian beliefs, even if indirectly so, can be inferred when there is evidence of: (i) an emphasis on developing interpretation from 'facts' rather than on testing among alternative possible explanations; (ii) reliance on serendipity rather than on rational analysis of the situation and the range of possible problems; (iii) persistent emulation of projects seen to have been successful, rather than the identification of underlying problems and their solution; (iv) continued reliance on inappropriate theory and thus inappropriate assumptions as to how the system under

investigation actually works; and (v) views that promote the belief that enough reductionist research is available around which to construct successful IPM programmes (e.g. Surtees, 1977), and which implies that knowledge is finite (see Chapter 2 for criticism of this view).

The various types of research issues that face entomologists who tackle pest-related problems need to be disentangled. This should help to enhance the prospects for solving problems. When, and how, should more fundamental principles in scientific method and in ecology be sought and employed? As a first step in assessing how the development of IPM is related to scientific method, in Chapter 4 the many components and aspects of IPM-related research are dismantled. This analysis is complementary to the outline of an IPM programme presented by Dent (2000). The emphasis then shifts to the ecological aspects of insect-related research (Chapters 5–9) before returning to the structure and operation of IPM in Chapter 10.

4

IPM: a diverse, interrelated suite of socioeconomic and scientific problem-solving activities

> Integrated Pest Management has become largely a delivery system with little attention to basic biological understanding of the crop/pest/ natural enemy system.
>
> J. R. CATE (1990, p. 29)

Introduction

The previous chapter analysed how society has been led to deal with pests at different stages in the development of modern pest management. Structured ways of thought, called paradigms, tend to direct efforts against pests and also direct the research needed to improve those efforts. Integrated pest management is accepted as the driving force behind current pest management practice and is seen as the guiding principle in pest management research (Dent, 2000; Kogan, 1998).

Contemporary analyses of IPM tend to concentrate on the successes of IPM and on other positive aspects. Although encouraging, this approach is not necessarily the best or most helpful in identifying and redressing the problems that still exist. A more introspective approach is needed if deficiencies are to be identified and targeted for improvement. One of the deficiencies noted in previous chapters has not been singled out for attention to any great extent in the IPM literature, namely the poorly articulated relationship between the practice of IPM and ecological theory. The aim of the current chapter is to scrutinise the research angles that relate to IPM and seek a realistic context for each of them. Research on pest or beneficial species for IPM purposes takes several distinct forms, depending on what problem area is being addressed. Each problem area has its own special features and each warrants consideration in its own right. At least some of the research that is conducted for IPM purposes falls squarely under the guidance of theories that fall outside the realm of IPM as traditionally defined or discussed. Therefore, few generalisations extend across all categories of IPM-related research on pest or beneficial insects. This point is clarified in the present chapter, and the relationships among the various 'kinds' of research conducted to improve IPM are teased apart and specified individually.

The analysis presented may seem to contradict the spirit of the epithet 'integrated pest management', as it represents more of a deconstruction than an integration. However, the term 'integrated' clearly refers to the application of control techniques rather than to the research that underpins those techniques or the research intended to integrate them. For example, Ehler & Bottrell (2000) describe the type of targeted research effort needed to provide the understanding necessary to ensure integration of techniques. Kogan (1998) confirms this viewpoint: 'By the 1960's there was substantial agreement that "integration" meant the harmonious use of multiple methods to control single pests'. This statement, like most others on this issue, does not mention research, although a major contribution of IPM has been the demonstration of the 'need to base all phases of the production system on sound ecological principles' (Kogan, 1999). When more direct mention is made of the research side of things, it seems to be treated as if it can take care of itself. This is perhaps a consequence of the biological research for IPM for long being treated almost synonymously with IPM practice. For example, Dent (2000, p. xii) writes of a need for individuals who can address the subject of IPM in an interdisciplinary way. That is, IPM research and application needs to be coordinated, managed and considered as a whole rather than as isolated disciplines. Although this statement is essentially correct and is an effort to include all the various needs of IPM, including those related to research, these latter must be dealt with explicitly if IPM success is to be enhanced.

The analysis of IPM-related entomological research that follows is an effort to redress this deficiency. The ultimate message from the analysis presented suggests that IPM research is of a diversity of types, all of which are needed if we are to improve success rates in IPM. Not all of the requisite research is likely to be interdisciplinary, although a good deal will be. Although strong interdisciplinary research requires much more than the sum of its parts (Petrie, 1976; Thorne, 1986), it also demands within-discipline strength as a foundation. Furthermore, a true interdisciplinary approach may even be needed for some of the 'reductionist' or 'specialist' research that undoubtedly will be required for a successful IPM outcome. Although such 'reductionism' is sometimes derided in the IPM literature (see Chapter 3), even the IPM successes in southeast Asian rice, with its attendant developments for extension and farmer participation in research, have benefited substantially from specialist reductionist research. For example, the view that *Nilaparvata* on *Leersia* plants can be ignored for IPM purposes could be confirmed only with research of that nature (Claridge *et al.*, 1985). Although this result could be seen as negative, it provides the depth of understanding critical to rational, scientific IPM.

Defining integrated pest management

Consensus is hard to find among authors as to what actually constitutes IPM (Bajwa & Kogan, 1996; Kogan, 1998), but the definitions available in the literature generally share several points, with some being points of omission. These common elements are expanded below.

1   Entrenched in most definitions is an economic basis for deciding when to take control or management action. The derivation of economic injury levels and economic thresholds remains a cornerstone of modern IPM, despite these concepts being applicable only to rapidly deployable techniques, which still means mainly insecticide application (Pedigo, 1995).

2   Generally an array of control tactics is advocated and it is the actions of these that are integrated for enhanced effect in dealing with pests. Biocontrol and chemical control seem to predominate at present, but resistant crop varieties have also been emphasised and genetically modified ones are likely to play an increasing role (Kogan, 1988, 1998). In practice, insecticides are generally preferred by producers, unless their use is precluded through policy changes being imposed on them (Matteson, 2000; Whitten, 1992), if a crisis occurs because of resistance or banning of compounds (e.g. Forrester *et al.*, 1993), or if pesticide cannot be delivered to the pest (e.g. many stem borers). Cost sometimes prohibits chemical application in low value crops and in situations in which resistance to cheaper insecticides has developed (e.g. Harris, 2001). Herbivorous biocontrol agents may be advocated against weeds that are difficult to control chemically or where the land is not worth the cost of the chemicals. In many situations reliance on chemicals has been the mainstay of an industry, usually when the crop is a high value one, because spraying is almost invariably considered cost effective. So,

in many instances, the aim of an IPM programme is simply the judicious management of a spray programme.

3   No explicit mention of problem solving in relation to pest organisms is made, although in more expanded treatments a list is presented of desirable information that should be acquired. Typically, advice of this type comprises no more than 'Know the ecology of the pest' (Flint & van den Bosch, 1981) or is so broad as again to be unhelpful: 'Consider the interactions amongst the whole range of organisms with beneficial, neutral and pest status' (Tait, 1987). This again seems to imply that one needs simply to go out and gather the facts that are required.

4   Integrated pest management is frequently written about, and thought of, as if it were a simple two-phase activity: conduct the research, institute the control. But 'IPM' is not as simple in structure, for it is an umbrella term that covers a diversity of activities, all equally necessary to achieving a successful outcome. The ultimate action of IPM invariably derives from a diversity of sources and is thus legitimately seen as a social activity.

The entomological components of IPM

To understand IPM one needs to dissect it into its constituent parts. In this chapter the insect-related aspects are considered, but similar treatments are needed for all components of the cropping system if we are to obtain efficiently the derivative knowledge required for the rational design of IPM programmes. From the fractions thus exposed, the various goals that remain hidden beneath the cover term can be established. From there, one can discuss what research is needed to achieve those goals and how best to conduct that type of research. One can thus ferret out the source of the theoretical inspiration for each such area of research, and even where such research effort is likely to lead. Finally, one can think about coordination of the activities for successful practice.

To illustrate what is meant by 'dissecting' the IPM concept, Table 4.1 lays out the entomological (or biological) information that would ideally be required if one wished to maximise the chances of success in IPM. This table is intended to expand a single box, that covering the definition of research needs, in Dent's (1991) 10-box flow chart (Fig. 4.1) that illustrates the specific aspects of the three phases in the development of an IPM programme. Included in Table 4.1 are aspects that have traditionally not been considered overly relevant to IPM, such as the more general theories of biology. The remainder of this book is devoted to detailing their relevance

Table 4.1. *The integrated pest management 'chain' is a schematic representation of the sources of information that may contribute to the design of an integrated pest management system. Special emphasis is placed on biological information, but other types of information are also included. Similar expansions could presumably be made for sociological, economic and psychological information*

	Components relevant to IPM	Examples of research generated
1	General theories of biology	• Developments in species theory and ecological theory
2	Hypothesis of genetic species status	• Does the species constitute a complex of sibling species? • Interpretation of the mating behaviour of different populations
3	Hypothesis of ecology of a particular pest species	• Host plant relationships, migration, diapause and patterns of abundance • Influence of predators under different circumstances
4	Potential means of reducing pest densities	• Synthesis of pheromones • Biological control • Development of application system (technology)
5	How effective is technology and what specifications should be set?	• Field trials
6	Set economic thresholds if relevant	• Relate pest densities to losses in yield
7	Design of IPM system, and what technologies should be combined?	• What technology can growers afford? • What suggested changes will be implemented? • How should the technologies be combined – field trials, models, personal experience?
8	Recommendations and extension	• What are the growers' perceptions of the problem? • Is further education required?

to IPM, mainly through detailed exploration of selected IPM-related topics. A few specific issues are covered to ensure sufficient depth to illustrate the significant general points.

The arrangement of Table 4.1 warrants initial comment, to eliminate possible unintended interpretations about its structure, especially about the order in which the components have been arranged. The order of the components is not indicative of the relative importance of each. Rather, the components represent the links of a chain, with each having equivalent status in the overall operation. The order itself has a different significance, and is based on logical dependence. To illustrate, if one takes

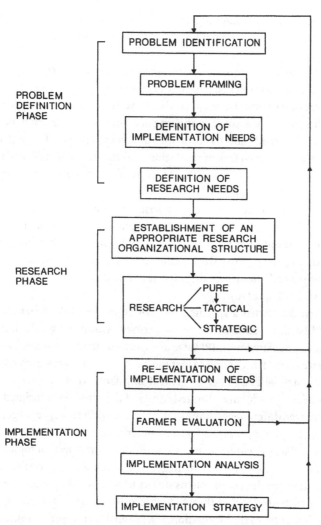

Figure 4.1. The three phases in the development of an integrated insect pest management programme. From Dent (1991). Reprinted with permission from CABI Publishing.

Component 2, which deals with the genetic species status of the pest organism in question (a topic which is discussed in detail in Chapter 6), the interpretation that one derives is dependent on the concept of species with which one works (Component 1). In turn, any ecological hypothesis about the pest (Component 3) will depend on the interpretation of species status in Component 2. The same logic holds for the more practical aspects lower in the table. This point is explored and illustrated more fully later in this chapter.

Logical relationships are one thing; the demands for action by society are quite another. In effect, all pest management programmes represent a degree of compromise between these two requirements. Surprisingly, conciliation between these conflicting demands does not seem to be high on the IPM agenda and has been mentioned by relatively few (e.g. Cate, 1990). However, such lack of resolution, and the frequent failure of those involved in Components 4–8 (Table 4.1) to play a role in demanding the appropriate research whilst ad hoc management measures are provisionally taken, denies a true scientific basis to a good proportion of IPM projects. Massive inefficiencies and their consequent financial losses undoubtedly result from a lack of attention to this point. Such failures in understanding and practice work to the detriment of IPM generally, as well as against a realistic view of the important role that science has in modern society (see Chapter 10).

When we discuss a specific IPM project (or even IPM generally), we should be clear about what specific activity, within the whole ambit of activities associated with IPM, we are speaking about. This applies even when we restrict discussion to just the biological side of research relevant to IPM. Little will be gained by generalising about 'IPM' as a whole when one is dealing with only a limited subset of the activities associated with it. Moreover, different generalisations, and even different types of generalisations, may be appropriate for the different aspects of IPM research. The inappropriate switching among decidedly different aspects or topics, even if inadvertent, confuses pest management discussion. Especially, definition of 'the' problem or problems suffers from a lack of appreciation that different activities within the whole enterprise usually require unique information or types of information, if a rational and scientific solution to the pest problem is to be achieved.

If one accepts the interpretation of IPM outlined above, the following points emerge. In general, they all require much deeper consideration than can be given them here.

Enhancing cooperation and communication

Ideally IPM is a social mechanism designed to obtain information that can be applied to achieve a defined end. Here the subsidiary goals are seldom well defined, despite each of these 'goals' really being a problem in its own right, as illustrated in Table 4.1. Currently the emphasis in IPM is not on problem solving but on application, so that the importance and process of problem identification (across all the subsidiary goals) attracts far too little attention. The enterprise suffers further because much of the research conducted for IPM purposes is of a fact-gathering nature and various shortcuts are frequently taken (see Chapters 2 and 3). Also, successful projects are far too often simply copied when research is aimed primarily at the design of IPM systems (see below and Chapter 8). In consequence, scientific principles are negated and the vast range of problems that warrants attention is seldom laid out for scrutiny and appropriate attention.

The earlier concern with scientific method (Chapter 2) and the nature and structure of IPM (Table 4.1) is an attempt to make the problem-recognition process, which is vital to the success of IPM, much more transparent. Recognising the diversity of inputs and endeavours that are relevant to IPM is a necessary first step in defining the structure of an overall plan designed to institute effective IPM. An over-arching plan that covers all areas of input is vital to IPM success because, in practice, different individuals are usually involved in different steps of the procedure. Although individual entomologists frequently do switch from one 'area' to another, Perkins' portrayal (1982, p. 168) that they span the whole range of activities is misleading, for he divided the research component of IPM too simply, as a dichotomy: 'entomology-as-technology' and 'entomology-as-science' (Perkins, 1982, p. 167). Such a categorisation obscures the subtle but meaningful differences that are found within each of his two compartments (see Table 4.1).

Few participants are au fait with the theoretical, historical and methodological background to each aspect that is relevant to IPM. The 'mimetic' and 'factual' approaches (Chapters 2 and 3) that are so common in IPM actually stand in the way of such an understanding. Furthermore, reliance on such faulty methods also undermines the sort of cooperation that serves science best, the type in which 'one finds oneself rather emotionally engaged in a controversy upon a scientific matter; but this ... is one's job, and one must learn to advocate powerfully, but to concede defeat gracefully, without making a personal issue of it. *In a sense, a well-fought controversy between two spirited champions is a form of cooperation*' (Ziman, 1968, p. 134, cited by Hull, 1988, p. 15) (emphasis added). Therefore, one would ideally like to see competing theories or alternative treatments tested at each step, where appropriate (see Chapter 2). Notice that competing for funds is not a consideration here. And contrast Ziman's view of cooperation with the one that is currently on the political agenda for science (see Chapter 10).

Appreciating the diversity of problems that face successful IPM

With regard to any pest species or situation that is considered a problem to agriculture, the cooperation and communication among individuals who may be operating within different 'components' of the chain (Table 4.1) is not assured. Prestige, authority and status may well intrude, which is yet another significant practical consequence of the paradigmatic nature of scientific knowledge. Furthermore, certain components, which may require particular pieces of information to ensure completion of the IPM 'chain', may attract no attention at all. Here the idea of logical dependence that is central to the design of Table 4.1 plays a crucial role. The weaker the interpretation at particular points in Table 4.1, the less likely that success will be achieved through the rational exploitation of understanding. Too many pest management endeavours thus rely on luck for a successful outcome.

The failure to consider each aspect of the IPM chain, as described above, appears to be consequential on several reinforcing factors that encompass, as common ground, too simplistic an attitude to IPM as a scientific endeavour. Foremost, though, appears to be the view that each pest problem can be defined solely with reference to the desired end product of satisfactory management of that pest. Instead of understanding the biology of the organism, as a means to application, we try to bend the organism to our aim. (Although many might disagree, and assert that such perversity is seldom practised, the examples covered in Chapters 5–9 help to vindicate the statement.) Add to this the emphasis given to fact-gathering as a cornerstone of scientific progress, as well as the belief that all 'IPM scientists' are equally qualified or experienced to cover all the requisite ground, and one has a situation in applied science that hardly falls under the umbrella of rational science. Some research may well be contracted out to an appropriate specialist, but when specialists are recognised as such on the basis of their technological skill, rather than on the basis of their understanding, additional hazards for scientific advance may well be encountered (see Chapter 2). Such 'techniques-centred' research is seldom as strong as 'problem-centred' research, and frequently it misses the mark of recognising and solving problems (Maslow, 1954; and see Chapter 6 for examples).

To facilitate constructive communication and cooperation across the span of activities detailed in Table 4.1, an overhaul of the basic interpretation of IPM will be needed. This will require a broader education for IPM purposes (Chapter 2), and a diversity of scientific opinion must be nurtured for maximum benefit. Competition in this regard is healthy for scientific advance; note how rapidly several independent physicists repeated the infamous cold fusion experiments to clarify the claims that were made. Not only did they rapidly demonstrate the problems with the original tests; they also identified new avenues of potentially profitable investigation (Youngson, 1998). By contrast, whole organism biology seldom, if ever, makes such rapid and decisive progress. The emphasis here on solving problems scientifically is not to belittle the participatory role for farmers in research and implementation activities (Dent, 2000, pp. 328–329). The research so far envisaged for growers is beneficial for the development and improvement of management systems, as well as for demonstrating the validity of management options. That sort of research simply cannot replace that of scientists in which the understanding and influence of theory is critically important. A balance between the two is clearly required, and needs to be built into the overall scheme. Ways in which IPM may proceed to overcome this problem are considered further in Chapter 10.

A rationale for selecting control methods?

How does one select an appropriate means of control for a particular pest or situation? The choice is wide: biocontrol, behavioural manipulation, host plant resistance, cultural control, genetic control, transgenic crops, and so on. Considerable pressure is applied by those who deal mainly with the socioeconomic side of IPM for 'early' decisions to be made with regard to control options. They are concerned that research effort should be directed towards those options most likely to be successfully implemented (see Tait, 1987, and Dent, 1991, pp. 440–441). In many ways this is understandable: 'Since the ease with which techniques can be implemented in any given situation will influence the likelihood of their being adopted, it is important that such matters are given early consideration in the definition phase of the programme' (Dent, 1991).

But care needs to be exercised, for if the words just quoted are taken literally then novel developments may well be stifled (if only indirectly). The statement seems to assume that we know all we are going to know or, at least, that we know all we need to know. But on what basis does one decide what needs to be known? How should one determine what control methods are likely to work against a particular pest species, especially when we are talking about taking decisions in ignorance of the research results that may make such a decision possible? For example, biocontrol may yet work against a given pest, despite several earlier failures with this method (see Chapter 6 for examples). Also, the statement assumes that any new methods that are developed will be 'hard' to implement, but that is not true. Once a breakthrough is made, technological innovation seems to follow quite readily, especially if a market exists. Failure to appreciate these points may well reduce the available options for dealing with particular pests.

Sometimes only one means of control represents an acceptable option and the decision is an easy one. For example, weed control on economically marginal pasture lends itself to biological control. In other cases, the choice may seem obvious; for example, when biocontrol success has been achieved elsewhere against the pest of interest, moves to import that natural enemy are frequently encouraged (e.g. Waterhouse & Norris, 1987b). Although such importations may well be warranted, developments of this nature should be made only on the basis of sound understanding (see Chapters 7 and 8). Anything less is more likely to inflate the proportion of unsuccessful biological control cases, or to generate case histories that will have to be unravelled at some future time before further advances can be made (e.g. Clarke, 1990; Clarke & Walter, 1993a, 1995; Jones, 1995).

In other cases, choice of an alternative technique to chemical methods is less clearcut. Presumably, the best choice can be made only in relation to an understanding of the pest organism's biology and ecology, but efforts to extract generalisations or principles have not been successful and are not in use. The most common means of connecting control method to ecology has been in relation to the concept, in evolutionary ecology, of the r- and K-selection spectrum (e.g. Conway, 1981; Dent, 2000; Way, 1977). Again, this is an ecological generalisation that is too seriously flawed on theoretical grounds to be practical in any way, for r and K are not directly and mechanistically related to one another (Stearns, 1977). Also, the theory is underpinned by assumptions of ongoing adaptation, or optimisation, in local populations (Nylin, 2001). This problem is dealt with in general in Chapters 5 and 6. The theory has also seen too many exceptions to be considered useful in any way (e.g. David, 1993; den Boer & Reddingius, 1996; Stearns, 1977).

The situation regarding choice of control methods for particular situations is far from clear, and there is little promise for developing prescriptive generalisations. This point is followed up in later chapters when sufficient ecological background has been considered.

Research conducted to enhance IPM

The urgency of practical needs usually means that IPM programmes are almost invariably designed and implemented before research can be oriented at building a sufficient understanding of the system, or local situation. Demands for research are frequently made whilst the IPM programme is under construction or is already running, seldom before. Consequently, the research demands for IPM tend to be those seen to be more immediate, with these being the components that lie towards the bottom of Table 4.1 (mainly Components 4–8).

To assist the design of IPM systems, models of the agroecosystem may well be developed. Such models are seen to be beneficial, in part because they inevitably suggest more research. Verbal models of IPM systems similarly yield research questions. How realistic and relevant are such suggestions for further research? Does the research suggested by IPM models coincide, for example, with the research suggested by each component of research listed in Table 4.1? Apparently the answer is no, and a conflict of interest is evident even between research questions generated by IPM models and those generated from biological theories for IPM purposes. Because the conflict is between the approaches to a problem, and not between competing scientific interpretations (as mentioned above), both science and the application of understanding are liable to suffer.

The questions generated by the models designed to mimic the IPM system under consideration are usually posed to fill information gaps. Almost always these are rather large gaps in somewhat superficial models, a situation that can be expressed in another way, as follows. The information gaps are identified with reference to the structure of an IPM model that was likely to have been developed, in the first place, on too little information. If this original information had a faulty underpinning, because of the assumptions that had to be made in developing the model, then 'filling in the gaps' may have less chance of improving the model, and thus understanding of the system, than would alternative research designed to investigate more fundamental aspects of the system. This is not to suggest that such models serve no purpose, or that they should not be developed (see Chapter 10). It is rather to warn against research questions that may be derived too superficially, for once a model has been developed it tends to hide its assumptions (although the basic idea of modelling is supposed to be exactly the opposite). The very existence of the model somehow encourages efforts to extend it or to fill gaps in its superstructure, and this process tends to discourage introspection of the model's foundations.

Despite the differences among the various categories in Table 4.1, reliance between the categories for understanding should still be strong. To illustrate, let us juxtapose two of the categories. The research generated under Component 3, which deals with the ecology of particular pest species, is compared with that generated in Components 6 and 7, which deal with economic thresholds and the design of IPM systems. Below, these two aspects are referred to, respectively, as Ecological Research (Component 3) and System Design (Components 6 and 7).

In considering a pest problem in relation to Ecological Research, evidence may suggest the need to clarify the species status of the pest itself or that of a natural enemy earmarked for biocontrol against the pest (see Chapter 6). Alternatively, the precise nature of the relationship of the pest species to its primary host plant species may need clarification (see Chapter 7). Two points are relevant here. First, when questions about the ecology of the relevant species are asked (whether it be a pest or beneficial), the researcher must not only determine what information is required, but also establish how the underlying components (1 and 2 in this case) influence the way in which the requisite information is best acquired. This is the approach canvassed in principle in Chapter 2, and some practical aspects that relate to IPM are explored more fully in Chapters 5 and 6. Second, results from such ecological research, if gathered appropriately, are of an enduring nature. That is, the results will hold their validity through time and across space, because species are expected to remain essentially the same ecologically, except for relatively minor changes such as insecticide resistance, which is coded at one locus and represents a response to intense selection pressure through high mortality rates (see Chapter 6 for a consideration of species stability).

The information that contributes to System Design in an IPM programme includes consideration of more than just the biology and ecology of the organism(s) in question. The ways in which the organisms influence the production system are incorporated here. Yield loss may need to be related to pest densities, the impact of combinations of technologies may have to be assessed, and so on. Again, two points emerge. One relates to the differences between this type of information and that collected under 'Ecological Research'. An appreciation of the relationship between the two should help in considering relative research budgets and time allocations for the two different research programmes. The second point relates to the problems that arise when System Design precedes Ecological Research and sets the research agenda, which is frequent in IPM. Both points are expanded below.

The first point, the difference between System Design research and Ecological Research, is best considered in terms of the differential dynamics between pest (and beneficial) organisms and agricultural systems. On the one hand, pest and beneficial species have rather slow evolutionary dynamics, as mentioned above and as explained more fully in Chapter 6. By contrast, the dynamics of agricultural systems are much more rapid, with change being the norm. New pest species invade, new crops are introduced into an area, new cultivars of well-established crops replace older ones, control methods change, agronomic practices vary among areas and among seasons, and so on. Where a species might once have been a minor pest, it may become a major pest; this happened with mites in orchards and bollworms in cotton when broad spectrum synthetic organics were introduced (Chapter 3), and it may also occur in response to climatic shifts (Hengeveld, 1990, p. 176). Conversely, major pests may be all but forgotten by subsequent generations of growers, as has happened when biocontrol has been persistently successful. Agricultural systems also include economic and sociological components that are absent from natural systems.

The differential in dynamics between pest and beneficial organisms, on the one hand, and agricultural systems on the other translates into a differential in the useful life of results obtained from research conducted in each of these components of the IPM chain. For example, economic threshold values may be established empirically for a particular crop because System Design may require it. Such research is intensive, time consuming and expensive, yet the results derived in one season may well be invalid for the next cultivar that is introduced, or may not be needed when an alternative control method is introduced. Because the system itself is inherently dynamic, and System Design questions are so context dependent, even high quality results are likely to fall by the wayside after some time. By contrast, the research mentioned earlier on the species status and host relationships of the pest organisms is of persistent value, for these represent the species' properties irrespective of the actual ecological setting. They will remain valid in relation to the crop at hand and are also likely to be useful at other times and in other contexts. This is not to belittle System Design results, but it does raise the question of just how much money and effort should be put into something that is difficult to measure accurately in the first place and then is in danger of being superseded within a relatively short time.

The second point, that of System Design preceding the Ecological Research stage, raises difficulties when the System Design then directs research of an ecological nature. An early System Design might require an alternative control technique at relatively short notice. This is not uncommon, and follows regularly enough in the wake of insecticide resistance. Consequently, field trials may be conducted on the use of alternative host species (trap crops or trap plants) to attract pestiferous herbivores away from the crop, for instance. The design of the research programme could readily be undermined if System Design requirements are conflated with Ecological Research requirements. The conclusions that derive from research that follows such confused design may well be undeservedly bolstered by scientists and industry if the demand for utility and facts is allowed to override scientific procedure (see Chapter 2).

Research on alternative host plants may, for example, proceed directly to relatively large-scale field trials, perhaps together with supporting laboratory tests. If the tests are developed uncritically, and without due attention to the scientific principles discussed in Chapter 2, serious complications may follow. For example, results from field trials may indeed show that many more pests are present on the alternative host than on the crop. The problem is that such trials are not readily replicated in a statistically acceptable way, for they are large, unwieldy and inconvenient. What was recorded may not, therefore, be representative. Results are also open to alternative interpretations, some as a consequence of the method used and some relating to the underlying assumptions. This caveat would hold even in the face of sufficient replication. For example, the favoured interpretation may hinge on the assumption that the insects on the alternative plant species actually do constitute a single species, an assumption that is frequently enough violated in pest management (see Chapter 6). Even if laboratory tests seem to verify the field outcomes caution is necessary, for such results can still be misleading (see Chapters 6–9).

Ideally, Ecological Research results should feed into System Design and thus improve it. Even if this has not been the temporal sequence followed in the development of the IPM system, we should still think in terms of the logical dependence specified by the IPM chain (Table 4.1).

Although the 'chain' has been set up in this chapter as an abstraction, its utility is to help to identify areas of logical dependence (among other things) and thus improve IPM research and, ultimately, IPM performance in the field. When System Design drives the Ecological Research that is needed for that design, rather than being driven by the appropriate ecological theory, the two approaches are much more likely to be conflated. Questions relevant to ecological understanding are thus likely to be generated, but without them being underpinned by ecological or evolutionary understanding. Consequently, misinterpretations are liable to misguide application because the logical dependence in the chain has been short-circuited.

In some ways the procedure suggested by the above analysis, of ensuring that all relevant components are analysed in their appropriate contexts for potential problems, indicates that research can be integrated for IPM purposes. This is a means of satisfying the need identified by Dent (1991, p. 519) of placing specialist knowledge, abilities and skills within a broader scientific framework, for better coordination of effort and resources. But still, it is not appropriate to think of this as 'integrated research'; rather it is a sequence of specific research exercises aimed at solving a single, multifaceted problem. It is a breakdown of the problem into its subcomponents, and each of these could well entail true interdisciplinary research.

Conclusions

The analysis of IPM presented above demonstrates that we should conceptualise the related problems in terms of an IPM 'chain' (Table 4.1) if we wish to underpin IPM with a scientific approach and high quality information. Anything less demands luck. Each agricultural situation that has been targeted for IPM could be seen as presenting a suite of potential problem areas. Thus the expertise relevant to IPM is diverse and resides in a diversity of scientists. Not everyone involved in IPM is likely to be competent across all areas of biology (and the related disciplines involved) that are relevant to the IPM chain. The overall task will best be achieved through the appropriate cooperation in identifying potential problem areas and investigating them scientifically. Integrated pest management is therefore best seen not only as an array of problems, but as an array of problems that stand in a particular relationship to one another. Who is competent to hold the strands of any particular IPM enterprise together?

Are we educating people for such a complex, multifaceted task? Such issues are considered in Chapter 10.

Finally, the term 'integrated' is considered in relation to the IPM chain of Table 4.1. The discussion thus far suggests that 'integrated' has a rather specific meaning in the context of IPM, namely that only the techniques developed for IPM are integrated, and the integration is aimed at maximising the negative impacts on pest populations (Kogan, 1998). Specific research should be conducted in this area, but usually is not; Ehler & Bottrell (2000) have outlined how this could best be achieved in terms of both horizontal and vertical integration. By contrast, research to extend knowledge or to apply knowledge for the development of technology is not integrated, because it answers highly specific and directed questions (although they are of relevance to the IPM enterprise). This last point is an important one because IPM research is often seen, rather generally and vaguely, to be different from research of a purely biological or ecological nature. However, IPM research is diverse, it is not 'integrated' (although it should be coordinated, and interdisciplinary where necessary), should always be scientific in its approach, and, to be successful, should rely on the recognition of the problems in the IPM chain that warrant a scientific solution.

Books on IPM deal with many of the aspects outlined above, though not in quite the same way. Most attention in the IPM literature is focused on Components 6–9 in the IPM chain of Table 4.1. By contrast, the ecological and evolutionary research required for IPM is seldom treated in any detail. The rest of this book is therefore dedicated to dealing with the ecological and evolutionary considerations that are directly relevant to IPM research, not only because they are seldom considered in this context but also because recent advances in evolution and ecology need to be worked into the culture of IPM. To begin the process, Chapter 5 considers the suggestions currently available as to what research in the evolution–ecology sphere is required for IPM, and extends perceptions by incorporating recent advances in these disciplines.

An ecological underpinning for IPM

It should be apparent to all students of applied entomology that while pest-management practices may often be agreeably simple in design and application, their successful development must be based on rigorous understanding of the natural laws that govern the abundance of insect pests and regulate their interactions with all other phases of the living environment.

R. L. METCALF & W. H. LUCKMAN (1994, p. XIII)

Whether we like it or not, ecologists have to start giving hard answers to some very hard questions. Without a sound theoretical base for our subject, we will be unable to meet the challenges of the next decade and beyond.

J. H. LAWTON (1989, p. 517)

Introduction

Overview of ecology, and chapter structure

Scientific progress is not reliant simply on the collection of facts. Even the entomological research conducted to support pest management practice is influenced in one way or another by underlying theory (Chapters 2 and 3). Sometimes that theory is recognisable only by searching out the unstated assumptions that underpin gathering and interpretation of the data. All the underlying theoretical structures that influence research on pests and beneficials therefore need to be located, scrutinised for the validity of their underlying assumptions, and examined for their practical contribution to IPM (Chapter 2). That a strong theoretical basis for IPM research is needed is perhaps best underscored by our persistent inability to understand the ecology of so many major pests, despite decades of research effort (much of it of a fact-gathering nature), and the low rates of IPM success achieved so far (Chapter 3). Chapter 4 advocates an approach to IPM that relies on a general appreciation of which different research aspects are required to support the rational development of an IPM programme, and how those areas of research relate to one another (Table 4.1). Each of these research areas is, inevitably, served by a particular theoretical backbone. Such a linkage to specific theory has not yet been attempted for IPM, as the research needs have never been so explicitly detailed, and because of the confusing nature of ecological theory (see below).

Ecological problem solving is seldom discussed in pest management texts. Many publications detail projects that have been successful, and some cover outlines of the information deemed desirable for developing various control techniques. Such accounts provide valuable background, and those of Pedigo (1999) and Dent (2000) are comprehensive and detailed. But the theory necessary to underpin the various areas of investigation and which is needed to help to crack those problems not tractable by emulating successful cases is still not dealt with explicitly. In any case, IPM does not provide that research direction (Chapter 4). This chapter explores the theoretical side of applied ecology and demonstrates that it is more subtle than implied in many publications. This is largely because of the fragmented and multifaceted nature of the discipline, which adds substantially to the complexity of the situation.

Some IPM writings do provide an idea of the type of ecological information considered relevant to the development of IPM programmes, but they are not consistent among themselves in what is really required of ecological research. Part of the problem is that each one may be thinking of a different component in the IPM 'chain' (Table 4.1) when making suggestions. The issue is further blurred, however, by alternative approaches being available in ecology. Research that follows one approach yields different information and interpretations from research designed according to an alternative approach (see below). Ecological theory, therefore, is even of itself a complex discipline, as detailed in this chapter. This adds immeasurably to the difficulties in unravelling the direction in which research for IPM should be pointed.

In the sections that follow, the various recommendations for ecological research in a pest management context are assessed. Each suggested approach suffers shortcomings that have already been identified in the primary ecological literature. These shortcomings are given priority here, because it is far easier to find approbative literature that outlines the advantages, methods and conclusions derived from such research than it is to locate even legitimate criticisms. This approach is not inherently negative; the argument is developed in this way so that a reasonable alternative approach to studying the ecology of pest and beneficial species can be developed, and this is also presented. This alternative is not new; it is practised to a considerable extent, but has not yet been fully formalised.

The ecology issue is tackled as follows. The following subsections outline the ecological hierarchy and review statements made in a range of IPM literature sources about the ecological information that is deemed necessary to underpin management programmes. The recommendations that are made relate quite neatly to the different levels of the 'ecological hierarchy', namely populations (and metapopulations), communities and ecosystems. Nevertheless, the overlap is considerable and some individuals call for research in more than one of these categories, even if their stated intentions seem to be the service of just one component of the IPM chain (e.g. Price & Waldbauer, 1994). The explanation for such duality (or pluralism) relates to ecological systems being viewed, by most ecologists, as hierarchical structures. This view implies that understanding at each level of the hierarchy is required if a complete picture of the system and the means to manipulate it is to be developed. By contrast, much of the research that is actually conducted in relation to IPM falls outside of these hierarchical categories and is best described under the heading 'species studies' (although this is not an inviolate rule).

Each subsequent section of the chapter relates to a particular level in the ecological hierarchy, in the order mentioned above. These are followed by a section that relates to understanding ecology from the perspective of species, which is termed autecology. In general terms, the suggestions for research that are made are assessed in relation to the particular components of research required for the IPM chain (Table 4.1), at least where such correlation between research and practice is clearly intended in the original. An important aim of this chapter is to resolve the discrepancy between the suggestions for ecological research, which relate to the ecological hierarchy, and the practice of IPM researchers, who tend to conduct work on species-related phenomena.

The ecological hierarchy

The lowest level at which ecologists theorise is generally the population. Although individuals make up populations, and would therefore represent a lower level for ecological generalisation, the explicit inclusion of individuals in ecological theories has generally been ad hoc and unsatisfactory (Walter & Zalucki, 1999). The more recent attempts (Lomnicki,

1988; Sutherland, 1996) have simply interpreted the behaviour of individuals in relation to the dictates of population models, rather than theory being developed from the perspective of the properties of individual organisms (Walter & Zalucki, 1999). Individuals, and their ecologically related behaviour, are more prominent in the research that concerns species, and so is considered there.

Groups of populations (of different species) that live in the same place (i.e. sympatrically) are seen to constitute the ecological community (Price & Waldbauer, 1994). Although community theory is independent of population theory in some ways, the two have been welded into a fairly coherent whole, which constitutes the demographic paradigm in ecology (Hengeveld & Walter, 1999; Walter & Hengeveld, 2000). However, in the current book community aspects are dealt with separately from population ones, in the third section of this chapter. In the last few decades, the ecosystem has come to be seen as representing an ecological level just above that of the community, with its separate status being conferred because the ecosystem incorporates the physico-chemical features of the environment (Price & Waldbauer, 1994). Demographic ecology tends to de-emphasise physical and chemical factors, with its focus on demographic influences like predation, competition and so on. Physical influences are, however, also included in broad terms in the form of the disturbances that disrupt the postulated equilibria of populations and communities. However, ecosystems are dealt with separately in the fourth section.

The fifth section considers species and individual organisms. These two components are dealt with last because they are usually not included in the ecological hierarchy (e.g. Eldredge, 1985, p. 165; Price & Waldbauer, 1994), except that species names are seen as labels for the various populations that are present. Although species are sometimes inserted into the hierarchy (e.g. Beeby, 1993), ecologists do not have well-developed theory to deal explicitly with species or with the individuals that make up species (Hengeveld & Walter, 1999; Walter & Hengeveld, 2000). 'Species-level' pursuits are, however, frequently seen as the material of the autecology paradigm. Autecology is given a different context by different ecologists. Some see it as one of several component pursuits open to ecologists (e.g. Gaston & Lawton, 1988), and it is frequently enough portrayed simply as natural history and therefore somewhat unscientific, albeit useful in applied ecology (e.g. Kareiva, 1994). Others consider that autecology constitutes an alternative paradigm in ecology (e.g. Cittadino, 1990; Hengeveld & Walter, 1999; Simberloff & Dayan, 1991; Walter & Hengeveld, 2000). The latter view is justified in the various sections that follow. Even if one opts for the view that autecology is just one of several legitimate components of ecological theory, the fact remains that it still has no generally recognised formal theoretical structure.

The following subsection demonstrates that each proposed level in the ecological hierarchy is seen, by different individuals, as the legitimate ecological steward for IPM research on pest and beneficial organisms. This justifies dealing independently with each of these levels in the subsequent sections of the chapter.

Suggested theory for IPM research on insects – an overview

Most ecological writings that claim a link to pest management (Clark et al., 1967; Flint & van den Bosch, 1981; Pedigo, 1999) call for research on the population dynamics of pest species that require IPM attention. Most of these authors also refer to ecosystems (e.g. Flint & van den Bosch, 1981; Pedigo, 1999), presumably because the agroecosystem is considered the 'unit' or 'aspect' that is subject to management, or that has to be changed for management purposes. However, Price & Waldbauer (1994) claim that a consideration of the various levels in an ecological hierarchy brings us 'by logical steps, to the only conclusion possible for understanding insect populations: the need for studying them at the ecosystem level, not forgetting that all components of the ecosystem provide insight into the basic mechanisms involved in insects and agricultural ecosystems'. Kogan (1998) has pointed out the discrepancy between most successful IPM programmes' having been implemented with little consideration of ecosystem processes but with reference to the ecological foundations provided by species and population ecology. Indeed, relatively few authors call directly and solely for an ecosystem-level understanding for pest management purposes. Those who do feel that agroecosystems can be characterised by a limited set of dynamic properties and 'hence can be employed in the design and evaluation of agricultural development projects, at all levels of intervention' (e.g. Conway, 1987, p. 96). Community ecology has also had relatively few protagonists for its incorporation into IPM theory. Mostly it has been recommended as a basis for improving pest management techniques, predominantly in biological control. Such calls have been made mainly by scientists who are primarily ecologists (e.g. Mills, 1994a,b) rather than pest management entomologists. The suggestion has, occasionally, also been justified by researchers with a primary interest in pest management (e.g. Liss et al., 1986).

In discussing the various influences of ecological theory on research for pest management purposes, no role has been specified for evolutionary interpretations. Evolutionary theory has long unified biological interpretation and provides the basis for interpreting the function of biological mechanisms, such as pheromone systems and polyphagous host relationships, and for understanding how they operate and how they should be investigated. The population and community levels of the ecological hierarchy are linked explicitly to evolutionary interpretation. The demographic principles that underpin these two areas of ecology are, not surprisingly, consistent with the demographic leanings identifiable in some areas of evolutionary biology. For example, population ecology approaches are frequently aligned with the view that sees natural selection as a process that optimises fitness continuously (or, at least, until halted by some form of constraint (e.g. Godfray, 1994)). Community ecology, on the other hand, sees species responding as units to ecological opportunities, for example to partition resources among one another (e.g. Ricklefs, 1987). These evolutionary views have all been criticised for emphasising relatively trivial variation, omitting more important patterns and invoking processes that could not operate in natural systems (Hengeveld & Walter, 1999; Masters *et al.*, 1984; Paterson, 1981; Rapport, 1991; Walter, 1991, 1995b; Walter & Hengeveld, 2000). For various reasons, these approaches to evolution and ecology have been criticised for their idealism (Hengeveld, 1988; Simberloff, 1980; Walter, 1993a). Autecological theory, by contrast, is logically consistent with evolutionary principles from which all identified vestiges of idealism have been removed, as holds for the Recognition Concept of species (Chapter 6). In discussing the ecological options for IPM research, an eye needs to be kept on these evolutionary connotations, an aspect that is pursued comprehensively only later (Chapter 6).

Population ecology, metapopulations and life systems

The assumptions that underpin current population ecology theory can be traced to a time when a perceived 'balance of nature' steered ecological interpretation and research direction (Worster, 1977). The basis of 'population regulation' was therefore sought. If one could understand such 'regulation' sufficiently well to influence it, management of populations could be achieved through the application of general scientific principles. The approach was enticing, but has not yielded the promised application, either in pest management or elsewhere. Successes in pest management have been achieved despite the 'laws' outlined by Turchin (2001), not because of them. Although this suggests problems with the underlying theory (Andrewartha, 1984), the approach still has many persuasive advocates. For this reason the other obvious problems with current population ecology theory are outlined below.

1   No consensus on what 'population regulation' actually means has emerged. The conceptual and practical problems are detailed by den Boer & Reddingius (1996), White (2001), Murray (1999), Chitty (1996), Wolda (1995), den Boer (1990a), Krebs (1995) and Murdoch (1994). Calls to ignore the meaning and concentrate on investigation (Turchin, 2001) may sound pragmatic but are likely to retard progress in developing sound theory in this area.

2   Too frequently the postulated 'population equilibrium', which is considered the consequence of regulation, is simply an average of densities estimated at different times. All such means do not actually represent anything real – they are human abstractions to which we should not attach particular biological significance (White, 2001). Nevertheless, the search for the processes postulated to be significant in the stabilisation of particular systems continues (e.g. Murdoch, 1994; Murdoch *et al.*, 1996b).

3   A debate central to population regulation involves the conceptual division of ecological influences into two categories.
    Density-dependent influences are biotic factors such as predation and competition that are theoretically capable of regulating populations because their impact increases proportionately to population density. Although density dependence is sometimes measurable in populations, it is too sporadic and spatially heterogeneous an influence to be seriously considered as a primary determinant of population numbers or any perceived stability in populations (Cronin & Strong, 1994; Dempster, 1983; den Boer & Reddingius, 1996; Stiling, 1987, 1988). An alternative suite of influences, mainly abiotic ones, is seen to act independently of density (although even this is questionable (Chitty, 1996, p. 170)). This conceptual dichotomy has sustained a long debate that remains unresolved, and population ecologists now tend to claim that both categories of influence are important (e.g. Lawton, 1991; Turchin, 2001). An alternative view is that the conceptual division and the discussion are centred on inappropriate underlying premises (Hengeveld & Walter, 1999; Walter & Hengeveld, 2000; Walter & Zalucki, 1999; White, 2001).

4   Several debilitating problems with the logistic and Lotka–Volterra population models, which formalise population and metapopulation

ecology theory (see Turchin, 1999), have long been publicised (e.g. Andrewartha & Birch, 1954; Heck, 1976). In short, logistic theory has restricted application and lacks the generality originally attributed to it. None of the criticisms have ever been satisfactorily addressed by proponents of the models, and now tend to be ignored.

5   As originally constituted, the theory included no spatial aspect. Indeed, typical studies claim their subject organisms do not move beyond the bounds of the study area (Southwood *et al.*, 1989; Varley *et al.*, 1973), but that is not correct. Even the winter moth, selected for population dynamics study because of its ostensibly sedentary life style, has first-instar larvae that redistribute themselves on an impressive scale (Holliday, 1977). Although metapopulation theory was designed to overcome this deficiency, its emphasis is still on the maintenance of an overall equilibrium or balance and it still relies on the logistic equation (see Hengeveld & Walter, 1999; Walter & Hengeveld, 2000).

6   The emphasis in current population dynamics theory is centred on numbers of organisms as a central influence of future population numbers. However, empirical studies indicate that the major influences on population numbers, even in studies purporting to show strong density dependence, are almost invariably such things as environmental factors, or nutritional availability and quality (e.g. Varley *et al.*, 1973; Webb & Moran, 1978; White, 1970a,b). Timing of climatic events is often crucial to providing conditions for population increase or decrease (e.g. Holliday, 1985; White, 1969).

7   The number of organisms in any place at any time is dictated not by numbers or postulated number-related influences but primarily by the interactions of individual organisms with the factors (physico-chemical and biotic) that influence them directly through their adaptive requirements and tolerances (Hengeveld & Walter, 1999; Walter, 1988b; Walter & Hengeveld, 2000; Walter & Zalucki, 1999). Some efforts have been made to adjust theory to correct for this misalignment (e.g. Hassell & May, 1985; Lomnicki, 1978; Price & Waldbauer, 1994; Sutherland, 1996), but these attempts have concentrated on the incorporation of variation among conspecific individuals into existing population dynamics models and they by-pass the species-specific adaptations that explain why species are ecologically idiosyncratic. This is not to deny other ecological influences, but such factors as predation, parasitism and competition are essentially secondary influences of any primary dynamic set up by the individual–environment interaction (Walter & Zalucki, 1999), with the last-mentioned being particularly weak because it is so sporadic in

nature (e.g. Connell, 1983). Ad hoc adjustment to theory is unsatisfactory when that theory is deficient at a more fundamental level, which seems the case with population ecology. Furthermore, emphasis on the variation among conspecifics does not reflect the fundamental properties of the individuals of a species, and consequently is inadequate as a basis for developing robust generalisations in ecology (Walter, 1995b).

8   If one focuses on individuals as a basis for understanding species' presence and abundance in a specified locality, the concept of population inevitably changes. The 'population' now becomes the end product of various interacting influences that affect the physiology, behaviour and survival of individual organisms. In other words, population ecologists have concentrated on an end product as their basis for defining principles or workable generalisations. The error is compounded when the population consequences are then taken as the cause (or selective advantage) of the individual behaviours. Logically, the population ecology approach is flawed for at least two reasons. First, a diversity of interacting influences is now known to contribute to changes in population numbers. Second, end products are effects, and scientific explanation is appropriately based on cause or mechanism (Hull, 1988, p. 44; Krebs, 1995; Paterson, 1985; Williams, 1966). Causal explanation is particularly important in biology, because adaptation underlies the operation of biological systems (Walter & Paterson, 1995). Adaptation, in other words, should form the basis of our generalisations in ecology because that is where we will detect repeatability, which makes for the most robust generalisations possible.

Efforts have already been made by population ecologists to incorporate adaptation, through the *life system* approach advocated initially by Clark *et al.* (1967). However, the connection between adaptation, environment and abundance is left rather vague. Adaptation is covered simply by the phrase 'inherited properties of individuals of subject species (genotypes)' (Clark *et al.*, 1967, pp. 6, 58). The inherent properties of individuals are seen to interact with the effective environment as *co-determinants of abundance* (Clark *et al.*, 1967, p. 58). The co-determinants achieve two things relevant to population dynamics: they control the life functions of individuals and thus reproduction and immigration, and they influence the circumstances of premature mortality, reduced reproduction and emigration. Nylin (2001) has recently pursued the issue further and relates to one another the theories of life systems, life history strategies and population dynamics. These bodies of theory intersect through their common link to life history traits and the numerical values estimated for them. These traits describe the life cycle of the organism rather than how the organism is shaped, how it works internally, or how it behaves. Although the morphological, physiological or behavioural traits influence or even determine the life history traits, the former are not the primary focus in life history theory because the latter quantify the transitions between different parts of the life cycle (Nylin, 2001). The properties of individual organisms are thus by-passed, although it is these properties that are the primary influences on the ecological interaction with the environment.

The life system approach is said to be one of the most common and successful approaches to field population studies (Jones & Kitching, 1981, p. 2). In many ways, aspects of that approach are incorporated, by default perhaps, in a large proportion of studies that set out to understand what is essentially a question of presence and abundance. The advantage to the life system approach is the effort to include animal movement in studies of abundance (e.g. Hughes, 1981; Kitching, 1981; Zalucki *et al.*, 1986). Consequently, interpretations of abundance relate to a more realistic scale, the species distribution (Zalucki *et al.*, 1994), although movement on a smaller scale has also been identified as a significant influence on presence and abundance at a locality (e.g. Jones, 1981; Zalucki & Kitching, 1982). However, the intellectual debt of these last-mentioned studies to the life system approach is not explicit, so other influences may well have been important in such a shift in emphasis.

The life system approach to population dynamics faces another major difficulty, its explicit failure to deliver generalisations (Jones & Kitching, 1981, p. 3). This shortcoming has several causes. First, the divergent inputs into the approach appear to represent an ad hoc combination with a basis in pragmatism rather than fundamental principle. Practical requirements for pest management purposes demanded information on particular species, and the ad hoc approach to gathering such information has simply been incorporated into population dynamics theory. Although population dynamics theory is fundamentally deficient (see above), its central presence in life system research maintains expectations that generalisations in ecology may be developed to mimic the achievements of physics and chemistry. As hinted at above, fundamental principles about biological systems must derive from a realistic understanding of evolution, and recent evolutionary theory reveals that population ecologists have been overly optimistic in this regard (see section on species in this chapter, below).

Because fundamental principles that relate to individual organisms, their species-specific properties and their spatial relationships are absent from the population dynamics and life system approaches, the associated theory remains unconsolidated. The whole is a rather expansive and amoeboid-like construct that is brought into service in too many ways and for far too many purposes. Ecologists therefore find it easy to slip from individuals to populations or species, in whatever combination, and in a very loose way. The relationship between each aspect remains unspecified and unclear, and the relationships of each to the spatial dynamics of the environment are left vague. Such lack of clarity in the general literature on what constitutes an alternative theory is unsatisfactory. Community theory also suffers such shortcomings (Walter & Paterson, 1995). Fundamental theory in ecology needs to be clarified and made stable enough to endure rigorous testing among alternatives and to have clearly specified internal relationships.

Any re-organisation of ecological theory will require the nature of adaptation to be incorporated realistically. In the life system approach, for example, little direction is available on how to deal with adaptation. For example, which inherited properties are the significant ones? Would such alleles as those for insecticide resistance be given top priority, or sole priority? And how would such complex inherited properties as host plant relationships be included in a life system model? In some cases the answer may appear to be straightforward, especially when one is dealing with monophagous species, because the individuals are adapted to the single host species on which they oviposit, feed and develop. Although such simplicity gives an impression of certainty, the other aspects of the organism's ecology require explanation, and invariably they are not simple.

The extent of the problem becomes clear when one considers species whose adaptations are not so clearly evident (if, indeed, any species' adaptations can be considered so transparent). Again, most polyphagous pests and beneficials provide good examples of species whose primary adaptations are not obvious. The real issue, then, is understanding adaptation. To do so we need to consider the *functional* aspect of the adaptation that is relevant and how it was *acquired*. To understand these features we need to know more about species and how species evolve, both of which are considered later (Chapter 6). At this point it is enough to remember that individuals acquire their characteristics (or adaptations) through their derivation from a particular gene pool, to use population genetics terms. The concept of 'species gene pool' (Ayala, 1982) links individuals to species most realistically (Paterson, 1993c, p. 2) and thus also provides the link between species and the ecological setting (Walter, 1995b; Walter *et al.*, 1984).

Before moving to species, we must consider ecological theory that is aimed at higher levels than populations – first communities and then ecosystems.

Community ecology

Community ecology was developed originally as an alternative to autecology (or 'species-level ecology') (Cittadino, 1990). Only subsequently did Schröter & Kirchner (1896, 1902) develop the classification of ecologists' activities that ultimately led to the hierarchical view that effectively apportioned ecological theory into complementary subdisciplines (McIntosh, 1985).

Modern community ecology has several interrelated aspects, covering community structure and guild structure, food web theory, succession, and so on (Price & Waldbauer, 1994). Again, virtually no restraint is exercised on what is included in community ecology, or what is excluded from it. Typically, community ecologists are expansive in claiming credit for ecological insights, despite the obvious derivation of those insights in other areas of ecology, as discussed by various authors (Lewin, 1986; Walter, 1993a; Walter & Paterson, 1995; Young, 1986). Most of the examples cited in support of using community ecology theory to improve pest management are ones that were discovered or developed by individuals who operated outside of the community ecology or ecosystem paradigm. For instance, Price & Waldbauer (1994) recommend use of island biogeography theory, a branch of community theory, as a prop for pest management research because 'we can profitably consider a crop field an island in a matrix composed of a mosaic of land types'. They continue: 'Its impact on the development of the whole insect community cannot be overrated, and we should keep in mind the possibilities available to us for manipulation of this matrix for the improvement of insect pest-management practices'. Their illustrative example involves '[o]ne enlightened use of wild plants' for maintaining populations of grape planthopper parasitoids in vineyards when the pest species is in diapause. Consultation of the original paper (Doutt & Nakata, 1973) demonstrates that these authors were working on their understanding of species ecology, not island biogeography theory. If the prime illustration of the utility of a theory comes from out-side of that theory, good reason should be provided for not simply employ-ing the theoretical background that generated the useful result in the first place.

The history of community ecology is, accordingly, exceedingly com-plex (see e.g., McIntosh, 1985). Nevertheless, the primary justifications of the validity of community ecology can be established, and were at least threefold.

1   Particular groups of species regularly occur together, and were originally seen to persist as a guild or community for long periods.
2   Malthus's doctrine specified that resources are finite, so the questions of how sharing of resources and the balance of nature were maintained became central issues in ecology, although those self-same questions had had definite theological origins (McIntosh, 1985; Walter, 1993a; Worster, 1977).
3   Competition for resources was seen as a ubiquitous driving force in nature (Darwin, 1859).

Community ecology has changed in response to various criticisms, to the extent that a 'new ecology' seems to appear with a periodicity of 10–15 years (McIntosh, 1987). However, the fundamental flaws remain.

1   'Communities' of species are temporary. This is especially evident when one considers time spans of thousands of years (e.g. Coope, 1978; Gleason, 1926, 1939; Pielou, 1991). This statement is also true, however, of periods less than that of the human lifetime, and such change is driven primarily by climatic fluctuations (e.g. Ford, 1982; Graham et al., 1996; Hengeveld, 1990; Huntley, 1991; Kitaysky & Golubova, 2000; Parmesan et al., 1999). The implications of such evidence for the theory of community ecology have been spelled out by Walter & Paterson (1994, 1995). Essentially, communities do not exist as entities in nature. They cannot, therefore, be specified objectively and unambiguously, so robust ecological generalisations cannot be made at the community level. Generalisations should be sought at a lower 'level' of biological organisation.
2   Interspecific competition is neither a ubiquitous ecological force nor a strong one (e.g. Andrewartha & Birch, 1954; Connell, 1980; Walter, 1988b; Wiens, 1977). Consequently, interspecific competition is not a universal evolutionary force, and adaptations attributed to the influence of competition are more realistically explained otherwise (e.g. Andrewartha & Birch, 1954; Heck, 1976; Simberloff, 1976, 1980,

1982, 1983; Walter, 1988b, 1991, 1995b; Walter *et al.*, 1984; Walter & Paterson, 1995). Although the points above have been long acknowledged, all that has really happened (in this 'new' pluralistic ecology) is that competition is now seen as only one of many ecological influences that may operate (e.g. Price *et al.*, 1984). Unsurprisingly, such a view has been part of ecology since its inception (e.g. Darwin, 1859) and even played a crucial role in the development of interpretations of coexistence theory (e.g. Hutchinson, 1948). Explaining coexistence is simply explaining why competition between similar species is so weak that it does not exclude one or other of them (Walter, 1988b). This tells us that predation, ecological disturbance and the other factors that have been incorporated into pluralistic community ecology have all been part of coexistence interpretations for a long time. As a theory, pluralism is methodologically weak because observations cannot be challenged (Simberloff, 1990), and because the ecological or evolutionary cause attributed to any observed phenomenon is tied to a 'principle' that is superficial and, therefore, unenlightening. Principles may well be simple, but they should never be so trivial as to state only that many influences are important in ecology. Such a statement warns that principles are being sought or abstracted at an inappropriate level.

3   The species included in 'communities' or 'guilds' are specified as members by the observer, on the basis of which species use a common resource. A typical example would be the species of insects that feed, as larvae, in fallen rainforest fruit (Atkinson, 1985). The group of species thus circumscribed is considered to have at least the potential to interact competitively among themselves (e.g. Abele *et al.*, 1984). Because most organisms use a range of resource types in any one category of resources (e.g. food or shelter), and since the food types consumed by one species are unlikely to overlap conveniently with the food types consumed by other guild members or community members, abstractions of communities based on resource use are inevitably overly restrictive caricatures of complex situations, even in such 'generalist' species as the drosophilid flies that feed on fallen rainforest fruits (van Klinken & Walter, 1996).

4   In community ecology, species are dealt with as interacting units, which is a form of typology (Hengeveld, 1988; Walter, 1993a, 1995b) deemed unsatisfactory by evolutionists (e.g. Mayr, 1982). Among other problems, such an approach tacitly invokes a form of natural selection called group selection. For group selection ever to have a significant evolutionary influence would require conditions that usually do not occur in nature (e.g. Maynard Smith, 1966; Williams, 1966).

How does one overcome the deficiencies outlined above? Logically, the only possibility that remains is to derive an objectively defined entity from among the dynamic complexities that make up natural systems. The fact that species are self-defining in terms of their sexual behaviour (see Chapter 6), respond to climate independently of other species, even those with similar requirements and those which co-occur, and use resources that do not map conveniently against those of co-occurring species suggests the level at which we may expect ecological generality is the species defined in population genetics terms. This line of reasoning is justified and pursued further after a consideration of ecosystem theory has been covered.

Ecosystem theory

Several influences led to the development of the ecosystem concept. Foremost, perhaps, was the distinction frequently drawn between the living and non-living attributes of nature, and the fact that the physical and chemical aspects of the environment, although acknowledged and studied, were not explicitly incorporated into population and community ecology theory. That was changed by Tansley (1935) and Lindeman (1942), who engineered the systems approach to dealing with the dynamics of nature. They were reacting to Clements' (1916) superorganism concept of community that was popular from its inception, and which still influences ecological practice (McIntosh, 1995; Simberloff, 1980; Walter & Paterson, 1994). Tansley (1935) thus saw himself developing an *alternative* concept to that of community, one that he considered *reductionist* in nature. Ironically, community ecology and ecosystem ecology have become quite strongly aligned (Price, 1984; Price & Waldbauer, 1994), although interpretation of ecosystem dynamics may well be more reliant on understanding the population dynamics of the species principally involved in the energy or chemical cycle of interest (Kokkinn & Davis, 1986; Peters, 1991, p. 17).

Ecosystem ecology was imported into agricultural entomology as a means of understanding the dynamics of a human-controlled situation, the agroecosystem, and then using the insights to alter the system in favourable ways (Croft, 1983). Of particular relevance to the agricultural situation is that ecosystem ecology, much like community ecology, set out to explain the dynamics of a subjectively designated entity. The entity was essentially spatial (McIntosh, 1985, p. 198), so it is not surprising that the relatively small bodies of water studied by limnologists featured prominently in early ecosystem studies. Central to the problem was quantification of transfer rates of matter, nutrients or energy within the spatial boundaries identified by the ecologist. However, relatively discrete boundaries are not typical in natural systems above the level of individual organisms. Even Forbes (1887), who pioneered lacustrine studies because of their perceived microcosmic nature, established that the lake border was wide open to nutrient transfer. In his study, vast amounts of pollen fed the lake (Bocking, 1990). To a large extent species are irrelevant to interpreting ecosystems as it is the energy transfer itself that is deemed important, not how it occurs (McIntosh, 1985, p. 198), but it has been the reduction of ecosystem function to units of energy transfer that has been elusive. 'Energy does not have the unifying power that was once attributed to it and, besides, treating the components of an ecosystem as non-varying participants in a mechanistic model may well obscure the very information that is most valuable to biologists' (Kokkinn & Davis, 1986).

That farms are also relatively small spatial units probably encouraged a belief in the utility of ecosystem theory. But two obvious features of insects seem to have been downplayed, quite consciously, by agroecosystem analysts. First, all insect species have at least one stage in the life cycle that is motile, capable of covering considerable distances, and doing so regularly enough to disrupt most other patterns of population change. Those who do offer solutions to this problem suggest only that 'we must define our ecosystem for study as a similarly extensive area' as that from which the invaders come (Price & Waldbauer, 1994, p. 37). Again, the migration feats of pest insects were discovered not by ecosystem-level studies, but by entomologists concerned with a particular pest species (e.g. Doutt & Nakata, 1973; Kisimoto, 1976; Rainey, 1989; Taylor, 1986). Redefining the spatial extent of the ecosystem to match the sphere of operation of invaders of one species seems realistic at first consideration. But it does beg the question of what other ecosystem properties should also be incorporated over the expanded area that is to be taken into consideration. If nothing more than the migrating insects and their ecological influences are to be included, we have effectively achieved nothing that is not already being taken into account by the entomologists concerned with those pest species. Use of this example again suggests that the justification of a 'higher-level' ecological approach is being made under false pretences, and that such approaches have had no real impact in pest management. Furthermore, the method of designating the scale of agroecosystems advocated by Price & Waldbauer

(1994) has a remarkable species-specificity to it. A species that migrates from across a continent suggests the agroecosystem is massive, whereas a species associated with the same crop, but from a nearby source, suggests an ecosystem of different dimensions. That both species may be pests on a single crop simultaneously further undermines the solution proposed by Price & Waldbauer (1994). Since the dimensions in each such case are dictated by a given species, the logical practical solution would be to deal with the pests on a species by species basis; no information would be lost, and unwarranted expectations would not be generated.

Second, the pests of agriculture do not behave like the other major components of agricultural systems. Most farm inputs are well under human control, including the cultivar that is grown. Perhaps this relative constancy, and the common perception that nature is not only there to be controlled, but is so controlled (see Worster, 1977), led analysts to ignore the fact that boundaries (whether farm, field or political) are irrelevant to insects. In any case, even recent pest management literature contains statements to the effect that 'there can be little doubt that the transformation of ecosystem to agroecosystem produces well defined systems of a cybernetic nature ... There is a strengthening of the bio-physical boundary of the system, a bund is created around the ricefield, for example, which makes the boundary less permeable. The basic ecological processes – competition, herbivory and predation – still remain, but these are now overlaid and regulated by the agricultural processes of cultivation, subsidy, control, harvesting and marketing. At least in cybernetic terms, an agroecosystem defined in this way is more similar to an individual organism than it is to a natural ecological system' (Conway, 1987, p. 96). That such claims were made, without justification, even after the migration feats of rice planthoppers in Asia had been made widely known (Kisimoto, 1976) is surprising.

The connection between an understanding of the ecology of specific pests and the proposed ecosystem approach to solving pest management problems is, therefore, extremely tenuous. In his summary of 'hard systems analysis', which is part of the planning of an IPM system, Tait (1987, his Table 7.1) includes all aspects of pest ecology under the rubric 'system description'. What information is needed to achieve such an end? Specific direction is not readily found in the literature, but the questions that are stressed in relation to agroecosystem research tend to be of two types.

First, 'How stable are agroecosystems?' ask Woolhouse & Harmsen (1987). The issue here is that system stability is considered a desirable attribute for agricultural purposes, and is an issue pursued by many concerned with theory that relates to biocontrol. Although any perceived stability of such systems is not a fundamental biological property, the idea has long persisted since arising in connection with a 'balance of nature'. And it has been linked functionally to species diversity in the system (e.g. Pimm, 1991). The belief that increased diversity leads to greater stability of systems persists (e.g. Price & Waldbauer, 1994), despite demonstrations of the fragile nature of this proposed principle or generalisation (Grimm, 1996; Grimm & Wissel, 1997).

Second, Conway (1987) has suggested that agroecosystems can be characterised by a limited set of dynamic properties that not only describe their essential behaviour but also can be used as criteria of 'agroecosystem performance'. Perhaps a typical question that might be developed under this mantle would be: 'What dynamic properties describe the essential behaviour of an agroecosystem?', but no hints as to how to achieve this result are presented. The problem seems to lie in the ecosystem concept being associated with ideas of balance and stability. Pest management actions are thus seen to be taken 'to restore, preserve, or augment checks and balances in the system' (Flint & van den Bosch, 1981, p. 108; see also Price & Waldbauer, 1994). But what is a 'check' or 'balance', and how does one go about measuring such aspects of the environment? No practical hints are given by Barfield & O'Neil (1984) when they claim: 'The boundaries of components of the system and the couplings among them must be identified before design and implementation of an IPM programme'.

Finally, research conducted at the ecosystem level returns results that suggest that different species respond individualistically to any alteration in the system, which is the same aspect that undermines the development of realistic generalisations in community ecology. For example, responses to elevated CO_2 levels tend to differ across species (Pitelka, 1994) and the overall response will depend on the relative numbers of individuals of each species that is present. Practical solutions for pest management have not been generated from ecosystem theory and again the idiosyncrasy of species is evidently part of the problem. Overall, it does seem time to bury the ecosystem concept (O'Neill, 2001).

Species, individuals and autecology

Three general points are developed in this section. The first details briefly the ways in which species are incorporated into aspects of the ecological hierarchy, as well as the inadequacies of this treatment. Second, the role of species studies in pest management is outlined, together with the way in which species names are used as a 'hook' on which to hang information relevant to their ecology and to pest management. Third, an alternative approach to ecology is outlined. Autecology works from the perspective of individuals, their ecologically important properties (which are derived asexually or sexually through their 'membership' of a species gene pool (Chapter 6)), and their interactions with the environment. 'Populations' and 'communities' thus represent the consequences of these most basic of ecological interactions. They are thus seen not to be entities or levels of ecological organisation, so no additional theory is needed to explain their presumed structure. And their postulated structure does not feed back inexorably to determine the actions and evolution of the constituent individuals and species.

Autecology was initially articulated over a century ago (see, e.g., Cittadino, 1990), but a full theoretical development has been proposed only recently. The utility of autecology theory for pest management is a point developed in the second part of this book, for autecology justifies the species-based approach already taken by so many involved in pest management, and it alone of all ecologically related considerations could be said to have driven most of the successes in pest management. In addition, it provides a theoretical basis for extending perceptions of ecology for pest management purposes, by helping to assess suggestions made regarding pest management by population and community ecologists, and by suggesting more appropriate alternatives.

Species in population, community and ecosystem ecology

The only way in which species are currently given prominence in ecological theory is when they are considered typologically or as units (Hengeveld, 1988; Walter, 1993a). For example, species *as units* are seen to fill a niche or to partition resources with other such units in a community, a view with distinctly group selectionist overtones (Walter, 1991). Species are similarly seen to play a role in a food web, food chain or ecosystem. The theoretical context for such views is community and ecosystem ecology, not species theory. Species theory explicitly rejects such an attitude (Paterson, 1986; Walter, 1995b), for reasons that will become clear in the following chapter.

Species in IPM-related studies

The study of species from the ecological perspective is the only approach that has yielded real dividends in pest management, for this has provided the various control options. Generalisations are not readily apparent when one considers the 'study of species' for pest management purposes, although the features of particular species, whether morphological, physiological or behavioural, are reported very commonly in the scientific literature. Frequently, this species-related work is referred to as 'natural history', even by those who approve of such studies (Rey & McCoy, 1979), which seems to undermine its scientific status. Aspects of the host relationships and of the sexual and other communication systems perhaps make up the bulk of such reports. The reports are frequently somewhat anecdotal in that they are seldom structured around any explicit theoretical generalisation, besides almost universally carrying the implication that the features documented are functional in one way or another. Not surprisingly, alternative generalisations about such functionality are available, and these are examined in the following chapter.

The 'species approach' remains without synthesis in ecology, which seems to be explicable only through the pre-eminence of demographic ecology. Consequently, information gathered about species seems to be considered somewhat peripheral to ecological theory, and this might explain the persistent and general decline of autecology since the beginning of the twentieth century. Indeed, emphasis on 'species' acquired a bad name as early as the nineteenth century because the natural historians who opposed evolutionary theory simply gathered observations about species and sought new species to describe. Because these so-called 'species mongers' were unscientific, any emphasis placed on species was viewed askance (Rehbock, 1983). Even species theory itself, which deals with species concepts and interpretations of species formation (speciation), has never quite recovered, as expanded upon in the following chapter.

A modern problem for species studies relates to the concept of functionality, mentioned above in this subsection. Functionality implicates natural selection. Many evolutionary and behavioural ecologists interpret natural selection acting as an optimising principle. Thus, parasitoid individuals of a given species, for example, are envisaged to compete among themselves to the extent that the most efficient of them passes on their optimising mechanisms disproportionately to the following generation, and thus 'relative efficiency' is seen to be the primary driving force in evolution (e.g. Godfray, 1994). Behavioural studies tend, therefore, to be channelled into investigations of relative efficiency, as this is considered to reveal the way in which behavioural mechanisms evolve. Such studies of ultimate causation are portrayed as more meaningful in relation to the development of generalisations than studies of the behavioural mechanisms themselves. These ethological aspects are dismissed as proximate mechanisms that are readily interpreted through descriptive studies (e.g. Waage & Hassell, 1982).

Because the conditions to which individual organisms are exposed are likely to differ among the various populations of a species, optimisation mechanisms are expected to differ among conspecific populations (e.g. Godfray & Hunter, 1994). As a consequence species and speciation processes are, at most, peripheral to evolutionary and behavioural ecologists. Some authors (e.g. Thompson, 1994) have written about species-defining characters such as the ones used in phylogenetic analyses, as opposed to locally acquired characters, but have not defined characters expected to fall into each category or why there should be a differential in this regard. One aspect that needs clarification, therefore, is the contradiction between the frequent reference to species in the ecological and pest management literature and the prominent view of demographic ecology that the significant evolutionary changes take place within the different local populations that are seen to span the geographical range of species. This is an issue that is tackled fully in Chapter 6, where species theory is detailed as a basis for extending autecological insights into IPM.

Autecology
An important, but little-appreciated, role for ecologists today is the development of a realistic theory of ecology that is consistent with the underlying principles of modern species theory (Walter, 1995b). The fundamental basis of ecology is provided by the inherited properties of individuals. Not all inherited properties are of equal relevance to interpretation of the ecology of individuals. Of primary concern are the complex behavioural and physiological adaptations that influence the conduct and well-being of individuals in their natural setting (Paterson, 1985, 1986; Paterson *et al.*, 1963).

Such complex adaptations, given a certain amount of pre-selection variation and a lesser amount of post-selection variation (Paterson, 1993b), tend to be species specific and species wide. These are the characteristics that influence the ecology of organisms, including pests and beneficial species. Since these characteristics are unique to each species, a point readily appreciated by those involved in pest management, species must be considered on an individual basis. The individuals of each species respond to the environment in a characteristic way, so their movements, local abundance and geographical representation (the species' distribution) is unique. Ecological generalisations, in the form of ecological theory, have to cope with this awkward property of organisms.

The environment of organisms comprises physico-chemical influences, which are primarily climatic or influenced by climate, and biotic conditions, which impinge mainly in the form of vegetation structure and the availability of host and vector organisms (Hengeveld, 1990). Individual organisms respond to particular subsets of the total environment, which defines their species-specific habitat requirements (Walter & Hengeveld, 2000). Organisms exist within an ecological landscape, which is the distribution of environmental factors across biogeographical space. It is essentially variable and ever-changing, and organisms respond to such heterogeneous dynamics principally by adjusting their position (motile organisms, including the propagules of sessile species) and/or physiology (mainly sessile forms) within the biogeographical landscape (Hengeveld, 1997). The result is the continuous shifting of individuals into more favourable localities and away from less favourable ones, so populations simply represent a statistical consequence of this individual behaviour (Fig. 5.1). The way in which these basic processes can be incorporated into a usable ecological theory has been developed by Hengeveld & Walter (1999) and Walter & Hengeveld (2000).

The primary message from the outline of autecology just presented is that we are concerned, in pest management, with individual organisms and that the individuals are referable to a particular species through the properties they carry. What is the precise nature of the relationship between individuals and species? How should we conceptualise the relationship? And how does one recognise and characterise the properties of a species that are of particular relevance to understanding its ecology, pest status and control? These questions are outwardly simple, but they delve deep into the subtleties that surround the understanding of species. Of particular relevance is that the individuals, whose behavioural and physiological properties we wish to work from, acquire those characteristics through their derivation from a particular species gene pool. That means that one has to appreciate what a species gene pool is, how that relates to the names given to species, and even how new species evolve. The origin of adaptations and new species is critical, for that tells us what we can expect of organisms in the field. Are they constantly undergoing adaptive change? If so, how does that affect their ecology and pest status? How does that, in turn, affect our interpretation and understanding of the ecology

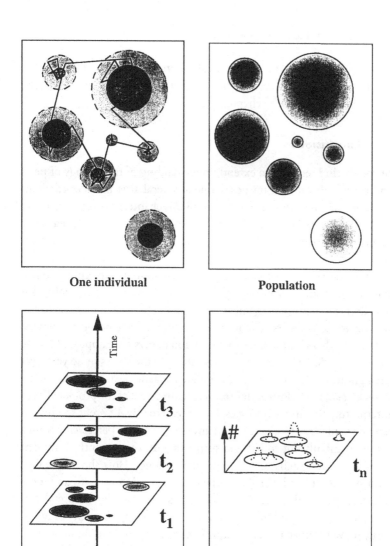

One individual   Population

Populations   Survival probability of populations

Figure 5.1. Diagrammatic representation of the 'metaphorical seascape' view of changing abundance and distribution of a species. Top left: A single individual moves across a landscape. When it encounters an area of suitable habitat it spends more time there (in this case by moving more slowly and turning more frequently). Top right: Consequently, the number of individuals (or population density) is greater in areas of suitable habitat, because the retention time of individuals is greater there. Bottom left: Suitable habitat shifts position through time, under the influence of environmental change. Consequently, populations shift with time as the individuals find localities in different areas that are suitable. Bottom right: Representation of the probability of survival of different populations at one time; the probability is an estimate of how long the environment remains favourable. Reprinted with permission from Kluwer Academic Publishers.

of those organisms? These are critical questions for understanding pest and beneficial organisms, and they suggest an entry into species theory, the topic of the following chapter.

Conclusions

In conducting research to extend understanding of the ecology of pests and beneficials for IPM purposes, one of several alternative research directions could be selected, as outlined in this chapter. A further difficulty is posed when various strands of ecological theory are intertwined, a frequent occurrence in publications that deal with ecological research for IPM.

This chapter has justified a species-based approach to ecology, but it cautions that such work is not going to be a matter of simply going out to collect the relevant facts. Substantial complexities exist in interpreting what characteristics are relevant to ecological understanding and how to investigate them. Remember that over 200 papers have appeared on the two pest species of *Helicoverpa* in Australia over the last 35 or 40 years, yet the significance of most results has still to be made clear (Matthews, 1999; Zalucki *et al.*, 1986). Furthermore, a substantial part of the problem lies in interpreting the limits of species, for we tend to mix different species under one name very frequently. And any such variation perceived within a species is usually not considered surprising, for the currently dominant view of evolution tells us to expect such changes. How should we deal with such complications? The problem is not straightforward and it entails conceptual subtleties. For these reasons, the following chapter delves deep into the background of species theory. It is a long chapter, but necessarily so. Without an understanding of each component of the subtleties, a logical interpretation of species is not possible, and as a result their ecologically relevant properties cannot be specified.

Chapter 6 is therefore designed to demonstrate just how complex are the theoretical and practical aspects that lie behind such outwardly simple statements as the one made by Flint & van den Bosch (1981): 'Identify the key pests; know their biology'. The implications for ecological research and IPM of the views developed in Chapter 6 are then followed up in Chapters 7–9, to illustrate how specific topics in ecology and pest management are influenced by a more realistic understanding of species.

Part 2

Specific directions in insect ecology research for IPM

6

Understanding species: good taxonomy, sexual species and pest management

It was a startling revelation that seven species with very distinct bio-logies had been confused under the taxon *Anopheles maculipennis*. The study of the *An. maculipennis* complex, more than any other, served to demonstrate the critical importance of cryptic (or sibling) species in ecology. 'Species sanitation' took on a new significance and a new com-plexity. The mysterious 'Anophelism without malaria', which had been such a feature of malaria in western Europe, was at last elucidated. It slowly became clear that the identification of ... species on morpholo-gical grounds is often not good enough. ...

H. E. PATERSON (1993a, p. 39)

And Einstein's theory, developed on what appeared to many of his contemporaries as a very frugal basis on which to found a physical theory, has revealed a richness of content that continues to baffle the imagination.

S. CHANDRASEKHAR (1990, p. 286)

Introduction

One of the most basic aspects in tackling any pest management problem is the initial identification of the species involved, whether it be pest or bene-ficial. A great number of expensive mistakes have followed identification failure. Fortunately, some of these have been well documented (Annecke & Moran, 1977; Clarke, 1990; Compere, 1961; Delucchi *et al.*, 1976), and a common call in the applied entomology literature is for 'good taxonomy' to underlie pest management. One of the main points made in this chapter is that good taxonomy on its own is frequently insufficient to solve the problem at hand.

In general, perceptions of species need to be extended a great deal if we are to approach the problem of identification in a more practical fashion. The major thrust of this chapter comprises a more expanded interpretation of the identification problem than is available elsewhere in the pest management literature. The expansion incorporates several nuances that are seldom clarified in relation to insect pests. In particular, species defined in terms of structural (or measurable) features need to be related to species defined in terms of their sexual behaviour. The strength of the latter approach is that it makes use of the way in which reproductive individuals themselves specify, in nature, their relationship to conspecifics. Species have therefore been described as self-defining (Lambert *et al.*, 1987), which is explained later.

Although the above summary may suggest that problem solving in this area is relatively straightforward, research on species and interpretation of empirical data suffers from at least three subtle difficulties, each of which is expanded in the body of the chapter. First, as in all scientific endeavours, the underlying theory affects the approach taken, methods used and final interpretation. Second, the study of sexual behaviour for the purpose of defining species usually has to be tackled indirectly because the appropriate direct observations are not possible on most insects in nature. Such observations and experiments are therefore likely to return outcomes that are ambiguous and demand further testing before resolution can be reached. To overcome this difficulty, method is given prominence within the chapter, but strictly in relation to theory. The third subtlety concerns the major consequences of the evolution of sex, both for the origins and interpretation of biological diversity, and for recognising species limits and classifying organisms. This explains why species that reproduce sexually have to be defined by criteria that are different from those used for entirely asexual organisms.

A study of the ecology of any species can be approached in various ways. One could seek what is believed to be additional facts about the organisms, as an empiricist would do (Chapter 2). Alternatively, understanding of the organism's ecology could be extended by testing alternative explanations of recorded patterns. In pest management research, the former prevails all too frequently, very likely to the detriment of progress in IPM. Ecological studies in pest management are further confounded by the agglomeration of the various approaches to ecology that are outlined in Chapter 5. Because the alternative approaches to ecological investigation are usually seen as legitimate complementary approaches, they are frequently all seen to be factual in some way and therefore entirely legitimate. Reasons are presented in Chapter 5 as to why ecological understanding should be built up from the study of individual organisms and their species-specific properties, which is essentially the autecological approach. Even if one does not agree that the autecological approach is a better alternative than other basic ecological theories, the question of species identity remains relevant to all ecologists.

In unravelling published interpretations of the ecology of any species, one should attack the problem through a process of *deconstruction*. The knowledge base should be meticulously pulled apart to find out what is claimed about the organisms of interest, how robust the data and associated interpretations are, and how the interpretations relate to current theoretical developments. Uncertainties are more readily exposed in this way and crucial questions can be identified. A crucial question is one whose answer ramifies through other areas of ecological interpretation regarding that species, and thus influences those interpretations. Issues of species identity are thus crucial to accurate ecological interpretations.

How easy is it to 'Identify the key pests' and 'Know their biology'?

The general problem: Good taxonomy in applied entomology
All ecologists and applied biologists would probably agree with Rosen (1978) that 'Correct identification and classification of organisms are essential for any intelligent interpretation of biological and ecological information', but few would be prepared to commit full attention to this central problem, despite repeated dire warnings in the scientific literature. A lax approach to identity threatens the rational basis that biology has provided for solving ecological problems. With regard to biological control, Sabrosky (1955) pointed out that some scientists wonder what difference the presence of undetected or misidentified species makes if control has been successful. His response to them is telling: 'Perhaps none at the moment, but we may have lost our best chance to know what we are doing, *to establish a sound understanding and scientific basis for our work*' (Sabrosky, 1955) (emphasis added). Clarke (1990) had good cause to cite these words in his critical analysis of what had been earlier referred to as a 'landmark' in classical biocontrol (see later).

The number of insect species is extraordinarily high, with insect taxonomy still requiring vast amounts of work (Oliver, 1988). Many groups of insects contain cryptic species. Also referred to as sibling species, they are virtually identical to one another in their morphology. However careful taxonomists may be in their morphological investigations, and even if forewarned about the presence of cryptic species (e.g. Annecke & Mynhardt, 1979; D. P. Annecke, pers. comm. 1978), they still may fail to detect or accept there is more than one species in a collection or sample.

Cryptic species are widespread in nature, including among insects of economic importance. For example, newly discovered cryptic species of heteropteran bugs, planthoppers and leafhoppers (Claridge et al., 1985; Claridge & Nixon, 1986; Gillham & De Vrijer, 1995), moths (Goyer et al., 1995; White & Lambert, 1994; Whittle et al., 1991), fruit flies (Baimai et al., 2000; Condon & Steck, 1997; Drew & Hancock, 1994), hessian flies (Makni et al., 2000), predatory beetles (Angus et al., 2000) and parasitic wasps (Atanassova et al., 1998; Fernando & Walter, 1997; Gokhman et al., 1998) are still frequently reported. Most species in a cryptic species complex are differentially associated with species of plant or insect hosts. In some cases, though, more than one cryptic species habitually feeds and breeds on the same host species (e.g. various planthopper species (Denno et al., 1987; Gillham & De Vrijer, 1995) and fruit flies (Clarke et al., 2001; Condon & Steck, 1997)). If a suite of sibling species is mistakenly perceived to be a single species, which undoubtedly still occurs regularly across all countries, an erroneous interpretation of the 'species' and its biology is developed. Furthermore, in parasitic and phytophagous associations, it is not unlikely that the host is also a complex, and this might explain some of the difficulties in certain weed biocontrol projects (e.g. lantana biocontrol).

Techniques for sampling and studying the 'species' may be developed and individuals from more than one species will be combined in ecological samples, behavioural observations or experiments. Disparate information from two vastly different species may thus be combined in studies of pest status, biocontrol potential, host specificity, geographical distribution, diapause, infection rates, activity cycles, the spread of insecticide resistance and so on. The severity of the problem for IPM becomes apparent when the ecology of pests that are difficult to manage is examined, as illustrated in the summaries of sugarcane black bug (Sweet, 2000) and cotton stainer (Schaefer & Ahmad, 2000) biology, ecology and pest status. The implications for pest management practice are serious, because each of the species within a cryptic species complex presents unique problems, demands and opportunities for pest management.

Despite the morphological similarity of one species to another, cryptic species are true species and no different in principle from species that express morphological distinctions. In particular, sibling species are not 'biotypes' or 'incipient species' that are in the process of evolving full species status, a point that is justified later. Each species in a 'complex' of cryptic species is almost invariably unique in aspects of their behaviour, physiology and ecology. For example, mosquitoes that are vectors for malaria parasites in Europe and Africa are frequently present in an area with mosquitoes that are virtually identical morphologically, or even indistinguishable at all stages of the life cycle, but which do not transmit *Plasmodium* and can therefore be ignored with respect to malaria control (Paterson, 1993a). If they were inadvertently included in control programmes then interpretation and management decisions would be compromised.

The economic consequences of such mistakes are considerable, as illustrated in example 1 below with reference to red scale insects on Californian citrus. Several other examples from different countries help to illustrate the ways in which identification problems manifest themselves. These examples illustrate the need to distinguish bad taxonomy per se from the undetected presence of cryptic species, because unresolved cryptic species complexes demand appropriately designed behavioural and population genetics investigations for their resolution. Research on cryptic species must be designed from a sound theoretical basis. A scientific rationale is available for this purpose and it also helps to pinpoint situations in which undetected cryptic species are likely to be present (Paterson, 1991) (see later).

Example 1: Parasitoids of citrus red scale in California
Since 1880 an aphelinid parasitoid (*Aphytis chrysomphali* (Mercet)) had been known to attack red scale insects (*Aonidiella aurantii* (Maskell)) in Californian citrus, after its adventitious establishment there, but it was generally considered ineffective. George Compere was sent to search in southern China, the native home of *Aonidiella*, for additional parasitoid species. His observations convinced him that red scale there was kept at low densities by a 'little yellow parasite', undoubtedly an *Aphytis* species. The species he observed was probably *A. lingnanensis* Compere, which was later to prove very effective in California. But L. O. Howard identified it as *A. mytilaspidis* (Le Baron), which was already present in California, but not on red scale insects. He also incorrectly identified *A. chrysomphali* in the

USA as *A. mytilaspidis*. However, Compere remained unconvinced: 'I don't care if all the entomologists in creation would coincide with Dr Howard that the yellow species here on red scale is the same as that in California, I would still be of the contrary opinion' (Compere, 1961, p. 204).

Howard warned Compere he was wasting time and the state's money in his efforts to introduce parasites into California that already existed there. This misconception was based on Howard's misidentification of *A. lingnanensis* and was possibly responsible for ending Compere's career as foreign collector and postponing related foreign work for several years (Compere, 1961). The pest management impasse was compounded by another misidentification, this time involving the host species. The red scale was originally mistakenly thought to belong to the genus *Chrysomphalus* and the search for natural enemies was thus misdirected at South America (Rosen, 1986). Further, red scale insect and yellow scale (*A. citrina* (Coquillett)) were not separated taxonomically from one another until 1937 (McKenzie, 1937), despite evident differences in injuriousness and physiology (Compere, 1961, p. 186). Meanwhile, host-specific parasitoids from yellow scale, the wrong species of *Aonidiella*, repeatedly failed to establish in California, which led to the erroneous conclusion in the 1930s that no effective parasitoids of *A. aurantii* existed in Asia (Rosen, 1986). The issue of red and yellow scale insect species status was debated publicly by two leading entomologists, C. V. Riley and B. M. Lelong. Riley was unmoved by the non-morphological evidence. In Compere's (1961, p. 187) view, he 'was attempting to defend an untenable position and was poorly prepared for the encounter with Lelong. When confronted with the evidence and the logic of Lelong, he gave ground, but only temporarily. He based his ultimate decision on the conventions of systematics and not on the facts of nature and, on the basis of morphological evidence, pronounced the red and yellow scales to be the same'. The account given by Compere is rather frightening in its timeless qualities.

With hindsight, the Californian citrus situation has been summarised as follows: 'A major reason for this 50-year failure was a series of misidentifications of the parasitoids and the scales, which resulted in searches being made in South America instead of the Far East, introductions of unsuitable parasitoids from similar species of scales, and failure to introduce the effective parasitoid species because they were misidentified as species already present in California' (Stehr, 1975, p. 174). Introductions of parasitoid species that subsequently proved effective in red scale insect control in California were 'painstakingly destroyed as an undesirable contaminant' (Rosen, 1978). Successful biocontrol of citrus red scale in California therefore has the dubious distinction of being one of the best examples of a long-standing failure turned into a success by the eventual introduction of appropriate natural enemies (Stehr, 1975). This example illustrates the consequences of bad taxonomy and the need to understand the limits of cryptic species. Indeed, understanding the limits of *Aphytis* species remains a significant problem for horticulture worldwide.

Example 2: Introduction of an exotic parasitoid of Karoo caterpillar into South Africa
The Karoo caterpillar (*Loxostege frustalis* Zeller) is an indigenous pyralid pest of South African sheep pasture. For various reasons, a supplement to parasitism was sought in North America (Annecke & Moran, 1977), where the congeneric *L. sticticalis* (Linnaeus) is a pest. Almost nine million individuals of the introduced braconid *Chelonus texanus* Cresson were released in South Africa over the decade ending mid-1952. To gauge the success of the programme *L. frustalis* caterpillars were field-collected during 1948 and 1949, and these 'collections yielded about 2% parasitism by the indigenous *C. curvimaculatus* Cameron, a species not then known to attack the Karoo caterpillar, and this species was taken for *C. texanus*, which it resembles' (Annecke & Moran, 1977). The misidentification was not corrected until early 1951. In the meantime, '[t]he recoveries of the misidentified *C. texanus* were a source of encouragement for the entomologists concerned, but the species never did establish in South Africa' (Annecke & Moran, 1977). Again, bad taxonomy and preconception were important influences.

Example 3: Taxonomic quality control in parasitoid cultures
The San José scale insect (*Quadraspidiotus perniciosus* Comstock) invaded apples and currants in southern Germany about 50 years ago. The major biocontrol thrust involved the mass production and release of various 'strains' of the aphelinid parasitoid *Encarsia perniciosi* (Tower). The parasitoid ultimately responsible for biocontrol was the so-called bisexual *E. perniciosi* strain. During the early years of the programme, the parasitoid cultures were unknowingly contaminated by the congeneric *E. fasciata* (Malenotti), a species native to southern Europe. The latter species apparently reproduced far faster than *E. perniciosi* in the insectary, and accounted for as many as four million *E. fasciata* being released in

the Heidelberg region in the period 1956–1958. However, *E. fasciata* fared poorly in the field and died out within a few years (Neuffer, 1990; Rosen, 1986; G. Neuffer *in litt.* vi 1996).

Biological control of San José scale in Germany was not permanently affected by the contamination, but similar events to those described have undoubtedly affected other projects and are likely to do so again. Contamination of cultures warrants quality control based on good taxonomy, but not only because of potential negative effects from field releases. Undetected contaminant species in cultures may also misguide the interpretation of host relationships and other adaptations, as effective quality control is dependent on understanding the species limits of the subject organisms.

A possible example of poor quality control compounded by inadequate understanding of the species status of the populations in question involves a braconid parasitoid of graminaceous stem-boring caterpillars, *Cotesia flavipes* (Cameron). Although the species is widespread in Asia and has been widely and successfully used in biocontrol (e.g. Alam *et al.*, 1971; Macedo *et al.*, 1993; Mohyuddin, 1991), its endemic distribution and natural host range cannot be defined precisely, partly because of early taxonomic confusion with *C. chilonis* (Matsumura) (see Polaszek & Walker, 1991) and partly because of frequent undocumented shipments even among Asian countries (Alam *et al.*, 1971; Polaszek & Walker, 1991). A dichotomy in the views of the species' host relationships also confuses the situation. On the one hand, it is portrayed as a parasitoid that principally attacks corn-boring *Chilo* species, and which has, without any rationale, been used successfully in the biocontrol of *Diatraea* species in sugarcane (e.g. Alam *et al.*, 1971; Bennett, 1971; Kfir, 1994). On the other, the species is reputed to exist in at least two strains, a 'Pakistan' strain that attacks primarily corn borers and an 'American' strain that parasitises sugarcane borers in North, Central and South America (Mohyuddin *et al.*, 1981).

The origin of the reputed American sugarcane borer strain is not altogether clear, but two options have been mentioned. Southeast Asian or Indian *C. flavipes* wasps belonging to 'sugarcane-adapted strains' are available (see Alam *et al.*, 1971; Mohyuddin, 1991, p. 20). That wasps mate across strains in the laboratory and produce viable offspring has been said to confirm that they belong to the same species (Baker *et al.*, 1992). An additional suggestion invokes adaptation of the '*Chilo* strain' in the laboratory. Whether these two options are considered mutually exclusive alternatives is not clear from the literature. The laboratory adaptation reputedly took place in Pakistan material that was shipped to Trinidad where a culture was propagated on the sugarcane borer *Diatraea saccharalis* (Fabricius). In less than two years that culture reputedly evolved into a '*Diatraea*' strain that mainly attacked borers in sugarcane and which was successfully established on *Diatraea* in the field in Brazil (Baker *et al.*, 1992; Greathead, 1994; Mohyuddin *et al.*, 1981, p. 580). Polaszek (*in litt.* x 1996) suggested, though, that 'strain' may be an inappropriate term and that *C. flavipes* simply has a broad host range, although at least some populations have biological characteristics not shared by assumed conspecific populations (Polaszek & Walker, 1991).

If host-specific populations do indeed exist under the name *C. flavipes*, a likely explanation is that endemic 'strains' in Asia attack borers in sugarcane. Such 'strains' probably represent distinct species even though they do interbreed in the laboratory (see Mohyuddin, 1991). The reputed case of adaptation in Trinidad is likely to be another example of earlier contamination of cultures or mixing of cultures to begin with and then the 'corn borer species' dying out on the unsuitable *Diatraea* hosts.

Claims such as that above warrant careful scrutiny and further critical research, not only because particular biocontrol programmes may be jeopardised through debilitating releases of contaminant species but also because inappropriate generalisations about biocontrol may be developed or supported (e.g. Alam *et al.*, 1971; Mohyuddin *et al.*, 1981).

Example 4: Successful control of *Salvinia* weed in Australia: the role of host specificity in cryptic species of insect herbivores

Salvinia molesta Mitchell is a free-floating freshwater fern of South American origin. In the many countries where it has been introduced it grows into dense undesirable mats that may cover extensive areas. In Australia, biological control was begun against *Salvinia* weed in 1978 (Room *et al.*, 1981). *Salvinia molesta* had been distinguished taxonomically from *S. auriculata* Aublet (Mitchell, 1972) only just prior to the start of the Australian programme. The original distribution of the weed species could thus be traced, to southeast Brazil (Forno & Harley, 1979). Prior to the Australian work, insects for biological control had been collected from *S. auriculata* much further to the north, outside the distribution of *S. molesta* (see Room *et al.*, 1981). In hindsight we see this as inconsistent with a primary doctrine of classical biological control, that potential control agents should be sought principally in the original area in which the target species was endemic (e.g. Greathead, 1994; Harley & Forno, 1992,

p. 13). The original efforts against *Salvinia* had been universally unsuccessful (see Room, 1986).

The realisation that *S. auriculata* actually comprised a complex of at least four species opened the way for a chain of fortuitous events that led to good biocontrol of *Salvinia* weed (see Room, 1986). The beetle that was eventually to devastate the fern mats is now known as *Cyrtobagous salviniae* Calder & Sands. When originally collected from *S. molesta* in Brazil, however, *C. salviniae* was considered to be a 'local race' of *C. singularis*. Locally adapted 'host races' have long been considered to have great potential for enhancing biocontrol success, a contention that is now being challenged (Clarke & Walter, 1995; see also Chapter 8). The 'local race of *C. singularis*' was therefore introduced into Australian quarantine (see Room *et al.*, 1981). Successful biocontrol followed the release of *C. salviniae* into Australia and Papua New Guinea, but it was only after release that the beetle was accepted, on morphological criteria, to be a host-specific cryptic species.

Successful control of *Salvinia* was contingent on good taxonomy, in this case involving the target organism. That the control agent was subsequently found to be an undescribed species suggests a fortuitous aspect to the project's success because the questionable principle of introducing 'local races' of a species was followed.

Example 5: A landmark case of classical biological control? Release of egg parasitoids against *Nezara viridula*

The egg parasitoid *Trissolcus basalis* (Wollaston) (Hymenoptera: Scelionidae) has been internationally acclaimed for the successful biocontrol of *Nezara viridula* (Linnaeus), at least in Australia and Hawaii (Caltagirone, 1981; Waterhouse & Sands, 2001). However, problems surfaced in the published literature regarding the origins of the successful 'Pakistan strain' (Simmonds & Greathead, 1977). How could a 'Pakistan strain' of *T. basalis* have been involved when the species had never been recorded from Pakistan, either before the project, or since? Furthermore, no direct evidence was ever presented to demonstrate that *N. viridula* was indeed under biological control (Clarke, 1990; Jones, 1995; Velasco, 1990), yet for some time the 'Pakistan strain' was recommended for biocontrol in other places (Waterhouse & Norris, 1987a).

The actual chain of events in the biocontrol of *N. viridula*, as well as the actual outcome, is still being pieced together, and it does involve bad taxonomy. At least three species of egg parasitoids were released into

Australia under the name *Trissolcus basalis*, with these being true *T. basalis* from Egypt, an Italian species known at the time of importation not to be *T. basalis*, and *T. crypticus* Clarke (Clarke, 1990; Clarke & Walter, 1993a). The last-named species includes the parasitoids referred to as the 'Pakistan strain of *T. basalis*'. The undetected introduction of yet other species is also a possibility.

Trissolcus crypticus is morphologically distinct from *T. basalis* (Clarke, 1993). Despite claims of its success against *N. viridula* (under the rubric 'Pakistan strain'), *T. crypticus* has never been recovered from the field in Australia (Clarke, 1993). Furthermore, *N. viridula* is still a serious pest in the major eastern Australian cropping areas (Clarke, 1992) and pecan orchards (Seymour & Sands, 1993). Perceptions of successful biological control may have been generated by changing agricultural practice reducing *N. viridula* abundance (Velasco, 1990). A reassessment of the Hawaiian situation has also indicated that *T. basalis* does not exert a major influence on *N. viridula* numbers (Jones, 1995), and no field-collected Hawaiian material has yielded *T. crypticus* so it presumably did not establish there (A. R. Clarke, pers. comm. 2001).

The *N. viridula* case demonstrates that inadequate taxonomy played a decisive role in undermining the biocontrol programme against *N. viridula*. Even if *T. crypticus* is found to have established in Australia, the early claims of success by the 'Pakistan strain' were scientifically irresponsible because no specimens collected in the field at that time were *T. crypticus* (Clarke, 1993). Indeed, the attribution of 'strain' status to *T. crypticus* implied that individuals could not be discriminated from any other strain of *T. basalis*, so it was inappropriate to attribute success to any one 'strain' over another.

General conclusions from the examples above

Each of the histories covered above is readily labelled as a case in which bad taxonomy per se reduced efficiency, or one in which a lack of appreciation of the widespread occurrence of cryptic species in various insect groups led to delays or other problems. An understanding of species theory (see below) suggests there will be many more cases of impeded pest management through such failures.

The negative effects of bad taxonomy can be overcome, to a considerable extent at least, through increased vigilance by taxonomists and non-taxonomists alike. By contrast, the problem of the potential existence of undetected cryptic species requires an appropriate theoretical

understanding and an appreciation of the potential roles and limitations of the various techniques that may be used in such circumstances.
Most such species problems cannot be solved confidently or scientifically
through the simple-minded application of technology, however sophisticated that technology may be (Paterson, 1991). Scientific problems of this
nature are primarily problems of logical deconstruction and the application of the most appropriate technique at each particular stage of the
investigation.

Although species identification problems are of the two types outlined
above (i.e. poor taxonomy and true cryptic species), most authors deal
with these situations under the single banner of 'good taxonomy', even
if they do acknowledge both aspects. For example, Stehr (1975, p. 175)
suggests: 'All this [see citrus red scale example above] serves to emphasise the necessity for sound taxonomy (along with sound biology and
ecology) as a fundamental base on which to build successful biological
control'. He does not elaborate any further, but Rosen (1978) puts forward
the following: 'Inadequate systematics of either pests or natural enemies
was the cause of prolonged delays in the ultimate success of biocontrol';
'Sibling species can be recognised as distinct only through biosystematic
research. With biparental forms, this can normally be achieved by reciprocal crossing tests, reproductive isolation being the ultimate criterion for
the determination of their systematic status'. Other authors who urge the
separation of cryptic species by means of behavioural tests are generally
unanimous that the ultimate criterion for judging species status should
be the degree of reproductive isolation between them (Rosen, 1986). This
contention is now examined, and is demonstrated to be unsound. A more
realistic alternative, which overcomes each of the problems identifiable in
the reproductive isolation approach, is available and is then discussed.

Proposed solutions to identification problems

We deal with species in two ways

The problem in specifying species limits is complicated by species being
conceptualised in two ways. First, species are treated primarily as nominal, or named, entities and this carries the implicit assumption that the
named entity accurately represents a natural entity. That assumption is
frequently inappropriate and thus results in inaccurate interpretations,
as expanded and illustrated later. Second, species are dealt with primarily as objective entities in nature. Their definition here derives from the

everyday observation that 'like' mates with 'like' in nature. For instance, in areas in which house sparrows (*Passer domesticus* Linnaeus) occur together with tree sparrows (*P. montanus* Linnaeus) the house sparrows invariably pair with house sparrows and the tree sparrows pair with tree sparrows. In population genetics terms, sexual organisms participate in positive assortative mating, which is why Lambert *et al.* (1987) could assert that sexual species are self-defining.

The nominal, taxonomic approach

Ideally, the system of naming species should reflect exactly the situation in nature. In many situations that is so, but in a significant number of cases their misalignment impedes accurate interpretation and, ultimately, effective management. For example, colour variation in Australian *Leptograpsus* crabs was initially interpreted as an adaptive polymorphism and as the initial stages in sympatric divergence (Shield, 1959), although the orange crabs were later shown to be of a separate species from the bluish ones (Campbell & Mahon, 1974). Note that both adaptive polymorphism and sympatric divergence are strictly intraspecific phenomena, so combining the two species as one was inappropriately used to support questionable evolutionary interpretations (Paterson, 1973). The independent existence of these two species was apparently not even suspected originally. Similar mistakes are undoubtedly still current. In entomology the *Cotesia flavipes* example dealt with above is a likely one, as are the adaptive polymorphisms claimed in cichlid and salmonoid fish by many authors (e.g. Logan *et al.*, 2000; Meyer, 1990) and even in birds (Knox, 1992). Cryptic species are being found in such groups where intraspecific variation was previously claimed (e.g. Blouw & Hagen, 1990; Waters *et al.*, 2001) and are likely to represent the real biological basis of the spate of sympatric speciation case histories that have been published in the last few years.

Species problems are frequently approached or dealt with from the perspective of the taxonomist's dual aims of demarcating entities on the basis of structural discontinuities and then fitting those entities to a system of nominal categories (species within genera within families). In many situations this approach presents no problems for the accurate demarcation of species. In others, understanding and insight suffer. A significant problem is that structural demarcation is too subjective because the observer simply seeks discontinuities. Furthermore, cases are known where structural discontinuities have not been found despite intense searching, and in some of these cases the chromosome banding patterns are also indistinguishable (i.e. are homosequential) across species (Coetzee, 1989). Morphological features are frequently tackled in this way, but the search for discontinuities may well be prominent even when non-morphological criteria are used, such as song structure or the relative mix of compounds in pheromone blends (e.g. Aldrich *et al.*, 1993, 1987). Location of such a discontinuity, even if it is statistically significant, does not necessarily imply it is functionally significant to the extent of defining species limits (e.g. Brézot *et al.*, 1994; Miklas *et al.*, 2000; Ryan *et al.*, 1995), and even molecular information is subject to this stricture.

In short, the nominal approach to species problems encourages observers to think in a particular way and that does not lead inevitably to a precise correlation with the situation in nature. For instance, morphological divergence is assumed to be inevitable between species because they are expected to continue their divergence with time (e.g. Ayala, 1982), and discontinuities that are measured are implicitly assumed to be biologically significant. The case of the green vegetable bug (or southern green stink bug: *Nezara viridula*) is enlightening. Because differences among populations in the ratio between two of the six chemical compounds that comprise the sex pheromone blend were portrayed as biologically significant, two 'strains' were designated on this basis (Aldrich *et al.*, 1987, 1993). However, no functional significance has even been proposed for these differences in pheromone blends, let alone demonstrated. Indeed, different blends exerted no differential effect on female bugs in a wind tunnel (although the bugs were walking and not flying) (Brézot *et al.*, 1994), and variation among individuals within a population spanned the range of ratios said to differentiate populations (Miklas *et al.*, 2000; Ryan *et al.*, 1995). In short, the taxonomic way of thinking may well impede a more realistic functional analysis of the behaviour of the organisms, the topic of the following subsection.

Approach based on sexual behaviour

The alternative approach to the strictly structural and nominal one (or taxonomic one) is to ask questions about sexual populations that mate assortatively in nature. Here we are concerned with a particular type of discontinuity, discontinuity in gene flow between the two populations (or samples) of interest. The differential may be discernible in any one of several aspects, such as habitat use, host plant use, time of mating, mating behaviour and so on. However, a differential in geographical distribution between the populations is insufficient on its own to be useful here. This issue of allopatric populations is a subtle one and is dealt with later, for it does have ramifications for pest management, when it would be useful to extrapolate an understanding of ecological characteristics or organisms in one area to those in other places (e.g. other agricultural areas or other countries).

The question about the two or more populations whose mating status is unclear concerns the amount of gene flow that takes place between them. In other words, are we dealing with one gene pool or two? Ultimately we are concerned with the mating behaviour of the constituent individuals in nature, because that will determine the possibilities for gene flow. Note the abstraction at this point. The term 'possibility' is used because individuals separated by a considerable area of unsuitable environment may well have every physiological and behavioural potential to mate with each other, but such an act is precluded simply by their allopatric distribution. Were such allopatric populations to be 'forced', by climate change for example, into a single area they would mate positively assortatively. Assessing the mating status of allopatric populations is notoriously difficult and is dealt with later, for additional basic information has still to be covered.

Dealing with species in two ways – consequences for IPM

If *Helicoverpa punctigera* is said to be a generalist species with a very wide host range, the implication is that the 'species taxon' *H. punctigera* has a wide host range. The species taxon *H. punctigera* has only ever been circumscribed and delimited on the basis of morphological criteria (Common, 1953; Paterson, 1991). Usually evolutionists and ecologists assume that the structurally defined 'taxonomic' species represents a group of individuals that mate assortatively in nature. Such a group represents a species in population genetics terms. That assumption is seldom rigorously tested and it is frequently not even recognised as an assumption. The question of scientific method thus enters the picture again and illustrates the importance of addressing fundamental questions and recognising underlying assumptions.

An important possibility for pest management is that more than one positively assortatively mating species may have been inadvertently forced into a single species taxon and given a single name. The problem for ecologists and pest managers is that ecological and IPM-related research may well be conducted on a particular species that is later recognized to comprise a species complex. The research results obtained earlier may, however, not be readily related to a particular species within the complex, unless meticulously labelled vouchers are amenable to subsequent molecular investigation. The potential devaluation, in this way, of hard-won and financially costly results should be avoided assiduously. If the new research is simply ignored future workers may be seriously misled, with the wastage of more valuable effort and money.

Although methods are available to investigate the possibility of cryptic species and are covered later, an inclination to downplay this option persists. Those who do tackle such situations are confronted by yet another dichotomy in interpretation. It involves the way in which the differences in mating behaviour between two species are perceived, interpreted and used to initiate research. Rosen's (1978) statement about reproductive isolation being the ultimate measure of species status was presented earlier and is expanded in the following subsection. Following this, the alternative to the reproductive isolation perspective is detailed and justified.

Reproductive isolation as species criterion

In reality, each sexual species is reproductively isolated from every other species. But 'reproductive isolation' is a theoretical interpretation of an observed pattern (Masters *et al.*, 1984), as described earlier for the concept of 'polyphagy' (Chapter 2). The central question in interpreting species status in terms of sexual behaviour is: 'What sets the limits to a species gene pool?' The theoretical nature of 'reproductive isolation' becomes evident at this point, because the answer to the question about species limits inevitably incorporates ideas about adaptation.

In the evolutionary literature, interpretations of the behaviour of the organisms that make up a species gene pool take one of two forms. One emphasises that gene pool 'X' is isolated reproductively from gene pool 'Y', and that mechanisms evolved specifically to keep them apart. The other stresses that the individuals in species gene pool 'X' mate assortatively with one another and that those in the second pool, 'Y', independently do the same. The two mate positively assortatively even when they occur together, because each has unique mechanisms, or adaptations, that ensure fertilisation. Under the second view reproductive isolation is simply a by-product of the evolutionary process that changed the mechanism of fertilisation in the first place (Paterson, 1985). Even though the distinction between the two interpretations may look semantic, it is critical for accurate interpretation and thus for IPM practice. Explanations that centre on by-products are weakened by the misplaced emphasis, for that immediately relegates the primary aspects of the whole process to the periphery of the explanation (see Chapter 5, p. 97). Were the nitrogen-fixing properties of legumes given an evolutionary explanation in terms of soil enrichment, to take an extreme case, significant aspects of the adaptive values and evolutionary significance of the bacteria–plant symbiosis would be lost.

The next task is to assess these two alternatives in relation to the behaviour of organisms in nature. That should yield the more realistic interpretation of the process of adaptation. The methodological implications of each approach are also important, for each suggests different criteria by which to measure and interpret species status. Some authors combine the two alternatives and suggest that both concepts yield the same interpretation of diversity (e.g. Claridge & de Vrijer, 1994, p. 222). Such a view is logically unacceptable and inaccurate, as demonstrated below.

The alternative definitions of sexual species

Only those concepts that deal explicitly with sexual species are covered here. Asexual species are dealt with separately in a later section. Although many demand a unified approach to sexual and asexual species studies (see Mayden, 1997; Templeton, 1989), particularly taxonomists and phylogeneticists, there is justification for their separate treatment in evolutionary or population genetics terms. In particular, species that are sexual (even if sexuality is only occasional, as in aphids and many flowering plants) contribute overwhelmingly to organic diversity (Hauser, 1987; Paterson & Macnamara, 1984), with only about 1000 obligate asexual species known (Foottit, 1997). Species diversification is thus a consequence of sexuality setting the 'limits' to which particular individuals can participate in sexual activity and thus which ones can recombine genetic material in the following generation. The evolution of new, different sexual behaviours thus sets up new gene pools or sexual species. Such a process is not open to asexuals, in which, by definition, gene pools do not occur. The clonal behaviour of asexuals not only precludes recombination between individuals, which constitutes the 'cement' of sexual species, but also inhibits change. However, mutation within a lineage does allow drift of characters but does seem to have limited impact on change. The call for asexuals and sexuals to be dealt with uniformly thus reflects an urge to categorise species nominally rather than to deal with species from the perspective of the processes that underlie their differences and similarities.

Several concepts cover sexual species, but I exclude all but two from consideration on the basis that the others are either taxonomic in nature or otherwise demand subjective delineation, including the phylogenetic and ecological species concepts. Their subjective nature and consequent lack of utility has been dealt with elsewhere (e.g. Masters & Reyner, 1996; Vrba, 1995).

The isolation concept of species

The *isolation concept* is the older of the two available concepts of species, and is almost invariably the only one portrayed in recent textbooks (e.g. Campbell *et al.*, 1999; Knox *et al.*, 2001; Purves *et al.*, 1998). Species are here interpreted in relation to one another, as suggested by the stress on reproductive isolation. They are considered to have evolved differences in their mating behaviour, with those differences functioning to keep the gene pools separate, hence their name of isolating mechanisms. Dobzhansky (1970, p. 357) formalised the isolation concept as follows: 'Species are ... systems of populations; the gene exchange between these systems is limited or prevented by a reproductive isolating mechanism or perhaps by a combination of several such mechanisms'. For example, the habitat of each species may differ, there may be seasonal or diurnal differences in mating period, their mating behaviour may differ, or there may be postzygotic 'mechanisms' such as those causing hybrid disadvantage (and including interspecific sterility). The emphasis of the isolation concept in specifying species status is clear: 'The major intrinsic attribute characterizing a species is its set of isolating mechanisms that keeps it distinct from other species' (Mayr & Provine, 1980, p. 34). That criterion is given specific prominence over the relationship of conspecific individuals to one another (e.g. Mayr, 1976, p. 358).

The recognition concept of species

The *recognition concept* was proposed and fully articulated by Paterson (1985, 1993c), and its significance as a valid and superior alternative to the isolation concept is quite widely appreciated (Carson, 1989; Lambert &

Spencer, 1995; Templeton, 1987). Under the recognition concept, species are not defined in relation to one another. Instead, any sexual species is definable solely on the basis of its fertilisation mechanism, which includes the set of processes that enables the sexes to recognise and locate each other in their usual habitat, and thus leads to successful fertilisation and formation of a zygote. The characters involved are diverse and include such aspects in the mating partners as the design features of the gametes, factors determining synchrony in the achievement of reproductive condition, the co-adapted signals and receivers of mating partners and their co-adapted organs of gamete delivery and reception (Paterson, 1988, p. 69). The *specific-mate recognition system* is that subcomplex of the fertilisation system that ensures motile organisms can find and recognise potential mates in their natural environment, even when present at low frequencies and despite the environment being heterogeneous and dynamic.

In formal terms, a species is 'that most inclusive population of individual biparental organisms which share a common fertilisation system' (Paterson, 1985, p. 25). This view is diametrically opposed to the isolation concept, principally in that the reproductive and genetic isolation of groups of individuals (i.e. species gene pools) is seen as an incidental consequence or by-product of individuals of the species involved having different adaptations for mate-finding and fertilisation.

Hybridisation
So far hybrids have not been mentioned, although some persistent hybrid zones exist between sexual populations. Such zones tend to be quite narrow, although they may extend linearly over hundreds of kilometres (Fig. 6.1), and hybridisation may take place at fairly high frequency among types (e.g. Moran & Shaw, 1977). In relation to the evolution of new species, hybrids have different significance under the alternative species concepts. Therefore, hybridisation is dealt with later, when speciation itself is covered.

Hybridisation also occurs occasionally between individuals from recognisably different gene pool pairs, and some authors use such observations to play down the non-arbitrary nature of species boundaries (e.g. Mallet, 1995; Schilthuizen, 2000). However, occasional hybridisation does not disrupt the overall pattern of species gene pools being discrete and is not relevant in this regard (e.g. Mayr, 1976), as indicated by recent studies on mosquitoes (Besansky et al., 1994) and geckos (Toda et al., 2001).

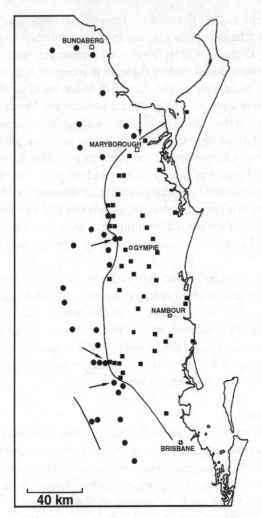

Figure 6.1. The line of contact between the southern 'Moreton race' of the acridine grasshopper *Caledia captiva* and the northwestern 'Torresian race' runs north–south. The arrows indicate where 'Torresian' populations (to the west of the contact line) were found with a low frequency (1–3%) of 'Moreton' chromosomes. Modified from Moran & Shaw (1977).

A functional approach to understanding species limits

To ensure the best chance of interpreting the limits to a species gene pool correctly, we must first of all know what a species is in adaptive terms. This is the standard approach to understanding biological phenomena;

once we understand function we can interpret operation (Paterson & Macnamara, 1984; Rosenberg, 1985). Recent calls to exclude functional terms from species concepts and to revert to the Darwinian view that species represent arbitrary designations (e.g. Mallet, 1995; Schilthuizen, 2000) ignore the ubiquity of the functional approach and provide no justification for such a departure.

A generalisation that holds for all sexual species is needed, one that captures the reason for different sexual species existing. Cast in population genetics terms, or in relation to gene flow in nature, a functional explanation is required for the observed pattern of positive assortative mating. From this interpretation of species, and only from this, an explanation can be derived of how a new species evolves. The change at speciation clearly reflects a change in the functional mechanism that leads to positive assortative mating in the field. Without understanding origins in this way, little will be understood about species because interpretations of host relationships and other ecological aspects must be compatible with the way in which the adaptations arose in the first place (Paterson, 1973; Walter, 1995b; Walter & Paterson, 1994).

Each of the two concepts of species reviewed above provides a mutually distinct interpretation of speciation and the acquisition of novel characters, or adaptations. This point has far-reaching consequences for the interpretation of the biology and ecology of species. The critical significance of theory in guiding the way the natural world is interpreted again becomes apparent. The chain of interpretation outlined above has direct importance in applied biology, but detailed examples are given later, when sufficient background has been covered.

Evolution of new species: overlap between the two concepts
Despite the differences between the isolation and recognition concepts of species, the two do have an important point in common: a new species can evolve only if a daughter population is separated geographically from the body of the species gene pool, or parent population. That is, speciation occurs allopatrically (Dobzhansky, 1951, 1970; Futuyma & Mayer, 1980; Mayr, 1963; Paterson, 1981). A new species cannot arise from individuals of the parental species if they are within the geographical area occupied by the parental species and are therefore potential participants in the same gene pool. That would be sympatric speciation, for which no unequivocal evidence exists, and, in particular, the mating behaviour and ecology of organisms supports the view that sympatric speciation is not

a plausible general pathway for the origin of new species (Paterson, 1981). Acceptance of sympatric speciation as a common method of speciation is growing (Berlocher & Feder, 2002; Itami *et al.*, 1998; Logan *et al.*, 2000) and the prime example is the apple maggot fly (*Rhagoletis pomonella*) in the USA (Bush, 1992, 1993, 1994; Feder *et al.*, 1998).

The apple maggot fly is native to North America. Although its main host is hawthorn, it is considered polyphagous and has been recorded on a range of plants among the Rosaceae, including *Crataegus*, *Pyrus*, *Cotoneaster*, *Prunus* and *Malus* (Bush, 1975). Its occurrence on introduced apples is considered noteworthy only because the 'switch' to apples apparently took place suddenly and after apples had been in North America for about 200 years (see Bush, 1992). The reliability of this historical information is now beyond test, but the sympatric speciation case does rest heavily on its validity (see p. 203).

The flies on apples have therefore been singled out as representing a newly adapted race of *R. pomonella*, on its way to full species status (Bush, 1994). So far, the only divergence between apple flies and hawthorn ones is a population-level difference in frequencies of several alleles (Bush, 1992). Such differences, on their own, are difficult to interpret (see later section on molecular techniques). In connection with behavioural evidence that the flies from each source do not respond preferentially to their fruit of provenance, and that mixing of populations takes place in the field, such differences could equally represent relatively minor differential selection pressures from the alternative host species, and the situation is readily explicable in behavioural terms (Marohasy, 1996).

That all reputed cases of sympatric speciation are inconclusive is readily revealed by the uncertainty of the protagonists with regard to the status of the relevant 'populations'. Authors frequently are uncertain as to whether they are dealing with two species or only one. If they are separate species, as frequently seems likely, then they are not sympatric races of a single species, and they are not on their way to 'full species status' in sympatry. Their origin requires a different perspective, part of which could well be the idiosyncratic shifts of the host plants and insects as climate changed and influenced the ecology and geographical distribution of the species (Walter & Paterson, 1994).

Lack of resolution about the species status of the relevant populations reflects a lack of clarity on how to define species. To understand how a new species evolves, one needs a clear and unambiguous understanding of what constitutes a species. Interpreting the origin of anything demands a prior understanding of the nature of that thing, despite Bush's (1993, 1994) emphatic denial of this point. The fallacy of Bush's argument is exposed in two ways. First, he assumes (but portrays as fact) that evolutionary change continues ineluctably in all populations, even extended ones (Bush, 1993, pp. 242ff.). Second, his definition of 'host race' (Bush, 1992, pp. 342ff.) shows that he has a concept of species in mind, one involving intersterility and reproductive isolation.

Bush's model of sympatric speciation assumes, further, that the principal ecological driving force behind divergence relates to resource use (see Bush, 1993, pp. 231, 238). Resources unused by the species – apples in this case – are seen to provide, in an unspecified way, an outlet or ecological opportunity for a species (hawthorn flies) that is perceived to be using its resources to the limit. Such anthropocentric views of organisms are idealistic, because they treat species typologically as units (Hengeveld, 1988; Walter, 1993a), and they are inconsistent with evidence on the ecological and evolutionary significance of interspecific competition (Connell, 1980; Masters & Rayner, 1993; Walter 1988b, 1995b; Walter *et al.*, 1984; Wiens, 1977), particularly that relating to phytophagous insects (Lawton & Strong, 1981; Strong *et al.*, 1984). In short, Bush's (1993) claim that he does not put the cart (species concept) before the horse (process of speciation) has been achieved by prejudging how evolutionary change takes place and only then mounting an interpretation of species formation, which is logically unacceptable. Such lack of clarity about species can lead only to lack of clarity in interpretations of species formation.

The determination to prove sympatric speciation has meant that several realistic alternative explanations of the situation have not been adequately investigated yet. Even the early criticisms raised by Futuyma & Mayer (1980), Jaenike (1981) and Paterson (1981) have not yet been addressed. Although new claims of sympatric speciation are rather frequent now, not one overcomes the difficulties outlined above. More than a century ago, Chamberlin (1897) warned us of precipitate explanation driving the acceptance of ill-investigated interpretations. His paper has been reprinted and frequently cited with approbation, but the practice seems clear in the wide acceptance of so many cases of sympatric speciation.

Interpreting speciation – the isolation concept

Under the isolation concept of species the principal specification about the daughter population is that it should be allopatric. The size of the population is largely irrelevant, and the model covers the simple halving of an extensively distributed species (dumbbell model) as well as the peripheral isolation of small 'satellite' populations (e.g. Ayala, 1982; Lynch, 1989). Even if the environment of the daughter population resembles that of the parental population, speciation is not expected to be inhibited.

In the geographically isolated daughter population some genetic change to mating behaviour is expected to take place, but no adaptive explanation is available or offered, or is even considered necessary. Therefore, pleiotropy in gene action is invoked; adaptive changes in characters other than sexual behaviour should lead to incidental change in mating characteristics through pleiotropy (e.g. Futuyma, 1987). At least two alternative scenarios come into play here.

The first involves hybridisation, usually with the parental population but even with other closely related species. For this to occur, the daughter population is thought to spread geographically after the initial phase of adaptation, and then to encounter the parental species. Hybridisation is thought to generate selection, through hybrid disadvantage, to reinforce differences in mating characters between the two populations (or incipient species) (e.g. Saetre et al., 1997). The formation of a new species thus occurs through hybridisation generating reinforcing selection to enhance such features as temporal difference in mating time, difference in mating behaviour, and so on. Such differences that evolve with the express purpose of holding two Mendelian populations apart are called isolating mechanisms, hence the name of the species concept.

The hybridisation option in the explanation of speciation is intuitively appealing, and has serviced evolutionary interpretation for a long time, but it suffers logical flaws. These are best illustrated with reference to Fig. 6.2, which illustrates a typical distribution of two closely related species. Quite obviously the area of sympatry is the only place within the entire distribution of either species in which hybridisation is possible. In other words, the overlap zone is the only place in which reinforcement could occur (Paterson, 1993c, pp. 14–15). Within the zones of allopatry, hybridisation cannot occur and reinforcement of isolation mechanisms is not possible. Characters selected in one area (e.g. in sympatry) are unlikely to spread to areas where positive selective forces do not sustain them (e.g. areas of allopatry). The genes coding for such characters would either be neutral in the zones of allopatry (because hybridisation does not occur there) or they could influence the organism negatively, through pleiotropic side-effects. Neutral characters could not be promoted by natural selection so they would not spread through the entire distribution of

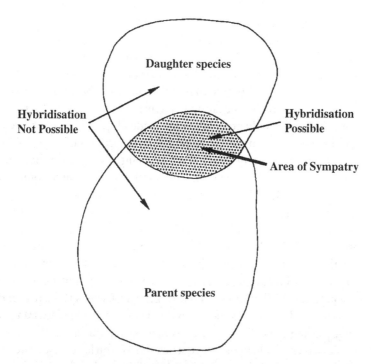

Figure 6.2. The geographical distribution of a daughter population that has undergone adaptive change in allopatry and has since made secondary contact with the parental population (i.e. the zone of sympatry) from which it originally derived.

the population. Negative influences, in contrast, would simply be selected out in the allopatric zone.

The second explanation of speciation suggests that the two populations achieve full reproductive isolation whilst in allopatry. Not surprisingly, this second alternative raises an immediate logical difficulty. How can natural selection mould a character in isolation of the situation in which it is to be of benefit? Such an argument is teleological, which means that natural selection would have to be working towards a future goal. This is simply not possible. Alternatively, the mating behaviour is not a mechanism to isolate. Isolation would therefore be a by-product or effect, which renders the whole alignment of the interpretation inappropriate. Thus a new, more logical start to the interpretation is warranted (and is offered by the recognition concept).

Observational data and analyses of mating behaviour demonstrate that adaptations associated with mating are often species wide (e.g. Henderson

& Lambert, 1982; Masters, 1991). Such a situation cannot readily be explained by the isolation concept, except by proposing improbable intermediate steps during species formation. For instance, one could insert the additional stipulation that the two species were completely sympatric at some stage in their history so that the mechanisms for isolation had reason to spread through the entire population. This may have been so for some species but could never have been so for all species pairs. It therefore could not be a general feature (or principle) of speciation.

The isolation concept also suffers from being entirely relational, which means that the origin and existence of a species is defined in relation to the species from which it is considered to be reproductively isolated (Paterson, 1985; Paterson & Macnamara, 1984). Reproductive isolation of a species, in other words, cannot be specified except in relation to another species, usually a closely related one. The isolation concept therefore cannot define those species that obviously evolved in isolation from their ancestral forms. Neither can it explain their origin. At least, it cannot do so in terms of reproductive behaviour or potential gene flow, which is the reason the concept was developed in the first place. The dodo and Madagascan partridge provide good examples (Paterson, 1985). Equivalent continental examples are also easily found (e.g. Ostrich and African hamerkop; Paterson, 1985). Some authors have tried to sidestep this logical flaw by asserting that all reproductive species concepts are relational because each species evolved from a parent species (e.g. Coyne, 1993), but such a claim misses the point and is clearly diversionary.

The isolation concept proposes that reproductive isolation evolves to maintain the genetic integrity of a gene pool. That implies the benefit is to the group, the gene pool as a whole, and this runs counter to the logical requirement in evolutionary interpretation of individual selection (see Chapter 5). The proposed benefit of such integrity lies in the more efficient use of environmental resources that is thought to follow, an old idea expressed again by Futuyma (1987) and Schilthuizen (2000), for example, despite earlier criticism by Paterson (1981, 1985, 1986). Under the isolation concept, natural selection operates to enhance diversity, which is teleological and contrary to the underlying requisites of individual selection.

Interpreting speciation – the recognition concept
The recognition concept of species was developed to overcome the deficiencies in the isolation concept outlined above. It is based upon different principles and is thus a logical alternative and not an adjunct to the

isolation concept (Paterson & Macnamara, 1984). Under the recognition concept evolutionary change is considered not to be easily accomplished in large populations, which is a far more stringent circumstance than that envisaged under the isolation concept. An important contributing factor here, but not the only one, is the nature and primary function of the mating behaviour that we observe in any sexual species. Figure 6.3 emphasises that sexual reproduction demands fertilisation as much as it requires reduction of chromosome numbers (meiosis), and it reiterates that fertilisation does not occur fortuitously. Fertilisation requires a mechanism or mechanisms to bring sperm and egg together. Such mechanisms are termed *fertilisation mechanisms* and together they constitute the *fertilisation system* (FS) (Paterson, 1985). In motile organisms such as insects the communication signals used prior to mating being achieved form a significant part, and often an obvious part, of the FS. These signals, their recognition and their induced responses together form the *specific-mate recognition system* (SMRS).

A subtle distinction is now required. Male and female individuals do not have a mechanism to ensure they mate with conspecifics or, conversely, to ensure they avoid mating with the wrong species (Paterson, 1980), and this is why the phrase 'species recognition' should not replace 'specific-mate recognition system'. The actual species status of organisms is perceived only by human observers. Individual organisms participate in sexual communication. Should the signals and responses of another individual match 'expectation' (or be 'recognised'), the interaction with that potential mating partner would continue. Thus, gene pools are demarcated incidentally by the nature of the FS and the pattern and features of their sexual communication (Paterson, 1985). Individuals have a mechanism to ensure they mate with an appropriate mating partner. That the partner is almost always a conspecific is a consequence of the behaviour, not a reason for it.

To achieve the end product of fertilisation, many subsidiary steps must be completed satisfactorily (Fig. 6.3). That several steps are necessary is dictated by the complex requirements of a heterogeneous and dynamic environment. Specifically, individual males and females will probably need to locate a mating partner over a distance. For this they may use a signalling mode that is necessarily different from the communication mode they use once in fairly close proximity to one another. Then there is a need for each sex to signal intention. In the green vegetable bug (*Nezara viridula*), for example, pheromones are claimed to bring the sexes together from

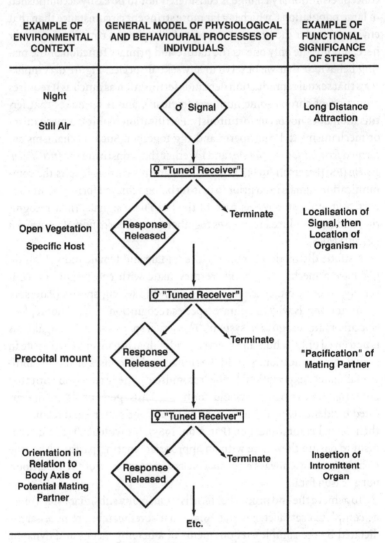

| EXAMPLE OF ENVIRONMENTAL CONTEXT | GENERAL MODEL OF PHYSIOLOGICAL AND BEHAVIOURAL PROCESSES OF INDIVIDUALS | EXAMPLE OF FUNCTIONAL SIGNIFICANCE OF STEPS |

Night

Still Air

♂ Signal

Long Distance Attraction

♀ "Tuned Receiver"

Terminate

Open Vegetation

Specific Host

Response Released

Localisation of Signal, then Location of Organism

♂ "Tuned Receiver"

Terminate

Precoital mount

Response Released

"Pacification" of Mating Partner

♀ "Tuned Receiver"

Orientation in Relation to Body Axis of Potential Mating Partner

Response Released

Terminate

Insertion of Intromittent Organ

Etc.

Figure 6.3. Diagrammatic representation of the sequence of events involved in the specific-mate recognition system of a biparental species of animal. Each event should be considered not only in its context in the chain of events that leads to fertilisation but also in relation to the nature of the environment and its functional role. The number of stages in the sequence is characteristic of the species concerned. Modified from Paterson (1985).

a distance (Harris & Todd, 1980), although particular plant species may play a role (Clarke & Walter, 1993b). Once on the host plant the bugs locate one another by means of substrate-borne vibrational duetting (Ota & Čokl, 1991; Ryan *et al.*, 1996; Ryan & Walter, 1992). Close-range communication may also entail physical stimulation, the male butting the female with his head (Borges *et al.*, 1987). The intensity and duration of each step in the courtship or close-range part of the interaction vary, but the signal and its sequence may be rigidly fixed to the point where an interrupted component cannot be re-initiated; a return to the beginning of the courtship sequence is then required (Stich, 1963). Stich (1963) also demonstrated that it is not the individual, as a whole, that constitutes the signal at any point in the sequence; a particular body part of the tipulid fly he studied served as the signal at any particular step.

The SMRS is undoubtedly more complex and intricate than portrayed above, but the need for multiple functional stages in the communication procedure becomes obvious. And the more stages that make up the system, the more aspects that may be subject to change. A functional change in only one of them may well be change enough that a new gene pool (or species) is demarcated, but the differences between species are virtually always much deeper than this. Adaptations that require several subsidiary steps to yield an appropriate outcome, as does the FS and SMRS, are complex adaptations. Each step on its own achieves little. Only in sequence and in the appropriate environment is fertilisation achieved, although a disrupted sequence need not always return to the initial starting point to be resumed (Paterson, 1985, 1986).

The complexities of the SMRS, and its specific functional qualities, have a critically important consequence. Any individual with an SMRS that is significantly different from the typical SMRS for the species is likely to be unsuccessful in mating (Paterson, 1993c). Such individuals would not contribute to the following generation, and their deviant genetic traits would be selected out of the population. Consequently the FS (or SMRS) is not readily subject to evolutionary change, and is considered to be subject to stabilising selection (Fig. 6.4) when the population is in its usual habitat or environment (Paterson, 1985, 1986). For the SMRS to change, the allopatric daughter population should be small during the period of adaptive adjustment. Novel combinations of genes are more readily spread throughout the population (or are genetically fixed in the population) when the population comprises relatively few individuals within a confined area. By contrast, individuals in more extensive populations are not

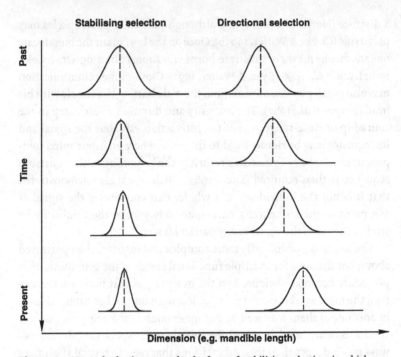

Figure 6.4. Natural selection may take the form of stabilising selection, in which case those individuals most different from the mean fail to survive, or they reproduce at a relatively low rate. The character thus remains stable at the mean value. Alternatively, natural selection may be directional because individuals different from the mean (in one or other 'direction') have enhanced survival or reproduction. The mean value of the character thus shifts.

restricted in this way. They are exposed to their usual environmental circumstances, genetic change is easily swamped, unusual recombinations are dismantled and stabilisation of characters is the norm.

Change is thus selected only if a small, isolated population is forced into a different environment from the one it usually occupies (Paterson, 1985, 1986). For instance, a few individuals may be displaced to new circumstances, such as on an island, by wind or water currents. Even land tortoises have reached distant islands, presumably by floating on ocean currents. Alternatively, some individuals may be cut off from their parental population and become trapped within a valley or some other pocket of suitable habitat. This may happen when vegetation types shift in response to climate change (Paterson, 1985, 1986). That populations do move like this has been extensively documented, in Pleistocene beetles for example (Coope, 1987). Of several beetle species found in England about

100 000 years ago, one is now found only on the Tibetan plateau, where aspects of the environment and habitat presumably match the requirements and tolerance of the species (Walter & Hengeveld, 2000).

If a small population does become trapped, and the climate continues to change in a given direction, then the environment and vegetation within the isolate will change. For example, warmer conditions mean that grassland will be replaced by forest. Conversely, when it becomes colder, woodland is usually replaced by grassland (Brain, 1981). The gradual change from one vegetation type to another, or one environment to another, brings populations under directional selection if they are trapped as outlined above (Paterson, 1986). They therefore go extinct or fortuitously undergo adaptation.

When small populations are trapped in an environment different from their usual one, average individuals are unlikely to survive or reproduce. If the population is to survive over time under such circumstances, the fertilisation system must function to some extent. Because the steps that make up the FS are adapted to the usual environment of the organisms, directional selection (Fig. 6.4) would adjust it to the new environment because variants in a particular direction from the average may mate successfully and produce at least some offspring (Paterson, 1985, 1986). Alternatively, or in addition, selection may operate on other complex adaptations that contribute to maintenance of the life cycle. For example, parasitic wasps that become isolated without their usual hosts may still be able to complete their life cycle if an alternative host is present and if it has sufficient in common with the original hosts. Under such circumstances a proportion of the wasps, however small, may interact with the new hosts and parasitise them (Walter, 1993c). Survival to adulthood of even a few individuals in each generation would allow directional selection to mould the host-finding mechanism of adults as well as the larval survival mechanisms. Such considerable genetic change is likely to influence the SMRS through pleiotropy. Because intersexual communication must ensure fertilisation, or 'closure of the life cycle' (Sinclair, 1988), does take place, the necessary co-adaptation between the male–female signal-response chain will dictate the pace of change (Paterson, 1985).

A newly adapted population is then free to expand out from its small epicentre, if suitable conditions have become more widely available. Recent phylogeographic evidence is consistent with such an interpretation of the origin of new species (e.g. Knowles et al., 1999). In parts of the new distribution, individuals may encounter individuals of the parental

population or members of other similar species. Should two such populations become sympatric, one of two outcomes could be expected.

1   If the SMRS of one population is now so different from that of the
       other, individuals will not hybridise, and a new reproductive entity or
       species now exists. That the daughter population has evolved to the
       status of a new species is incidental, because it achieved species status
       merely as a by-product of the adaptive change that had been forced on
       the small population. Evolution does not proceed to ensure that new
       species will exist. Such a view is teleological, despite it being quite
       common in the literature – for example, when species are said to evolve
       to use the environment more efficiently (see discussion of sympatric
       speciation above).

2   If the SMRS of the daughter population is similar to that of the parent
       population there will be mating between individuals of the two
       populations. In terms of their mating behaviour in nature, a new
       species has not evolved, although differences between them may be
       discerned. Reinforcing selection will not enhance any slight
       differences in mating behaviour between the two populations
       (Lambert et al., 1984; Paterson, 1978; Spencer et al., 1986, 1987). If there
       is any hybrid disadvantage, natural selection will remove the genetic
       elements that cause the disadvantage. If the cause of the disadvantage
       is too great to be removed simply, the rarer population goes extinct in
       the overlap zone (Paterson, 1978), as observed in some species that are
       rare and of conservation concern (Wolf et al., 2001).

After the newly adapted population expands out of the restricted
area in which it speciated, most natural selection that affects the population is likely to be stabilising selection, and species are therefore expected to show a considerable degree of stability. In particular, the SMRS
is predicted to be stable across space (e.g. D. melanogaster (Henderson &
Lambert, 1982) and various primates (Masters, 1991)) and through time
(e.g. antelope; Vrba, 1980). Furthermore, the other complex adaptations
carried by the individuals that make up a species' gene pool are also
expected to be stabilised, mainly because survival in a heterogeneous
and dynamic environment is likely to demand the appropriate functioning of those mechanisms. The stabilising influences that operate on the
SMRS are likely to contribute, in part, to the stability of the other complex adaptations, because any pleiotropic influence that affects the SMRS
is likely to be selected out. This view of organisms in nature provides
an alternative explanation for observations of stability, for example in

biological control case histories (Holt & Hochberg, 1997), that otherwise cannot be accounted for without invoking the negative explanation of constraints.

Implications of the recognition concept of species for IPM

The recognition concept of species has several implications for pest management. These are all logical deductions from the concept, but are increasingly supported through the results of empirical tests. Before expanding, in the three subsections below, on those aspects most relevant to pest management, let us consider a point common to all three subsections.

The recognition concept differs fundamentally from the isolation concept in the way in which adaptation is interpreted. The isolation concept, at least as originally constituted by Dobzhansky (1951, p. 208), demanded that isolating mechanisms are refined after the period of allopatry. In other words speciation is 'completed' through evolutionary change continuing after the allopatric event. That implies we should expect post-speciation change, and we should expect such change to features as complex as the mating sequence. Such changes should be observable in different parts of the species' distributional range (e.g. Bush, 1994; Fox & Morrow, 1981; Mallet, 1995; Mopper & Strauss, 1998; Ricklefs, 1987, 1989). This now seems to be a central premise of 'isolationist' evolutionary interpretations as well as the 'Darwinian selectionist' view of species that now underpins reputed cases of sympatric speciation (Mallet, 1995; Schilthuizen, 2000). It is bolstered as well by views of natural selection being an optimising process that selects for efficiency of resource use (Nylin, 2001; Schilthuizen, 2000; Travis, 1989).

An obvious reason for the stress on post-speciation adaptation relates to the anticipated role of hybridisation in the evolution of reproductive isolation. However, an earlier and more pervasive influence can be traced to a significant intellectual constraint that Darwin faced. The theological view of the fixity of species that dominated pre-Darwinian biology could presumably be overcome only by emphasising the observable variation among individuals, as opposed to their shared complex adaptations, and by Darwin stressing his postulate of ongoing change in species. A good idea of the strength of the constraint he faced is transmitted in the historical works of Ellegård (1958) and Ospovat (1981).

Darwin portrayed complex features of organisms as emerging gradually and in small incremental steps. New species were seen to emerge in essentially the same way. Such views are extrapolated to complex characters being spread to fixation across the entire geographical distribution of species. This is not acceptable as some populations are distributed across thousands of kilometres and others are spatially separated from one another. Nevertheless, such views persist, although consensus on *Homo sapiens* having had a single origin seems now to be emerging.

Of particular relevance to understanding the origin of complex adaptations is that the subcomponents of such an adaptation may have no benefit to an organism outside of the context of the complex adaptation as a whole and in the environment in which that complex adaptation is functional (see Paterson, 1985, 1989). Interpretations of adaptation that assume complexity is readily spread to all members of a species' gene pool do not reflect observed pattern. Even relatively minor allelic substitutions are not spread to fixation in this particular way; at least, no empirical case of such fixation has been claimed.

The recognition concept overcomes the logical problems outlined above and is not prejudiced by notions of the 'benefits' of reproductive isolation. Hybridisation is therefore relegated to a place, in theory, that is commensurate with its status as an ineffectual by-product. The emphasis thus shifts from minor variants among individuals to the primary role of complex mechanisms in the survival and propagation of individuals of any species, but specifically in relation to the usual environment inhabited by that species. Complex adaptations do not readily change, which suggests a different approach to understanding adaptations such as host relationships and reputed cases of host shifts. This is a topic covered more fully in subsequent chapters, principally in relation to interpreting the ecology of polyphagous pests (Chapter 7) and in using theory to improve biological control practice (Chapters 8 and 9). Finally, the recognition concept provides reasons why optimality models and life history theory based on optimising or competitive selection will not lead to as good an understanding of organisms or their ecology and evolution as we would like.

Understanding cryptic species: using theory to anticipate their occurrence

The recognition concept was developed only when cryptic species had been understood to be true sexual species, and not populations in the process of evolving full morphological divergence, as anticipated under the isolation concept (e.g. Ayala, 1982, p. 195). Consequently the recognition concept explicitly incorporates statements about the status of cryptic species, their relationship with morphologically similar species, and the situations in which they are likely to evolve (Lambert & Paterson, 1982; Paterson, 1991). The recognition concept thus primes us on when to anticipate species problems or species complexes.

The recognition concept emphasises the adaptive relationship of the fertilisation system to the normal habitat of the species. The isolation concept, by contrast, emphasises the importance of reproductive isolation from closely related species in interpreting mating features, which explains the current emphasis on hybridisation in evolutionary studies. Consider the normal habitat of organisms in fairly general terms. The calls of crickets, birds, cicadas, frogs and so on are adapted to work effectively in their normal habitat. Forest organisms tend to have calls made up of pure tones, which transmit readily through dense vegetation. By contrast, grassland species tend to have pulsed sounds or trills because pure tones are too easily disrupted by air currents, and so on (Masters, 1991; Morton, 1975; Paterson, 1989).

Among the Lepidoptera, moths are principally nocturnal and butterflies diurnal. A large number of butterfly species signal visually with their wings. Optical signals transmit effectively in the open habitats in which brightly coloured butterflies are found, whether above the forest canopy, in sunny forest clearings or in open woodland or grassland. Humans, as visual organisms, detect such signals with facility, except for those involving UV wavelengths, and we can thus readily interpret species limits of most of these organisms (Paterson, 1991). Contrast this situation with that in moths. Visual signals are inappropriate for nocturnal organisms, unless they produce light signals. In place of the long-distance visual attraction of butterflies, many moths use highly volatile chemicals at a time when turbulence in the atmosphere is very low. The sense of smell in humans is not particularly well developed, so we do not detect these sexual signals. Because night-flying moths are generally dingy or cryptically coloured for daytime protection, we frequently have difficulty recognising species limits (Paterson, 1991). In other groups that use signals that are not readily detected by humans we also encounter such difficulties. Small insects such as many microhymenoptera use tactile communication for close-range communication, for example.

Theory helps to predict in which types of situations or in what type of organisms sibling species might be expected (Paterson, 1991), as outlined above. Consider what that means when preserved specimens are examined and decisions are made as to their species status, for taxonomic purposes. Many animals that use visual signals will present no problems when their species limits are being designated, because they are so readily categorised according to the physical features used for signalling. By contrast, animals that use signals that are cryptic to humans may be difficult to separate accurately from one another on physical appearance. Many moths resemble one another in outward appearance, as do mosquitoes that are night-fliers (*Culex* and *Anopheles*) (Paterson, 1991). These organisms use chemicals, sound and tactile signals in various combinations to communicate between the sexes. To illustrate, some parasitic microhymenoptera have a complex behavioural interaction associated with mating and which seems to be diagnostic for species, as in the genus *Coccophagus* (e.g. Abeeluck & Walter, 1997; Walter, 1993b). Other parasitoids do not behave in this way; the interaction is brief and reveals little about species status, as in the genera *Aphytis* (Fernando & Walter, 1997; Gordh & DeBach, 1978) and *Trichogramma* (Pinto et al., 1991). These cases reveal the need to assess how the organisms themselves recognise appropriate mating partners. Appropriate approaches and techniques are discussed later.

Consider now the example that was used to start Chapter 2, that of the polyphagous bollworms. Both *Helicoverpa armigera* and *H. punctigera* have been defined only in morphological terms (Common, 1953; Paterson, 1991). The species limits designated to each may well be accurate, *but we do not know*. Similar uncertainty attends situations in which taxonomists, using structural criteria, and usually morphological ones, are explicitly uncertain about species limits. Frequently, but not always, such cases involve populations with allopatric distributions. Species problems are also a possibility when exceptionally generalised habits or ecologies are assigned to a particular species, particularly if the organisms signal sexually in a mode that humans do not readily detect unaided (Paterson, 1991). Broadly polyphagous 'species' may, in fact, be made up of more than one species each of which is somewhat host specific on a subset of the range of hosts attributed to the 'generalist species' (Paterson, 1991). Such situations are increasingly described, across a range of taxa.

To deal effectively with such pest problems the hypothesis proposed by the taxonomists should be investigated. But it needs to be tested on the terms imposed by the organisms themselves. Approaches to such problems are dealt with in the following subsection.

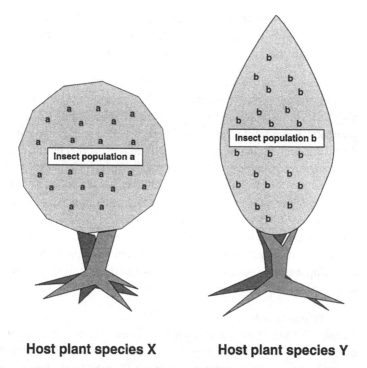

Host plant species X   **Host plant species Y**

Figure 6.5. The distribution of herbivorous insects on two host plant species. All individuals are classified within a single species, but their status in relation to gene flow is unclear. They have therefore been designated population 'a' and 'b' as an aid to setting up hypotheses about their species status, for later testing.

The reproductive limits of sexual species – experimental approaches

Problems of species limits usually present themselves in a form that is, in one way or another, analogous to the situation depicted in Fig. 6.5. In the case illustrated, insects are present on host plant species X and they resemble morphologically those on host plant Y. Such a problem is usually associated with organisms that use a communication mode that is not readily perceptible to human sense, as outlined above. The central question to be addressed is whether mating takes place at random, in nature, between individuals observed on host plant X and those on plant Y. Alternatively, mating may be positively assortative. The equivalent question, in population genetics terms, is whether free gene flow takes place between the insects from the alternative host plants. In general, such questions are tackled by first setting up two *a priori* groups. Should there be more than two populations, or samples, the equivalent number of *a priori* groups has to be set up on the basis of one or another of the following discontinuities.

1   Biological differences across space or through time
Biological differences in different parts of the geographical distribution of the species of interest frequently provide a basis for setting up an appropriate test. For example, if parasitoids of a particular species in one locality do not attack hosts that the species usually attacks in other places, the situation warrants an investigation for the presence of cryptic species. Alternatively, moths of a particular taxonomically defined species may be caught in a pheromone trap in one area, whereas identical traps do not work elsewhere despite the presence there of that species (e.g. Goyer *et al.*, 1995). Another example: two herring populations spawn off Kiel, Germany, one in spring and one in autumn. They mix in winter, but separate for the next spawning season (Sinclair & Solemdal, 1988), so these are undoubtedly different species in population genetics terms.

2   Taxonomic difficulties
Taxonomic uncertainty about the species limits of various taxa is not uncommon and may well indicate the existence of a species complex. For example, taxonomists were not certain whether the Jarrah leaf miner, a small incurvariid caterpillar, was comprised of one or more species. By addressing questions about the reproductive status of the various host-associated types, the situation has been partly resolved; at least two species are encompassed by this taxon, possibly more (Mahon *et al.*, 1982).

3   Behavioural and other discontinuities in a locality
Difficulties may be experienced in interpreting aspects of the behaviour, physiology or ecology of a 'species' within a particular area. For example, *Anopheles gambiae* mosquitoes, which transmit malaria in Africa, were controlled effectively by spraying dwellings with various insecticides. Although malaria transmission declined almost altogether, '*An. gambiae*' individuals were still present in the area, but now only outside houses and feeding on cattle rather than on people. Because evolutionary theory, even now, anticipates ongoing adaptation, initial claims heralded 'behavioural resistance' to the insecticide. Once the *An. gambiae* species complex had been unravelled, the appropriate interpretation became clear; *An. gambiae* is an endophilic and anthropophilic species whereas *An. quadriannulatus* is exophilic and zoophilic. The latter species was not affected by benzene

hexachloride applications inside houses, persisted in the area and was assumed to be *An. gambiae* with a new behaviour (all from Paterson, 1993a). Despite this revelation, similar claims of the evolution of 'behavioural resistance' are still made for Pacific mosquitoes. Other circumstances warrant serious inspection for the inadvertent amalgamation of different species under one name. Broad polyphagy and adaptive polymorphism have been covered earlier, but claims of alternative reproductive tactics and so on are also likely candidates.

To test the species status of different populations, a technique must be selected, of which several exist (each outlined in a separate point below). Not only must the most appropriate technique for the situation be selected, but a thorough understanding is needed of the limitations of each available technique. Such insight about techniques derives not only from a thorough understanding of the principles on which the technique is based, but also from an appreciation of the functional nature of the fertilisation system of the organisms concerned. Only occasionally can the species status of populations be tested directly (see 1, below). Furthermore, the recognition concept explains how indirect tests are prone to yield asymmetries in experimental outcomes, for it is normal that one outcome may produce an unequivocal conclusion, but the alternative outcome from the same test may well be ambiguous and thus inconclusive.

1   Behavioural observations in nature
Direct observation of organisms in nature constitutes the most direct approach, but is practical only when organisms can be observed without disturbance, their behaviour can be manipulated in the field (by the playback of sound for example) or testing whether mating is random in a zone of overlap is possible. In general, these methods are appropriate with larger organisms that are readily observable in nature. Birds, individually marked if necessary, are obvious candidates, but the method could possibly be used on larger day-flying insects such as butterflies and dragonflies. The advantage of such observations lies in their natural setting. All functional steps of the SMRS are thus played out in the appropriate context.

Tests of whether mating is random or positively assortative are possible even with small organisms in overlap zones if sufficient information is available to score the different types and hybrids accurately. This method was used to confirm for the first time that *Anopheles gambiae* comprised a complex of independent cryptic species, by capturing wild, mated females and isolating them to obtain egg batches.

Offspring were scored for sex ratios, testicular sperm (adult male offspring) and asynapsis of larval polytene chromosomes (they are asynaptic in hybrids). All families had normal sperm, all chromosomes were synapsed and there was no sign of distorted sex ratios, despite the different 'types' being present in a ratio of 5:4:8. The strong evidence for positive assortative mating in the wild confirmed the coexistence of three species in the one locality, with these being *An. quadriannulatus*, *An. gambiae sensu stricto* and *An. arabiensis* (Paterson, 1964).

2   Cross-mating experiments

Reciprocal cross-mating experiments are frequently portrayed as an ultimate solution to species problems (e.g. Mayr, 1963, p. 50; Rosen, 1986). The results they yield should, however, be interpreted with due attention to the fundamental principles of sexual species theory. Any interpretation offered will rely inevitably on theory derived from basic assumptions. Thus, any commitment to the 'facts of the matter' or any evasion of concepts of species simply reveals an impoverished understanding of the situation and a failure to appreciate the subtleties involved. Results and interpretation will almost inevitably be compromised.

Cross-mating experiments are set up as illustrated in Fig. 6.6. The range of possible outcomes is detailed in Table 6.1. The first outcome (Table 6.1) would imply that the experimental arena is inappropriate for the organisms under investigation. A suitable arena is needed or an alternative technique should be considered (see below). The only outcome that yields an unequivocal result is the second one, because successful matings in the control and no mating in the test crosses dictates that the two samples represent two distinct species. That conclusion holds even

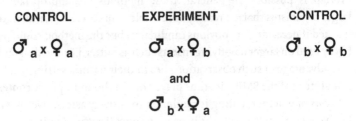

Figure 6.6. Cross-mating experiments are designed to test the species status of individuals from different hosts (as in Fig. 6.5), localities or samples. Five outcomes are possible, but only one is unambiguous (see Table 6.1).

Table 6.1. *The crossing experiment depicted in Fig. 6.6 could yield one of five general outcomes, as detailed. The interpretation of each outcome is given in terms of the isolation concept of species (IC) and the recognition concept (RC), and is explained fully in the text*

	Outcome number				
	1	2	3	4	5
Controls	No mating	Mate	Mate	Mate	Mate
Experiments	Mate or no mating	No mating	Mate	Mate	Some mating
Offspring from crosses	Viable or non-viable	–	Non-viable	Viable	Viable or non-viable
Interpretation (IC)	Not interpretable	Two species	Two species	One species	–
Interpretation (RC)	Not interpretable	Two species	Not interpretable	Not interpretable	–

if representative individuals are not distinguishable morphologically, as documented in host-associated populations of parasitic wasps (Fernando & Walter, 1997) and planthoppers (den Hollander, 1995). Should two such populations exist sympatrically they would persist there as independent gene pools, or species, as do the two 'populations' of Kiel herrings (Sinclair & Solemdal, 1988).

Outcomes 3 and 4 in Table 6.1 yield definite conclusions only under the isolation concept of species. When reproductive isolation is specified as the defining criterion of sexual species, the particular means by which such a result is achieved demands no specifications in terms of mechanism. Although sterility could never have evolved for the purpose of reproductive isolation, as Darwin realised (see Mallet, 1995; Paterson, 1988), it is still often enumerated as an isolating *mechanism* (Templeton, 1989). Outcome 3 is actually even problematic, for sterile 'matings' are typical of certain relatively common intraspecific crosses, such as cytoplasmic incompatibility (Rousset & Raymond, 1991) and self-incompatibility in plants. The latter even involves different flowers on the same individual plant, so their conspecificity is beyond question (Paterson, 1988). Sterility therefore cannot serve as the defining criterion for species, so outcome 3 does yield equivocal results.

Should the experimental individuals mate and produce viable offspring (outcome 4), the reproductive isolation criterion suggests the two populations are conspecific. This conclusion is, however, also question-able, as shown by the way in which the recognition concept deals with the situation. Under the recognition concept, reproductive isolation is viewed as a by-product of different sexual populations having evolved, independently, differences in their SMRS. The SMRS is a complex adap-tation made up of a sequence of functional steps to ensure fertilisation takes place (Fig. 6.3). Early steps in the sequence usually function in dis-tance attraction. Individuals of one species may have a different long-distance attractant from those of a second species. Cryptic species among the moths provide good examples (e.g. Goyer *et al.*, 1995). In the field, pheromone released by individuals of either species would not attract in-dividuals of the second species. When crossing tests are conducted, the ini-tial step (or steps) of the SMRS may be obviated. If those steps that oper-ate after the distance attraction are similar to one another across the two species, the heterospecific individuals would mate in the cage, and possi-bly produce viable offspring (e.g. *A. gambiae* mosquitoes (hybrid females are viable) (Hunt *et al.*, 1998) and *Ribaudodelphax* planthoppers (De Winter, 1995)), an outcome that provides yet another reason for rejecting steril-ity as a gauge of species status (Paterson, 1988). Nevertheless, sterility is still frequently used implicitly in delimiting species (e.g. De Barro & Hart, 2000; Pinto & Stouthamer, 1994).

The final outcome (5) in Table 6.1 has been obtained in some cross-mating tests, for example with parasitic wasps by Rao & DeBach (1969). These may be situations in which the organisms perhaps did not mate on initial exposure to one another, but were left for so long that attempts by males to mount females were eventually 'permitted' (Fernando & Walter, 1997). This outcome does not represent circumstances in the field at all realistically. An appreciation that recognition of potential mating part-ners is the more realistic defining criterion of species demands that atten-tion be given to the duration of the experiment. The experimental crosses should not run longer than the time it takes for the control pairs to mate (Fernando & Walter, 1997). Interpretation of work on trichogrammatid species, which are claimed to show much intraspecific variation in mating success across populations (Pinto & Stouthamer, 1994), may be affected in this way if cryptic species are inadvertently included in the samples.

3   Molecular genetic techniques

Molecular techniques provide a range of methods for assessing whether mating takes place at random between organisms that have different origins or derivations. The techniques are modern and popular, but each has particular strengths and weaknesses that should be appreciated. They are therefore best applied in relation to appropriate theory and matched to specific questions. Again, with each technique that can be used to resolve species problems certain outcomes are conclusive but others are not necessarily so. Some molecular techniques are not appropriate for investigating species limits, but are useful for the development of easily used markers for the rapid identification of species whose limits have been recognised by an appropriate technique, although this distinction is not always clarified in the literature. Unknown individuals, regardless of life stage, can thus be assigned to their appropriate species once the limits have been defined by other means. Several PCR-based tests have been developed for this purpose and have facilitated identification outside of the laboratory (see Armstrong *et al.*, 1997; Cook, 1996; Kambhampati *et al.*, 1992; Paskewitz & Collins, 1990; Perring *et al.*, 1993; Wilkerson *et al.*, 1993). Such species-specific markers that are developed will not necessarily detect any additional species that may be discovered. For example, the newly discovered *Anopheles gambiae* complex species discovered by Hunt *et al.* (1998) was cryptic to the PCR-based identification procedure that is routinely used for species in this complex.

Electrophoresis was the first of the molecular techniques to be developed (see Richardson *et al.*, 1986, for review) and has been used in population genetics and for the investigation of species limits for several decades. The more recently developed DNA analyses have the advantages over allozyme electrophoresis that (i) both sexes and all developmental stages can be identified by the same technique, (ii) material suitable for analysis does not demand such stringent preservation and storage conditions, and (iii) DNA suitable for analysis can often be recovered from material preserved simply or even naturally (see Collins & Paskewitz, 1996).

Electrophoresis is also used, in one form or another, to help to visualise the molecular characteristics of individuals subjected to most of the more recently developed genetic techniques. For this reason, and because enzyme (or allozyme) electrophoresis remains the method of choice in species studies (Avise, 1994; Loxdale & Lushai, 1998), the principles of electrophoresis are outlined initially in some detail. Explicit details, procedures and relevant recipes are available elsewhere (Richardson *et al.*, 1986; Symondson & Hemingway, 1997). Briefly, a potential difference is set up, by means of an electric current, between two points in an ionised solution. The solution is usually supported within a stable and permeable medium, like starch gel or cellulose acetate. The potential difference causes charged particles, such as proteins, to move.

The charged protein particles targeted in electrophoresis are the soluble enzymes. Once a current has been run for the requisite time (20–120 minutes, depending on substrate, running conditions and so on), the target enzyme must then be singled out from the conglomeration of all charged molecules that have shifted between the electrodes. Enzyme-specific substrates are therefore added to the entire 'running medium' and the specific products are then stained so they can be visualised. The distance they moved can then be assessed.

In each electrophoresis run, individuals from each population or sample to be compared must be included. The distance travelled by the enzyme will vary with conditions, and the comparative distances run by enzymes coded by alternative alleles are small. Interpretations of the resultant pattern (or zymogram) requires an understanding of enzyme structure and Mendelian genetics; reviews of both topics are available (e.g. Richardson et al., 1986). The technique assesses which populations exchange genes in nature, so species limits can be meaningfully investigated only if the samples are derived from areas of sympatry. Most molecular markers *are useful only after species limits have been established by some other means.* They can be used on individuals from allopatric populations, but they will not necessarily detect the presence of unrecognised cryptic species (e.g. Hunt et al., 1998).

In reading a zymogram, one can take a strictly structural approach to defining species limits, or population genetics principles can be used. The former yields only limited insights, however, as it does with other techniques used to establish species limits. Again, we see the situation in which one type of result yields an unequivocal conclusion and another is inconclusive. Structural interpretation is reliable if one has set up two *a priori* groups, suspected of representing different species, and one electromorph of the screened enzyme is found in one group whereas the alternative electromorph occurs only in the second. The locus in each group is said to be monomorphic, and such fixed differences are sufficient to put the matter beyond doubt (Richardson et al., 1986). By contrast, the lack of fixed differences reveals nothing about species limits because such a result could be obtained whether one is dealing with one species or more.

One can look at the results outlined above from the population genetics perspective. Interpretation is not affected, but this perspective illustrates how one can go beyond fixed differences to understand species limits. A lack of heterozygotes at a locus, even in a sample of only five individuals from each population (Adams *et al.*, 1987), yields certain evidence that one has two species. In short, no recombination of genetic material takes place between representatives of the two populations.

In searching for fixed differences between samples, the optimum approach is to screen as many enzyme systems as possible. As pointed out above, relatively few individuals are required. Should fixed differences not be located, an alternative approach is still open for probing species limits. If a locus is not fixed for a particular allele, that locus is said to be polymorphic. In a large sample of individuals drawn from a Mendelian population, the distribution of the various allelic morphs (or electromorphs on the electrophoresis substrate) at any one polymorphic locus should be in Hardy–Weinberg equilibrium. The same should be true of a sample from the alternative population. However, the frequency distribution of allelic morphs in one population is likely to be different from that in the alternative population because of the differential influences of ecological conditions and historical events. Such a comparison is readily made statistically. Outcomes in which allele frequencies do not differ among populations are insufficient to demonstrate that the two samples derive from a single panmictic population.

Even if a significant difference in allele frequencies is demonstrated between two samples, derived sympatrically, interpretation is confounded by various possible influences. For example, the differences between host plant-associated samples of incurvariid moth larvae were presented as evidence for cryptic species (Mahon *et al.*, 1982), whereas similar data on apple maggot flies are presented simply as an intraspecific differential in host associations (Feder *et al.*, 1988). Presumably, the safest way to deal with such data is to acknowledge that they are simply an indication that positive assortative mating may be taking place within a taxon (taxonomic species). In other words, the results are parsimoniously explained by the two species hypothesis, but other influences could, at least in theory, generate such a pattern of allele frequencies. For example, adults of different genotype could respond differentially to the host species, or larvae of particular genotypes could survive better on one host than the other.

Of the more recently introduced molecular techniques (Loxdale & Lushai, 1998), microsatellite loci have become popular for investigations in which individuals need to be distinguished (as in parentage studies) or in which high levels of heterozygosity and Mendelian inheritance are needed for population genetics studies (Lambert & Millar, 1995;

Rosenbaum & Delhard, 1998; Schlötterer & Pemberton, 1998). Microsatellites have proved revealing in species studies where allozymes yielded little, and are envisaged to be used more for such purposes (Waters *et al.*, 2001). For studies of species limits, the same population genetics principles covered above for electrophoresis are relevant to analyses involving microsatellites, so studies of allopatric populations remain difficult, and information related to the structure of the SMRS should take precedence. It is worth remembering that the acquisition of species status (in population genetics terms) does not necessarily correlate with measurable genetic divergence, because the measures of molecular variation that we use have 'little, if anything to do with what species are' (Lambert & Millar, 1995). Again, the theory is as important as the technique.

4   Cytogenetics

The chromosomes of some organisms are amenable to much closer inspection than is possible for most other organisms. The giant polytene chromosomes of certain Diptera, including many anopheline mosquitoes, vinegar flies, blackflies, sciarid fungus gnats and chironomid midges, are best known in this respect. They are found particularly in larval salivary glands and in the ovarian nurse cells. Various staining techniques allow visualisation of specific components of the chromosomes, in the form of bands.

Species-specific rearrangements of the chromosomes are useful for accurate identification of cryptic species, and have been used in this way for over half a century. Homozygosity for alternative inversions in sympatry indicates cryptic species, because heterozygotes are disadvantaged and one of the inverted states would have disappeared if the two types mated at random in nature. Chromosomal banding patterns are usually read structurally, with consistent differences inferred to represent different species. By contrast, homosequential banding may occur despite species differences (see Collins & Paskewitz, 1996; Green & Hunt, 1980; Narang *et al.*, 1993), so again an asymmetry in experimental outcomes is evident.

5   Analysis of mating signal structure and function

Species complexes are often investigated by analysing one or other of their mating signals, be they chemical, acoustic or visual. Appreciation that the SMRS is a complex adaptation explains why many studies of sexual signals have more in common with the structural approach to defining sexual species than with the behavioural approach, despite dealing with pheromones or sounds, for example. Description of a difference in signal structure, even if it is statistically significant, may not necessarily reflect a functionally significant feature. The 'pheromone strains' of *Nezara viridula* (Aldrich *et al.*, 1993, 1987) may provide a good example. Functionally significant differences in pheromone blends may exist among *N. viridula* populations, but to date no convincing evidence is yet available (Brézot *et al.*, 1994; Miklas *et al.*, 2000; Ryan *et al.*, 1995). Any differences detected among populations need appropriate tests for any functional significance, as outlined in Baker's (1993) readable account.

Attention to detail is critical, and the signal has to be fractionated and recombined (in chemical communication language) to test the sensory and behavioural responses of other individuals (Baker, 1993). Do they recognise the signal? Do they respond to it? Appropriate tests need to be conducted to test for such responses, and tests will vary with function. When investigating pheromones, for example, tests for volatile distance attractants will necessarily differ from those conducted on relatively non-volatile cuticular lipids, which are close-range pheromones (Rungrojwanich & Walter, 2000). In functional terms that relate to the behaviour of individual organisms, the concept of recognition takes precedence over isolation. Isolation of the group is the consequence, or by-product, of the behaviour of individuals.

Allopatric populations provide special difficulties, because sympatry is a convenient 'template' against which to make judgements about the mating behaviour and thus species status of individuals. The only alternative is to assess the species status of allopatric populations indirectly, by investigating the responses of individuals to signals of the alternative population. If the populations have different signals from one another and individuals do not respond to signals of the alternative population, the existence of two species has been demonstrated. Tests can sometimes be conducted in the field. The caterpillars of the tortricid *Merophyas divulsana* (Walker) are pests of lucerne in part of Australia but are not so in other parts, despite being present. The deployment of a synthetic pheromone blend from the pest individuals attracted moths in the 'home' area, but not where lucerne is not attacked by *Merophyas* caterpillars, thus demonstrating the existence of two species (Whittle *et al.*, 1991). Similar tests can be conducted for other modes of communication; this has been achieved through song playback experiments with birds in the field, for example, and insects in the laboratory (e.g. De Winter & Rollenhagen, 1990; Ratcliffe & Grant, 1985).

Such an approach is clearly needed when dealing with populations that mate in crossing tests and produce viable offspring, as in the *Trichogramma* studies described by Pinto & Stouthamer (1994). This approach has strength in that it overcomes deficiencies noticed in the quantification of reproductive isolation by Pinto & Stouthamer (1994); the SMRS of one species can be analysed, quantified and defined without reference to that of another species. In other words, the mating signals and mating behaviour are seen positively as characters in their own right (e.g. Rungrojwanich & Walter, 2000), although the information can still be used comparatively.

The isolation concept does not encourage research on the mating system beyond that component judged to be responsible for isolating the populations of interest, which might help to explain why investigators focus on such features to locate postulated 'speciation genes' (Coyne, 1992). By contrast, an appreciation of the complex nature of the fertilisation system, and the various functions carried out at each step, does provide a basis for comparison among populations, even allopatric ones.

Intraspecific categories, related to species defined reproductively

The isolation concept of species holds that evolutionary change takes place inevitably in different populations of a given species (e.g. Ayala, 1982, p. 182). Those evolutionists who advocate structural definitions of species share similar views (e.g. Mallet, 1995). That change is seen to be differential among populations (or groups of populations) but significant for speciation only when such populations are cut off from one another, to the point that gene flow is significantly impeded. A hierarchical view of the various stages that a population is envisaged to pass through on its way to full species status was therefore developed. Ayala (1982, p. 193) plotted the degrees of genetic differentiation against the categories of evolutionary divergence in the *Drosophila willistoni* group of species. Each new species is said to evolve through the sequence from local populations, subspecies, incipient species, sibling species and finally to morphologically different species. Local populations are variously also referred to as strains, biotypes, forms and races (Pinto & Stouthamer, 1994; Steiner, 1993). This view remains current, even to the extent that it is used against positive assortative mating as a defining criterion for species (Mallet, 1995). The recognition concept warns against such an interpretation.

The use of informal subspecific categories encourages the view that significant evolutionary change is inevitable in different populations and that adaptive change and speciation sweeps across extensive geographical distributions. They cannot, however, be defined in any non-arbitrary way, and the terms have the potential for inconsistent use among observers, perhaps even by the same observer. Unlike species, these entities are not self-defining. Nevertheless, such terms are frequently used for organisms that (i) may be genetically different, but in which the extent and significance of the difference has not been assessed, (ii) are assumed to be genetically different because they have been reared from different hosts, for example, and (iii) that simply derive from different geographical areas. Many 'biotypes' that are obviously different from one another and that, for example, perform differently on different host plant or host insect species are likely to represent cryptic species. Many early parasitoid 'biotypes' turned out to involve cryptic species (Clarke & Walter, 1995), and the only neutral way in which to deal with populations that have not been investigated for their species status is to refer to them provisionally as host-associated populations until the issue has been resolved. The ease with which genetic variation can be documented and the view that ongoing adaptation is characteristic of populations seems to be sustaining an approach that allows the subjective designation of a population's reproductive status, when the only significant criterion in assessing the status of such groupings should be the assessment of their mating status in nature.

The intentions in specifying subspecific categories are undoubtedly good. The epithets attached frequently provide useful information on ecology, such as the host species attacked and the provenance of introduced material. That could be crucial to any efforts, including future ones, aimed at interpreting and improving particular biocontrol situations. However, use of the terms 'race' and 'strain' has great potential to mislead, and the associated information is conventionally, and conveniently, given in other ways.

With regard to sexual species that have been arbitrarily categorised into 'races', 'strains' or 'biotypes', several questions need to be addressed before the ecology of the various populations can even begin to be interpreted. For example, do two host-associated species exist? If not, why does mating among 'races' not 'homogenise' any differential in host association when both races have been released in the same general area? To disregard precision in such circumstances is likely to undermine efforts in pest management, for example by encouraging biocontrol release attempts of such organisms 'in case they work'. Instead, a more rational scientific underpinning for applied entomology should be sought. When host-associated 'races' have been investigated, they have almost invariably been shown to be distinct species (as exemplified earlier in this chapter). The morphological similarities between the two species, which led to them being designated 'races' in the first place, justify them being called cryptic species.

Species represent distinct gene pools, so if a population is demonstrated to be a separate gene pool it should be recognised as such by calling it a distinct species. To illustrate, insecticide resistance is best understood in terms of population genetics, which is gene exchange within a gene pool. Studying the population genetics of insecticide resistance, to prepare a resistance management strategy, for example, is seriously undermined if distinct gene pools are ignored by calling them deceptive names such as biotype when they are distinct species with separate gene pools.

Misunderstanding not only attends each situation in which a population or sample is arbitrarily designated with an intraspecific categorisation; generalisations, concepts and practice are also likely to be distorted. Many authors accept that the introduction for biocontrol of 'strains' will inevitably be good because increased genetic diversity must increase the chances of obtaining the most appropriate natural enemy (e.g. DeBach, 1969). However, this assumption does not stand theoretical scrutiny on the basis of principles derived from population genetics theory and the recognition concept, and empirical justification is also lacking (DeBach, 1969).

One subspecific category is useful, that of subspecies, because the subspecific status of populations can be determined in relation to their mating behaviour (same as one another) and geographical distribution (different from one another) (Ford, 1974). Such a categorisation also helps us to cope taxonomically with those evolutionary situations in which adaptation took place in a small isolated population, but in which change to the SMRS was limited or did not take place. In nature we thus have populations (or gene pools) of organisms that are differentially adapted to slightly different habitats or environments. Should such habitats become contiguous, perhaps through ongoing climate change, mating between individuals from the alternative gene pools would take place, and homogenisation of any differences would follow, through elimination of the rarer gene arrangement or allele, within the zone of overlap. In cases that involve hybrid sterility, elimination of the rarer form would ensue within the overlap zone (Paterson, 1978). The frequency of 'hybridisation' in the area of overlap would vary among specific cases, as natural selection does not select for a fixed endpoint with regard to species status (Paterson, 1993c). In effect, different degrees of differentiation of the SMRS could be expected in allopatry. That regular hybridisation takes place at relatively low frequency in zones of overlap (and at different frequencies for different pairs of gene pools, or species) does not necessarily indicate that the two populations are 'on a continuous route to sympatric speciation via natural selection' (Drès & Mallet, 2002). A non-teleological alternative is that they underwent adaptive change in allopatry, have become sympatric after a distributional shift, and hybridise at low frequency as a consequence of shared elements of their SMRS. Such hybridisation is insignificant in the ecology and evolution of the organisms and will continue (see above).

In summary, a rational approach to pest management demands that the reproductive status of populations be understood if they are to be manipulated as effectively as possible. The cost of ignoring this aspect will be that pest management will have to continue relying on luck, a luxury that the application of science really cannot afford (see Chapter 10).

Asexual organisms

Asexual organisms are also referred to as uniparental or thelytokous. They are quite common among natural enemies, especially among the parasitic Hymenoptera (e.g. DeBach, 1969). Organisms whose asexual condition is transient and dependent on environmental conditions can still be defined in sexual terms, however infrequent sex may be. For example, low latitude populations of the aphid *Myzus persicae* are typically asexual (Blackman, 1974), but are still potentially part of the species gene pool. Similarly, the sexual *Rhopalosiphum maidis* (Fitch) aphids in their presumed area of geographical origin help to define the asexuals elsewhere in its broad distribution (Blackman, 2000; Remaudière & Naumann-Etienne, 1991). Some strictly uniparental species do produce occasional males, which are apparently induced by temperature extremes, as in *Ooencyrtus submetallicus* (Howard) (Wilson & Woolcock, 1960). Antibiotic and thermal elimination of endosymbiotic *Wolbachia* micro-organisms may even return forms to sexual reproduction permanently (Louis *et al.*, 1993; Stouthamer *et al.*, 1990). However, the males of some forms are non-functional (e.g. Wilson & Woolcock, 1960).

The asexuality of insects is, in almost all cases, known to be a derived condition. That is, their evolution was contingent on the sexuality of their ancestral forms. Only a few asexual groups are 'species' rich and ostensibly asexual (Judson & Normark, 1996; Norton & Palmer, 1991). Only two aphid genera fall into this category, although several acarine groups have been listed. But when can one be certain that sufficient information exists for us to declare a group entirely asexual? Genetic analysis on some entirely 'asexual' organisms has overturned views on the sexuality of certain parthenogenetic groups (Hurst et al., 1992), and may do so with others. In any case, interpretation of the origin of asexuals can be complicated. For example, did the species in the aphid tribe Tramini, all of which are asexual, evolve from one another, as generally implied (Moran, 1992) or accepted (Judson & Normark, 1996)? Could they not have independently lost their sexuality under the influence of their similar ecological circumstances, or could they have evolved independently from one or more undiscovered or extinct 'core sexual species' (Hurst et al., 1992)? In other asexual groups with many taxa, for example the weevils, the taxa represent various grades of polyploidy (White, 1970), and have not necessarily undergone adaptive change. Clearly, dealing with asexuals requires considerable care.

The switch of most forms to asexuality not only reflects a response to ecological contingency but has also incidentally ensured their evolutionary stability, as follows. Most asexual insects are automictic (see Crozier, 1975; White, 1973). The egg cells of automicts undergo a reduction division. To restore diploidy the future egg has to fuse with one of the other haploid products of meiosis. Some crossing over and recombination does take place, but without the need for a sexual partner. Although such organisms are freed from the constraints of a co-adapted fertilisation mechanism, two features ensure stability of lineages, at least in terms of the complex adaptations of the organisms involved. The limited recombination is insufficient to disrupt complex adaptations, and any extreme variants are likely to be selected out because the life cycle of the organisms is still played out in a dynamic and heterogeneous environment. Thus, parasitoids such as *Aphytis chilensis* (Howard), *A. chrysomphali* (Mercet), *Ventura canescens* (Gravenhorst) and the San José scale form of *Encarsia perniciosi* are consistent in their morphology and host associations across geographical space (DeBach, 1969). They do not adapt locally, although they may house considerable genetic variation (Foottit, 1997), so they seldom give rise to new behavioural or ecological types.

Because asexuals do not form Mendelian populations, the concept of species limits does not apply to them. They can be defined only structurally, either in terms of their morphology or in relation to discontinuities in behaviour or ecology (see DeBach, 1969; Foottit, 1997).

Closing comments: Understanding species and interpreting adaptation

In summary, the common call of applied entomologists for 'good taxonomy' should be re-interpreted. What is needed, in fact, is a sound understanding of species theory, and sound application of the theory by phrasing the appropriate questions, the collection of appropriate samples and the use of appropriate techniques. The place of good taxonomy is to underpin such an approach; without sound taxonomy all may be lost.

Understanding species is relevant not only to understanding species limits and cryptic species. Different species concepts influence fundamentally the interpretation of such ecological features as host relationships of herbivores and parasitoids, prey requirements of predators, habitat associations, searching behaviour of parasitoids and predators, and so on. Consequently, an understanding of species from a fundamental perspective influences, in turn, interpretation of the ecology of pest and beneficial organisms. Before illustration of this point in relation to the host relationships of polyphagous pests and parasitoids (in Chapter 7), and examination of the theoretical underpinning of biological control programmes (in Chapters 8 and 9), the different approaches to such ecological investigations that are suggested by the isolation concept relative to the recognition concept are outlined in general terms.

The isolation concept predicts that different populations of a species will almost invariably have adapted or be adapting to local conditions. Therefore, such local populations are expected to have considerably different adaptations from other conspecific populations. For example, herbivorous insects are expected to evolve different host associations should local circumstances elsewhere impose selection pressures on the herbivores (e.g. Fox & Morrow, 1981; Mopper & Strauss, 1998). Because adaptive change is seen to take place so readily, numerous 'examples' of such change are uncritically added to the literature, thus falsely bolstering the original view. Failure to understand organisms and adaptation undoubtedly contributes quite considerably to failure rates in applied entomology.

The idea that local adaptation is relatively easily accomplished is frequently bolstered with reference to those situations in which adaptive
change has been observed, as with insecticide resistance. But caution is
warranted here, because such changes are relatively minor, in that they
usually involve only a single allelic substitution. But even then, disad-
vantageous pleiotropic effects usually become obvious, to the extent that
removal of the selection pressure results in relatively rapid reductions in
frequency of the resistance allele (e.g. Muggleton, 1983). Even without obvious fitness reduction such alleles may be lost (Sayyed, 2001). Local adaptive change to features that are part of a complex mechanism, such as host-
finding behaviour and its associated physiology, is therefore less likely.
From the understanding that derives from the recognition concept as to
how adaptations arise it becomes clear that such features are not as readily
altered as is usually portrayed in the ecological literature. Adaptive change
to complex adaptations requires the special conditions of a small population being confined by environmental circumstances to conditions different from usual. For local adaptation to be claimed, as in postulated cases of
host shifts to novel plants, requires the phenomenon to be demonstrated
beyond doubt, not simply claimed uncritically. Acceptance of this stricture will enhance understanding, and thus improve the application of understanding to achieve desired results.

To illustrate: *Salvinia* was controlled in Australia through the fortuitous introduction of *Cyrtobagous* beetles from Brazil, for they were
thought to be conspecific with those already present in Australia. Would
they have been considered for introduction if other herbivore species had
been found simultaneously in Brazil with the *Cyrtobagous* on *S. molesta*?
An effective biocontrol agent might have been overlooked through insufficient understanding of the species concerned. Fortunately the beetles
were introduced, biocontrol was achieved and the species status of the organisms eventually resolved. We have no such detailed records of projects
that have failed through a lack of understanding, which is unfortunate as
they would be revealing.

A better indication of the inadequacies of the older views of species and
adaptation is derived from a consideration of so-called 'generalist' species,
such as polyphagous herbivores and natural enemies. Modern developments in adaptation theory suggest alternative ways in which to deal with
such species, a topic covered in the following chapter with extensive reference to empirical studies.

7

Polyphagous pests, parasitoids and predators: trophic relations, ecology and management implications

[We] demonstrate the existence of significant amounts of variation in feeding and oviposition behaviour below the species level for both host preference, or degree of specialization to particular host plants, and host suitability. The study of such variation contributes greatly to the understanding of resource use and other ecological processes, and of adaptation to stressed environments such as are caused by agricultural practices.

L. M. SCHOONHOVEN ET AL. (1998, p. 219)

Introduction

The range of principles and interpretations covered in the earlier chapters are applied in Chapters 7–9 to the investigation and interpretation of the ecology of pest and beneficial species, with emphasis on those aspects relevant to pest management. Chapters 7–9 are illustrative rather than exhaustive in that they cover only a minute subset of all ecological understanding and application relevant to IPM.

Chapter 7 deals with polyphagous species, which are those associated with many host types. Such species are considered to be generalists (e.g. Bernays & Chapman, 1994; Briese *et al.*, 1994; Futuyma, 1991; Leather, 1991; Schoonhoven *et al.*, 1998; Ward & Spalding, 1993) and are seen to provide a strong ecological contrast with host-specific herbivores and parasitoids, which are referred to as monophagous. Polyphagous species present special problems for ecological understanding and therefore for pest management, although many host-specific species are notorious pests (e.g. rice planthopper and olive fly). Herbivorous pests, parasitoids and predators are used to illustrate these difficulties and to help to plot a way around them.

Chapters 8 and 9 focus entirely on natural enemies, and their use in biological control against insect pests. These two chapters are designed, with Chapter 7, to illustrate how a strong focus on the most robust scientific, evolutionary and ecological principles available can alter outlook, influence research direction and improve management practice. Some evolutionary principles and processes are generally accepted as flawed, but are still fairly extensively applied in ecological interpretation relevant to the subject matters of Chapters 7–9. For example, group selection, teleology and typology are invoked frequently, albeit perhaps tacitly. They thus influence the questions that drive current empirical research in this area, and also the interpretations offered. The interpretations generally accepted in the discipline areas of herbivore/host, parasitoid/host and predator/prey also warrant some introspection, and the principal alternative approach to understanding adaptation, which was developed earlier in Chapters 5 and 6, is applied to these discipline areas to illustrate the strength of that approach. Only a relatively narrow area relevant to pest management is dealt with in Chapters 7–9, but the treatment does illustrate how such an approach will return benefits if applied elsewhere in research on pest and beneficial species.

The examination of scientific method in Chapter 2 drew on the illustrative case of a polyphagous insect pest, the native Australian bollworm *Helicoverpa punctigera*. Inspection of the available information about the host relationships of *H. punctigera* revealed that logically it could support more than one interpretation of the host relationships of this species. The 'available information' in this case is the raw data on host use that has been collected in the field, where most information on this species has been gathered. Stated in this way, 'available information' does not infer inevitably that the species is polyphagous, or to what extent it may be polyphagous. To gather information that extends interpretation and understanding of the host relationships of species recorded on a diversity of food sources requires further attention to patterns of host species use in the field, as well as to the behavioural and physiological processes that underpin those host relationships.

An understanding of host relationships can be built only on an understanding of adaptation. The two alternative views of adaptation were covered in Chapter 6. This dichotomy signals different approaches to the investigation and interpretation of host and prey relationships. The inspection of polyphagy presented in the present chapter reveals a reasonable alternative to the current interpretations of the host relationships of polyphagous species. This option has remained hidden because of the general stress on 'ongoing adaptation' and intraspecific variation that is evident in the literature and which is, in turn, related to demographic ecology and the isolation concept of species. The text that heads this chapter is an accurate portrayal of the current emphasis in the interpretation and investigation of host relationships. Bernays & Chapman (1994, p. 2), for example, specify that 'behaviour, and especially variation in behaviour, is central to our understanding of the major evolutionary questions' related to insectan interactions with their host plants. The alternative presented in this chapter emphasises the importance of testing interpretations of intraspecific variation and finding the most appropriate functional interpretation for patterns of host use. Emphasis is thus removed from an uncritical acceptance of the power and ubiquity of optimising selection acting inevitably on the intraspecific variation that is so readily documented, and instead is refocused on the central mechanisms of organisms and the usual outcomes of those mechanisms in the typical environment of the species in question. The alterations to interpretations of the host relationships of polyphagous species that follow from such an approach have significant implications for the way in which pests should best be targeted in management schedules, and also for the way in which natural enemies should best be deployed in biological control programmes.

The chapter is organised as follows. First, recent insights about pattern and process in herbivore host relationships help considerably to clarify what requires explanation and which directions are likely to be profitable. This leads into a summary of current interpretations of the adaptive value of polyphagous host and prey relationships. Many observations contrast in some way with current expectations, so these are outlined in a separate section to help to circumscribe what requires explanation. These observations feed into a section that provides a general alternative approach to the understanding of host relationships that are considered polyphagous or generalist. This allows a re-examination of what details need to be known about the host relationships of polyphagous species if we are to understand their ecology. Naturally, the implications for pest management that follow are different from those generated by the research questions emphasised by current interpretations, and these are outlined.

Innovations in understanding host relationships

The evolution of the host and prey relations of insectan herbivores, parasitoids and predators is usually portrayed as the outcome of a coevolutionary arms race (e.g. Thompson, 1994), a metaphor with its origins in the Cold War era of the 1950s and 1960s (Ehrlich & Raven, 1964; Fraenkel, 1959). The phylogeny of herbivorous insects is probably much better known than that of parasitoids. Evolutionary analyses of their host relationships therefore tend to have more of a phylogenetic influence, which has proved revealing and is likely to reflect on the interpretation of aspects of parasitoid evolution. One of the most significant revelations in recent understanding of the evolution and ecology of phytophagous insects has not yet influenced much research and interpretation, although its influence will inevitably be felt. The pattern of host relationships of insect herbivores does not match the expectations of coevolutionary theory. Jermy (1984) analysed the general pattern of the phylogeny of insect herbivores relative to the phylogenetic pattern of their host plants. In principle, four kinds of relationships are expected (Fig. 7.1). Classic coevolutionary theory cannot explain the evolution of patterns A and D in Fig. 7.1, yet these are the two that include, by a long way, most herbivorous insect species (Jermy, 1984).

Jermy's analysis of the basic premises of coevolutionary theory indicates that coevolutionary theory is unlikely to explain the relatively few remaining cases. A phylogenetic perspective of an apparently tightly coevolved mutualism between an insect herbivore and its host plant, yucca moth on yucca, demonstrates that almost all of the adaptations of the yucca moth, which outwardly seem to have evolved to suit it specifically to exploit yucca seeds, were already present in the ancestral moths that exploit plants that are quite different from yucca (Davis *et al.*, 1992; Thompson, 1994). The move to yucca seems to have been made possible by preadaptation plus the usual adaptive change associated with an evolutionary shift to a new host species. The points outlined above imply that the postulated coevolutionary arms race is a poor model, although it is still popular.

Changes in perception at such a fundamental level lead inevitably to further questions. Re-examination of the role originally proposed by entomologists for secondary plant metabolites indicates that many of them are not primarily defensive against particular insect herbivores. They have a range of possible roles that have long been appreciated by botanists, including response to various stresses (such as extremes in light levels,

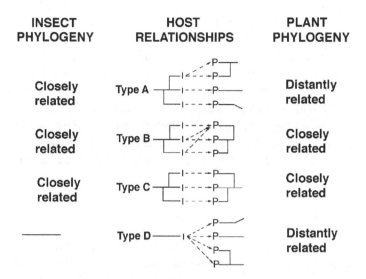

INSECT
PHYLOGENY

HOST
RELATIONSHIPS

PLANT
PHYLOGENY

Closely
related

Type A

Distantly
related

Closely
related

Type B

Closely
related

Closely
related

Type C

Closely
related

——————

Type D

Distantly
related

I = phytophagous insect species
P = plant species
— = trophic relations

Figure 7.1. Diagrammatic contrast between the host associations of closely re-
lated herbivorous insects and the phylogenetic relationships of their host plants.
All herbivorous insects fall roughly into the four categories represented. After
Jermy (1984). Reprinted by permission of the publisher.

temperatures, moisture, heavy metal concentration), injury (including
herbivory) and invasion by other organisms (including fungi, bacteria,
nematodes and insects) (Jermy, 1984; Kutchan, 2001; Seigler, 1998). Even
if not primarily defensive against insects, they may have serious effects on
insect herbivores that do not have appropriate adaptations to cope with
them.

The secondary metabolites do, nevertheless, play a major role in the
sensory physiology and behaviour associated with the location and recognition of oviposition substrates and feeding sites by herbivorous insects
(Bernays & Chapman, 1994; Jermy, 1984; Schoonhoven *et al.*, 1998). Interactions at this level play a predominant role in dictating insect–host plant
relationships, and it is not surprising that many ecologists have shifted
emphasis to investigation of the behaviour of individual organisms to understand specialisation and generalisation (Bernays & Chapman, 1994).
This approach has led to the production of the most useful general models
of how consumers interact with the various resources they use. Such behavioural models have been developed mainly by those who target weed

and insect pests for biocontrol, and the models are covered in Chapter 9. Surprisingly, then, the relevance of adult behaviour to interpretations of adaptive change in host relationships has taken a long time to be more generally appreciated. Specifically, in the evolution of host relationships, a behavioural change in adult phytophagous insects has inevitably to precede any adaptive change of the immature stages to the new host plant (Bernays & Chapman, 1994, p. 2). Diversification of herbivores is thus led by the physiology and behaviour associated with finding a host for oviposition, so this is the aspect that needs to be looked at when adaptation and speciation are considered.

Current interpretations of polyphagy

Polyphagous species are widely considered to be generalist users of an array of resource types, with species usually being the unit of resource measurement. Unlike specialists, which target a single type of resource, they are considered somewhat inefficient but more flexible to externally imposed conditions (e.g. Carriere & Roitberg, 1994). This interpretation is underpinned by several evolutionary assumptions that require consideration and critical assessment.

Chapter 6 detailed the most widely accepted perception of natural selection. The driving force comes from the organisms themselves generating an environment that is essentially competitive. Greater efficiency is considered to be positively selected because that is seen to translate into enhanced fitness, and usually it is efficiency in resource acquisition that is emphasised. The physical environment is not ignored, but physical factors tend to be treated as unidimensional features of the environment, as indicated by their axes in diagrammatic representations spanning such continua as 'stable vs. unstable' and 'persistent vs. ephemeral'. Organisms that inhabit the more 'unsatisfactory' conditions are seen to exploit a competitive vacuum whereas in the more 'satisfactory' situations competition is the primary driving force. Differential adaptation is expected across the local populations that span the distribution of species (and is expected to lead ultimately to allopatric speciation) and even within local populations (where sympatric speciation is expected). Such evolutionary change is expected even in 'complex adaptations' like the specific-mate recognition system and host plant relationships. In general, change is expected to proceed until constraints to further change are encountered.

The specific theory that tends to drive research aimed at understanding polyphagous species is as follows. Because fitness benefits are emphasised, polyphagous species are considered to have obvious advantages over specialists under most conditions, even though specialists may be more efficient at converting a particular resource. The expected fitness benefits of polyphagy derive from incorporating more resource categories into the diet. This, in theory, provides more opportunities for the organisms to maximise their fitness by using alternative foods, and this is driven by differences (both spatial and temporal) in host abundance, diversity and quality (e.g. Bossart, 1998). Frequent claims are made that generalists benefit from having a broad geographical range, many generations each year, an almost certain availability of food at any time, and the possibility of balancing their diet by taking food from alternative sources. Generalists are therefore said to be in a position to build up their populations on a sequence of alternative hosts, and this increase in 'ecological amplitude' (Futuyma & Peterson, 1985) is often portrayed as the adaptive advantage (Van Valen, 1965), despite the group selectionist overtones.

A primary evolutionary incentive seen for generalist species is therefore the incorporation into their diet of unused resources (e.g. Futuyma, 1987), a feature that underpins models of sympatric speciation (e.g. Bush, 1994). The means open to organisms to exploit such resources, which represent 'empty niches' in demographic ecology terms, reflect the end points of the generalist–specialist continuum. Species may pre-empt an empty niche and become specialists, or they forego competitive superiority and become generalists. The expectation is that populations of a generalist species 'expand into' situations to take advantage of available resources. Therefore, it is not unexpected to find species that are generalists, nor resource-based diversity of populations within a species.

Specialisation, by contrast, is expected to incur the costs of (i) a high risk of extinction when the host becomes rare, (ii) a smaller available niche space, (iii) exposure to a single immune or defence system that can evolve to eliminate the parasite (Red Queen hypothesis), and (iv) an increased risk of mortality when using an unsuitable host (e.g. Sasal et al., 1999). These points are either teleological or typological (see Chapter 6), and even though the argument can be recast in terms of individual selection – 'in every generation of a host-specialized species, some individuals must fail to find suitable hosts' (Futuyma, 1991) – acceptance of the argument seems to rely mainly on intuition, or 'all other things being equal', which they are not. Nevertheless, the issue of what countervailing forces of selection, or trade-offs (Bossart, 1998; Carriere & Roitberg, 1994), favour specialisation drives a substantial amount of enquiry.

The expectations outlined above drive the following research agenda. Measurement of the fitness consequences of using alternative resource types is considered crucial, not only to understanding polyphagy but also for predicting future evolutionary trajectories and subsequent specialisation of the organisms in question (Carriere, 1992; Carriere & Roitberg, 1994). Therefore ecological correlates with polyphagy are explored in various ways, with suggestions of large larval size, larval overwintering, wood-feeding, ephemeral food resources, poor dispersal ability, environmental heterogeneity and unpredictability of resource availability being significant (e.g. Cates, 1981; Kassen, 2002; Novotny, 1994; Ward & Spalding, 1993). Polyphagous species are also seen as most likely to be deterred from feeding, whereas monophagous or oligophagous species are considered most likely to be stimulated to feed by allelochemicals in a potential host plant (Martinat & Barbosa, 1987). Almost inevitably, causation is implied in terms of selective advantage or 'trade-offs', so this approach also assumes adaptation is an ineluctable process and it overrides the details of species-specific characters such as their mechanism for locating hosts and their life cycle requirements.

Specific differences are detectable among the interpretations of polyphagy that are offered by different scientists, but by and large they are all underpinned by the same set of premises of how evolution works and how contemporary resource use drives the ecology and evolution of organisms. The views of adaptation canvassed by the isolation concept and by those who downplay the relevance of species (e.g. Bush, 1994; Mallet, 1995) are logically consistent with and cope comfortably with interpretations of the 'generalist' ecologies of polyphagous crop pests and beneficial species.

Evidence and interpretations that counter current theory

Several lines of observational evidence and a few re-interpretations suggest that the approach outlined above needs scrutiny, even at the most fundamental level.

1   Specialists predominate among phytophagous insects to an overwhelming degree (Jermy, 1984). Such a discrepancy between

theoretical expectation and empirical pattern has tended to focus attention on specialist species, to search for the factors involved in the maintenance of specialisation as this is more difficult to account for under theories of competitive selection and ongoing local adaptation (e.g. Bernays & Chapman, 1994, pp. 258ff.). Generalists may be represented to a lesser extent even than currently acknowledged. In various ways, as detailed in the following points, the tag of 'generalist' is invoked prematurely.

2   Definitions of polyphagy vary widely (Bernays & Chapman, 1994; Schoonhoven et al., 1998). Some specify the consumer species should use many species of hosts, hosts in several genera or even families or orders (Mitter et al., 1993; Novotny, 1994), which emphasises the arbitrariness of the concept (Bernays & Chapman, 1994). This subjectivity can be contrasted with the non-arbitrariness of species boundaries (e.g. Mayr, 1976) and species being self-defining (Lambert et al., 1987). Development of theory related to polyphagous species will have to transcend this issue if robust generalisations about host and prey relationships are to be developed.

3   The terms 'generalist' and 'polyphage' tend to draw attention away from the behaviour of the insects by emphasising their pattern of resource use. The functional context given to different plants relates to the assumption that organisms optimise their fitness under each set of circumstances they encounter. Host 'preference' is therefore expected to correlate with subsequent 'performance' on preferred plants, whether that performance be egg maturation by adults that also feed on larval foodplants or larval performance after hatching on the plant. Organisms are expected to 'do the best they can'. The basic neurosensory process of 'recognition', which itself is seen as an unsolved challenge for neuro-ethologists (Huber, 1985; Schneider, 1987), is pushed aside in favour of the more complex process of 'preference', and the functional interpretation hinges on the generalised idea of maximising fitness. Such ideas about optimisation draw attention away from the mechanistic basis that underpins the host association and thus the environmental and physiological context of the association with the particular plant species.

Labelling a species 'specialist' or 'generalist' therefore does not necessarily imply any ecological understanding. At best, such terms are preliminary approximations, but in initially labelling something in this way a particular direction on subsequent research is likely to be imposed, even if unintentionally. Because such direction has not been reasoned from first principles it may not be the most appropriate line of thought for generating understanding.

4   Scrutiny of reputed generalist species across an array of taxa, both
insect and others (Barbieri *et al.*, 1995; Bidochka *et al.*, 2001; Coulson,
1990; Duffy, 1996; Gordon & Watson, 1986; Green *et al.*, 1972; Groth,
1988; Kirsch & Poole, 1967; Knowlton & Jackson, 1994; Tatarenkov &
Johannesson, 1998, 1999), often reveals unrecognised cryptic species
(see Chapter 6), each with a restricted set of requirements (e.g. host
range) that accounts for part of the broad set of tolerances attributed to
the generalist or polyphagous 'species'. Insect 'species' that might
originally have been said to have been polyphagous or generalist, but
comprise a complex of species each with a rather narrower range of
hosts, even to the point of monophagy in all component species in a
few such complexes, include various fruit flies (Condon & Steck, 1997;
Drew & Hancock, 1994), lepidopterous pests (Goyer *et al.*, 1995; Pashley
et al., 1992; Whittle *et al.*, 1991), cerambycid beetles (Berkov, 2002), bugs
(den Bieman, 1987; Wood, 1993), two-spotted spider mites (*Tetranychus
urticae*) (Navajas *et al.*, 2000; Tsagkarakou *et al.*, 1998), probably
European corn borer (Linn *et al.*, 1997; Roelofs *et al.*, 1985), possibly
Gonipterus scutellatus (Clarke *et al.*, 1998) and perhaps even *Leptinotarsus
decemlineata* (see Hsiao, 1978; Lu *et al.*, 2001).

Almost inevitably, the whiteflies covered by the taxon *Bemisia tabaci*
belong to several unrecognised species. The difficulty in
understanding *B. tabaci* ecology is compounded by arbitrary decisions
about species status within this taxon. Consider the statement: 'If the
recent precedent to recognize new species of *Bemisia* is followed by
others, an excessive number of *Bemisia* species could be erected for
which there are no (practical) distinguishing morphological
characters . . . This unfortunate trend would essentially reverse that
which was initiated in 1957, in which an attempt was made to simplify
the taxonomic status of the group by synonymization of at least 19
entities into the *B. tabaci* epithet . . . In view of recent advances in our
understanding of the *B. tabaci* species complex, the creation of new
epithets based upon limited information should be discouraged'
(Brown *et al.*, 1996). Advocating the amalgamation of 'biotypes' may
well be as inappropriate as separating them into species. *The issue is not
whether consistent morphological or molecular correlates exist, but whether
cryptic species are present.* The existence of at least two species in *B. tabaci
sensu lato* has already been adequately demonstrated with cross-mating
tests and enzyme electrophoresis (Perring *et al.*, 1993), and further
appropriately designed tests need to be conducted on the other
so-called 'biotypes' as soon as possible. Without such tests providing a
basis for interpretation, the true host associations of these insects will

Table 7.1. *The various 'biotypes' recorded within the taxon* Bemisia tabaci, *with comments on their host plant range and virus transmission capacity. This table is summarised from that of Brown et al. (1995) and the relevant authorities can be found in that publication. 'Biotype B' is the species B. argentifolii of Perring et al. (1993). Undoubtedly this list is incomplete, but it does illustrate the need to screen for cryptic species in this taxon*

'Biotype'	Locality	Host plant use	Virus transmission[a]
A	USA (Arizona)	Polyphagous	GV (OW and NW) LIYV
B	USA (Arizona)	Polyphagous	GV (OW and NW) LIYV (poorly)
E	Africa (Benin)	*Asystasia* spp.	AGMV
H	Africa (Nigeria)	Sweet potato	TYLCV-Ye
J	Africa (Nigeria)	Polyphagous	TYLCV-Ye
N	Central America (Puerto Rico)	*Jatropha gossypifolia*	JMV
'Non-cassava'	South America (Brazil)	Polyphagous, but not cassava	GV (NW)
'Cassava'	Africa (Ivory Coast)	Cassava, eggplant	ACMV
'Okra'	Africa (Ivory Coast)	Polyphagous, not cassava	GV (OW) Not ACMV
'Sida'	Central America (Puerto Rico)	Polyphagous	GV (NW) Not JMV

[a] GV (OW) and GV (NW), Old and New World geminiviruses, respectively; LIYV, lettuce infectious yellow virus; AGMV, *Asystasia* golden mosaic virus; TYLCV-Ye, tomato yellow leaf curl virus, Yemen strain; JMV, *Jatropha* mosaic virus; ACMV, African cassava mosaic virus.

undoubtedly remain obscure. A good indication of the potential species diversity within the taxon is available (see Table 7.1). The arbitrary designation of 'biotypes' and the associated reluctance to confront the species issue is more likely to retard understanding than advance it (see Chapter 6).

Unfortunately, research on species limits is not as simple as it may seem, despite the importance of correct interpretation in these situations. The uncertainty with which authors treat species and interpret any differences, without supporting tests, as speciation in progress is strong testimony to the lack of clarity in such interpretation. The widespread belief that ongoing adaptation is an inevitable process ensures that authors feel secure in extrapolating what they believe to be local phenomena (e.g. claim of a local host shift) to being an essential part of a speciation trajectory. 'Relaxing' the rules in this way has at least two tacit consequences that undermine understanding. First, the assumption that natural selection selects for speciation is teleological, although it is frequently assumed in work on herbivorous insects. Speciation is a by-product of adaptation under specific environmental circumstances (Chapter 6); natural selection cannot act on a population to ensure it forms a new species, and teleology of this nature should not be allowed to influence investigation or interpretation. Second, cryptic species are not dealt with effectively in conceptual treatments of host or prey relationships, although Futuyma & Peterson (1985) did warn that many so-called biotypes and races were likely to be cryptic species. When cryptic species are mentioned (e.g. Schoonhoven *et al.*, 1998), they tend to be dealt with explicitly as an intraspecific phenomenon. That is, cryptic species are still seen as representing the transient condition of populations along their inevitable trajectory to full species status, as tabulated so explicitly by Ayala (1982), and despite the inherent teleology in that view.

Studies, concepts and interpretations most susceptible to being undermined by the unrecognised presence of cryptic species include many of the reputed cases of:

(i) sympatric speciation,

(ii) local adaptation, including the widely cited notion of phytophagous insects being 'species-wide' generalists while practising 'ecological monophagy' in different local areas of their distribution (Bernays & Minkenberg, 1997; Cates, 1981; Fox & Morrow, 1981; Scriber, 1986), and

(iii) 'disjunct oligophagy', where an insect feeds on a small number of plants from different families with 'no obvious connection between the host plant types' (Bernays & Chapman, 1994, p. 6).

Most cases in these three categories have been researched and interpreted as intraspecific phenomena despite the insects on the alternative plant types not having been rigorously investigated for the presence of host-associated species. Consequently, their current widespread acceptance might well be a contemporary illustration of Chamberlin's precipitate explanation (see Chapter 2) at work. This point does not imply that all cases of polyphagy are expected to be explicable in terms of relatively host-specific cryptic species complexes; only that the relatively low proportion of polyphagous species is likely to be reduced further when investigated appropriately. Consider the situation with leafhoppers, when 'statements such as "polyphagous on grasses" is frequently the only information given . . . and relatively narrow host specificity is found whenever more detailed observations are made' (Novotny, 1994). Leafhoppers are not unusual in this respect, as the examples cited elsewhere in this chapter demonstrate.

5   Current treatments of 'generalist' vs. 'specialist' life histories are categorisations seen to reflect ecological opportunities at opposite ends of a continuum. Treatment of organisms in this way results in both the herbivore and its host plants being treated in typological fashion, despite typology being inimical to the development of realistic generalisations in evolution and ecology (see Chapter 6). From the perspective of the herbivore, the continuum is considered to reflect the adaptive strategies that different species can adopt. Although appealing, the dichotomy misrepresents the way in which organisms interact with their environment, for environments do not present themselves in such simple unidimensional fashion. Use of dichotomies in this way misrepresents the way in which adaptation takes place. These issues are considered in more detail in the following chapter, but the lack of utility in this approach is foreshadowed here to explain why it is not used further in this chapter.

Host plants are also treated typologically. The different host plants recorded for a herbivore species are, for example, accorded equivalent rank in a listing. Whereas no one is likely to claim that each such species has equivalent status, various views expressed frequently in the literature are clearly under the influence of typology, as follows.

(i)   Unused resources (= empty niche) are seen as the driving force for local adaptation and consequent specialisation (e.g. Ward & Spalding, 1993), which underpins belief in sympatric speciation (e.g. Bush, 1994; Feder et al., 1998).

(ii)   'Taxonomic conservatism' in host relationships, which is the association of all species within a herbivore taxon with the plant species in a single taxon (Types B and C in Fig. 7.1), is still portrayed as a consistent pattern across insect herbivores and their host plants (despite the analysis of Jermy, 1984), and is explained with reference to toxins or constraints on genetic variation (Futuyma et al., 1993, 1995; Janz et al., 2001).

(iii)   The belief is implicit that hosts further down the 'list of preferences' will inevitably be used when the 'preferred' hosts are not available. Although this may be the case when the major host species are considered, it may not be inevitable with relatively minor hosts, which may make up the bulk of the host lists of polyphagous species (see point (v) below). Herbivorous insects have the ability to move, and host use is dictated by behavioural and physiological mechanisms, the physiological status of the herbivores, and environmental circumstances.

(iv) A herbivore that lays eggs with equal frequency across several host species may well be responding to them in this way because they give off the same or similar signals. To that herbivore, and to natural selection, the species of host plant is irrelevant, even if survival of offspring is significantly less on one or more of them. In such cases natural selection simply cannot remove a substandard host plant from use by the herbivore. Such herbivores are 'chemical specialists' (see below), and referring to them as oligophagous does not confer any real information other than that the species may be found on several host species.

(v) In comparisons of polyphagy or oligophagy as a strategy relative to monophagy, the alternative hosts of species in the former two categories are automatically given equivalent status to one another. This implies that each is functional in exactly the same way as the others, but that is likely to be incorrect more often than not, as further demonstrated by examples used later in this chapter.

6   The population dynamics of polyphagous species do not necessarily reflect an ability to utilise alternative hosts in the way portrayed for generalists. Their dynamics do not even follow similar patterns in different parts of their distribution. For example, outbreak populations of autumnal moths in northern Fennoscandia occur where plant diversity is relatively low (Tammaru *et al.*, 1995). They do not make up their performance (or fulfil the same potential) in lower latitudes by using alternatives. Similarly, *Helicoverpa punctigera* populations persist in regions where their primary hosts occur (Walter & Benfield, 1994), and do not necessarily establish permanent populations outside of these areas. After periodic incursions into Brisbane by this species, larvae survive on common weeds like *Ageratum houstonianum*, but the species does not persist there (pers. obs.). This point has been generalised for polyphagous herbivores with reference to quantitative data on the generalist thrips *Frankliniella schultzei* and its potential for invading new areas (Milne & Walter, 2000), and is considered also for generalist predators in Chapter 9.

7   Enhanced fitness is said to accrue to polyphagous species through relatively higher 'efficiency' in locating hosts, but how this efficiency is related mechanistically to the behavioural and physiological processes that relate to resource use remains unspecified. Indeed, the relationship seems to be unspecifiable in terms of the underlying physico-chemical, biochemical, physiological and behavioural processes involved. Nevertheless, suggestions are made frequently enough about behavioural plasticity and even the acquisition of

additional behaviours for each host species added to the resource list of a particular herbivore species (e.g. Mitter *et al.*, 1993). Some species deemed to be generalist or oligophagous do feed from a range of host plant species that may be quite substantial. For example, 'the migratory locust, *Locusta migratoria*, will eat many, perhaps hundreds, of different grasses because all of these grasses possess features in common that are used by the insect in selecting food' (Bernays & Chapman, 1994, p. 5). Such species have been called 'chemical specialists' as they have specialist neurosensory responses to a range of species. In terms of 'ecological strategies' and the 'specialist–generalist continuum', the term generalist or polyphage for such species is largely meaningless, and the issue of their evolution is not as straightforward as implied by labels of polyphagy or generalist. Did selection act on *L. migratoria* for it to have a broad host range or did selection impose a particular interaction based on specific neurosensory and behavioural features that, incidentally, conferred a broad host range? That the latter is more likely is justified in the following section.

The points detailed above add weight to the initial claim of this chapter, that the concepts of polyphagy and ecological generalist have been too readily accepted as accurate portrayals of the way of life of various insectan and other consumers. The interpretation is self-fulfilling in that the underlying premises are not tested in specific circumstances. That the general pattern of herbivorous insect–host relationships is one of specialisation encourages the exploration of alternative interpretations for the specialist–generalist dichotomous continuum. This is addressed in the next section. Following this, polyphagous herbivores and predators are re-examined, the specific case of the Australian pest *Helicoverpa* species is expanded to illustrate the species-specific ecologies of these so-called generalists, and the implications of the interpretations derived from this chapter for IPM are discussed.

An alternative concept for polyphagous species

The recognition concept of species was built up in a logical progression starting with the fundamental properties of sexual reproduction (see Chapter 6). Sexual organisms have a complex mechanism to maximise the chances of achieving fertilisation in their usual habitat. Such an association between males, females and the environment imposes stabilising selection on species. Individual organisms track suitable conditions (Chapter 5), so populations adjust spatially to changing environments and therefore remain under stabilising selection within their usual habitat (Chapter 6). Emphasis is thus placed squarely on the complexity of the major adaptations of organisms and thence on the consequent stability of those adaptations. Adaptive change takes place only if forced under particular circumstances, when a small population is trapped in a changing environment. Under such confinement, strong directional selection is imposed on the entire population in its geographical isolation from the parent population.

The premises of the recognition concept suggest a coevolutionary arms race is not at all likely, and this is increasingly borne out by various analyses of host associations and the processes that underpin the documented patterns. At one end of the spectrum are the demonstrations (e.g. Jermy, 1984; Smiley, 1985) that most herbivorous insect associations are inconsistent with the coevolutionary process in any of its several forms. Most associations are more consistent with the evolution of host plant relationships being 'sequential': insect diversification has followed the diversification of plants, rather than driving the diversification of secondary metabolites as defences against specific herbivores. At the opposite end of the spectrum is the documented pattern of an ovipositing herbivore's 'preference hierarchy' for plant species remaining the same across different populations of the same species, even if those populations had not been exposed to the same suite of hosts for extended periods. *Papilio zelichaon* in the north-western USA provides a good example (see Fig. 7.2). The lack of adaptive change to predators and parasitoids released for biocontrol purposes also supports the model just outlined, although Holt & Hochberg (1997) did not include it as a possible explanation for the observed stability.

The above outline specifies that an understanding of the evolutionary origin of host relationships is required, as that is when the host-associated characters are 'set'. Those characters dictate, in turn, where individuals of the species can survive and reproduce, and thus influence their abundance in a locality (Walter & Hengeveld, 2000; Walter & Zalucki, 1999). In practice, different populations of a species are expected to be similar to one another in their major adaptations, and this should hold, too, for so-called generalist species. 'Soft' selection for efficiency or optimisation is not expected to impose change in local populations (Chapter 6), which explains the emphasis on understanding the origins of adaptations and species rather than on the maintenance of characters through studies of relative fitness (Walter, 1993c). Emphasising maintenance at the expense

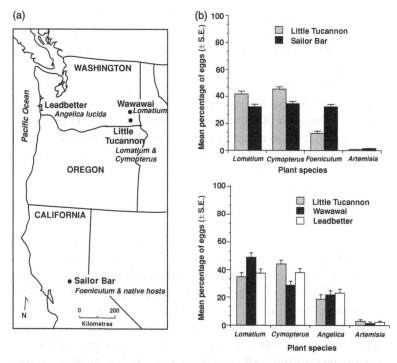

Figure 7.2. Map of the northwestern USA to show (a) the four populations of *Papilio zelichaon* that were tested for host preference and (b) the distribution of the host plants that were tested relative to the four butterfly populations. The Leadbetter population occurs well away from the host plants used in eastern Washington. It feeds almost exclusively on *Angelica lucida*, and may have done so since the retreat of the ice cap 10000 years ago. The Sailor Bar population feeds predominantly on the introduced *Foeniculum vulgare* as well as on other native hosts. Despite these different host associations, the four populations show similar oviposition preferences. After Thompson (1993, 1994). Reproduced with permission from Evolution and by permission of Chicago University Press.

of origin is the prerogative generated by interpretations of natural selection being a competitive process.

The view of species given above sees polyphagy or generalist feeding from a different perspective than the one that is traditionally accepted. The new features evolved by all species are specific adaptations against particular environmental conditions, and each species can therefore be seen to be a specialist in one way or another. Note that the recognition concept does not explain specialisation or cases of 'taxonomic conservation' (insect species within a taxon all associated with a particular plant taxon; see Ehrlich & Raven, 1964) with reference to the constraints on genetic variation invoked by Futuyma *et al.* (1993), Janz *et al.* (2001) and others.

Rather, the influences of preadaptation at speciation, the selection pressures that prevail at the time of adaptation, and post-speciation stabilisation provide a more likely starting point, especially when one considers the vast amount of genetic variation within a gene pool and the fact that a relatively small sample of individuals represents that diversity quite precisely. The usual stasis of species (Chapter 6) also shifts emphasis from the issue of maintenance of characters to an understanding of origin.

Host specialisation is expected because the mechanism of interaction of the consumer is geared to the localisation and exploitation of a food source with which its offspring will spend perhaps the greatest part of their pre-imaginal development. Although not all plants are toxic to a particular herbivore, immature insects tend to be able to develop effectively on a relatively small subset of hosts, at least within a single locality. Extensive host lists developed for a species cannot represent accurately the availability of hosts to particular organisms in particular situations. Furthermore, oviposition, phenology, pupation and other aspects of the life cycle are also keyed to features of the host plant and the environment (or a subset of the environment) in which it lives. To ensure synchrony with their host plant, females of *Platycotis vittata* membracid bugs, for example, have a summer reproductive diapause, and *Enchenopa* membracids have a host-associated egg dormancy (Wood, 1993). The continuity of the life cycle within the usual environmental setting occupied by the organisms is the minimum requirement for the persistence of organisms, and our concept of life cycle is an abstract generalisation of complex 'spatial and temporal ecological mechanisms' (Sinclair, 1988, p. 143). The demand for synchrony with a host species, or any other environmental variable, suggests host specialisation should be expected. Indeed, it is consistent with the specialisation seen across all other organisms, from sponge-inhabiting shrimps (Duffy, 1996), independent gene pools (species) of Atlantic herring (Sinclair, 1988; Sinclair & Solemdal, 1988), African antelope (Vrba, 1995), to mycorrhizal fungi (Bever *et al.*, 2001), and so on. The significance of considering the entire life cycle in interpreting the evolution, ecology and biogeography of species is little appreciated, but has been dealt with in general terms by Sinclair (1988).

In host-specialist herbivorous insects, not even all stages of the life cycle can necessarily run equally effectively on a single host, at least as the host is affected by climatic variation (e.g. Bjorkman, 2000). This does not justify invoking 'constraints' in evolutionary interpretation, for constraints explain what we see by reference to why we think it has not reached our expectation, based on the assumptions of competitive or optimising selection (Chapter 6). Rather, Bjorkman's example illustrates that the organism cannot 'fit' (see 'Autecology' in Chapter 5) the totality of its environment, and that ensuring completion of the life cycle on even a single host is not as straightforward as assumed by optimisation models. To look for explanations of specialisation in terms solely of so-called 'ecological factors' (Bernays & Graham, 1988; Dyer & Floyd, 1993) or plant chemistry, or some combination of these (as outlined by Courtney & Kibota, 1990), is to miss the point. The species-specific life cycle of the organism is geared to a suite of conditions that represent those at the time it was forced to adapt to novel environmental circumstances in a small, localised population under intense directional selection. This explains why (i) complex adaptations are fixed across a species' gene pool and why they persist through geological time (Chapter 6), (ii) species generally dwindle at the edges of their distribution rather than compensate ecologically through local adaptation (Hengeveld, 1990; Walter & Hengeveld, 2000), and (iii) a clear correlation cannot be found between chemically mediated suitability of hosts and actual host lists (Courtney & Kibota, 1990). The stability of adaptations of species provides a single, functional explanation for the observations covered in the third point, and thus obviates the current explanations that are negative (based on constraints) or multifaceted (Courtney & Kibota, 1990; Futuyma *et al.*, 1993, 1995; Holt & Hochberg, 1997).

The model of adaptation and speciation outlined above justifies why the host-searching mechanism is specifiable independently of an ultimate, evolutionary explanation of host specificity, as anticipated by Courtney & Kibota (1990). The mechanism to be investigated and interpreted essentially 'selects itself' when the observer is dealing with a herbivore that oviposits and undergoes larval development on a single plant species, as in studies of host plant recognition in the monophagous weevil *Ceutorhynchus inaffectatus* on its cruciferous host (Larsen *et al.*, 1992) and monarch butterflies on milkweeds (Oyeyele & Zalucki, 1990; Zalucki *et al.*, 1990). Herbivore species that use several to many hosts present the problem of knowing which hosts to begin with, a central issue in the investigation of such species (Milne & Walter, 2000) and one that is addressed later in this chapter.

Faced with the realisation that 'specialisation' is far more common than 'generalisation', ecologists have tended to turn immediately to the question of what the selective advantage of host specificity could be (e.g. Jermy, 1984). To understand why host specialisation in one 'population' of the geometrid *Alsophila pometaria* evolved from the more polyphagous *A. pometaria* 'population' is said to require data on the 'selective advantages and disadvantages of relatively polyphagous and oligophagous genotypes within populations' (Futuyma, 1991). The alternative explanation of host specialisation offered by the recognition concept (and outlined in the previous section) suggests a perspective in which the host or prey range recorded for a consumer species is not selected as a 'package'. Rather, host specialisation is a consequence of specific adaptations of the host location mechanism as it was selected within the context of the entire life cycle and the context of a particular subset of environmental conditions. Host specialisation, for example, is thus a consequence of specific adaptations and is not a cause of them as sometimes claimed (e.g. Sasal *et al.*, 1999). A potential reduction of fitness cannot, therefore, influence what organisms actually do in nature, as demonstrated for weevils on thistles by Louda (1998). The use of 'substandard' hosts is a consequence of those hosts fitting to some degree the pattern that influences host recognition, with the threshold for acceptance influenced perhaps by physiological condition of the consumer organism. Comparative fitness studies therefore reveal less about evolutionary progress than about an aspect of the ecological consequences of the species-specific neurobiological and behavioural mechanisms.

This view suggests that the concept of 'generalist' is not very useful. It combines into a single category various organisms that have in common only that their diet comprises organisms of diverse taxonomy, and this obscures the underlying ecological interactions that are significant to each situation. Although spiders are treated as generalist predators, they do specialise in constructing a web of a particular type. Usually it is made in particular situations within a particular biotope or vegetation type, and at least some species tend to place them where prey is more available (Foelix, 1996; Harwood *et al.*, 2001). A trap of this nature ensnares prey organisms of a particular size. Depending on the orientation of the trap and its height above ground it may trap predominantly flying insects or ground-dwelling arthropods, for example. The array of prey species seems to be secondary in information content to aspects of the way of life and ecology of the prey that intersect with the trapping situation of the spider species of interest. Only then can the interaction with the major prey species be meaningfully interpreted. For example, at least some species concentrate on particular prey species rather than take even congeneric prey at the relative rates at which they occur in the web (e.g. Harwood *et al.*, 2001). Some prey species are even detrimental to these so-called generalists, a situation not necessarily improved by a mixed diet (Bilde & Toft, 2001). Many generalist predators can therefore be considered as specialised to a 'way of capture'. Others are associated with specific aspects of the environment, as extensive field sampling of phytoseiid mites has revealed. These so-called generalist predators are strictly associated with specific host plant species (Beard & Walter, 2001).

The idea of true generalist species adapted to take advantage of a diversity of conditions or resource opportunities is inconsistent with the principles of the recognition concept. By contrast, underlying the diversity of food types of all generalists can be discerned a suite of primary requirements, behaviours, habitat associations and host/prey relationships. To understand polyphagy requires homing in on these primary host relationships (Milne & Walter, 2000; Velasco & Walter, 1992; Walter & Benfield, 1994; Wint, 1983) but this is seldom done. Most research on polyphagous species addresses instead issues of relative fitness across a few arbitrarily selected 'alternative' hosts, oviposition choice and its relationship to subsequent larval performance, and the relationship of polyphagy to dichotomous sets of ecological circumstances. In this way most studies tend to isolate monophagy or polyphagy and treat it as a character in isolation from the overall suite of life cycle requirements. The categorisation is thus treated as if it is subject, perpetually, to directional selection and ongoing adaptive change.

To illustrate just how much the seemingly innocuous concept of polyphagy or generalist actually hides, the ecology of two Australian pest species is detailed as far as is possible. The comparative outline of the primary aspects of their host associations provides a platform for summarising, in the subsequent subsection, the various problems that use of the concept 'polyphagy' carries when efficiency or optimisation of resource use is inferred to be its functional basis.

Host plant relationships of Australian pest *Helicoverpa* species

Both *Helicoverpa punctigera* and *H. armigera* are notorious pests in Australia. Whereas the former is endemic, the origins of *H. armigera* in Australia are not altogether clear as it has a wide distribution into temperate regions on either side of the equator, but outside of the New World (Matthews, 1999). Both species have been reported from numerous plant species. For example, by 1986 eggs and larvae of *H. punctigera* had been reported from at least 127 species in at least 39 families (Zalucki *et al.*, 1986), and that list has grown substantially since (Matthews, 1999). Almost all the early records are of larvae or eggs collected in unspecified environmental circumstances (see Chapter 2). That implies that the ecological pattern to be explained has been portrayed simply by the single term 'polyphagous', which carries with it the standard adaptive implications specified in the section 'Current interpretations of polyphagy' earlier in the present chapter.

The pest *Helicoverpa* species have recently been deemed to be 'highly polyphagous', and that view has been extended to infer interpretation of the behaviour and physiology of the individual insects. The females are said to be flexible in their host selection behaviour or to have a 'general mechanism' for host location and oviposition, to allow oviposition on a wide range of plant species (Fitt, 1989, 1991). Such a view encourages particular lines of research. To explain the mechanism of oviposition, for example, various combinations of the following points have been suggested. The moths are attracted primarily to flowers, they prefer monocultures, moths respond differentially to various plant heights, and they are mainly repelled by allelochemicals. Whereas each of these features may be accurate, their influence on host-searching behaviour must be contingent on circumstances, because exceptions are regular when sampling in the field (Walter & Benfield, 1994). For example, seedling cotton may attract considerable oviposition, flowers of many species in monoculture are not attacked, and repellents and deterrents on their own cannot explain the patterns observed. In other words, the underlying mechanism of host location and host acceptance remains hidden, obscured by the view that a 'general' mechanism drives these organisms. By contrast, if organisms are adapted to specific environmental circumstances, then specific mechanisms that place organisms rather precisely within the appropriate context can be expected, and a different line of research is suggested.

The host associations of *H. punctigera*, outlined in Chapter 2, clearly need explanation, but the ecological pattern that defines the host relationships of this species remains unclear. Specifically, which plant species are to be considered significant as hosts? And in assigning host status to a plant species, does it have equivalent status to all other host species? Therefore, in simply documenting the ecological pattern to be investigated, judgement needs to be exercised, and the basis of that judgement demands consideration.

A more explicit statement of the ecological pattern to be explained is required; the traditional approach outlined above tends to pre-judge the situation because of its basis in pattern analysis and expectations of optimisation. Further, each host species is accorded more or less functional equivalence. Should a host species be absent, then the next best one in the area is expected to fulfil its role in maintaining or generating populations. The emphasis is thus firmly on the exploitation of resources by organisms. Nevertheless, host plant species are seen to differ in their position in the preference hierarchy of the species, and in their impact on larval performance. Polyphagy is thus also seen to have important ecological consequences for pests, mainly in providing great potential for population persistence in an area and for rapid population increase before invasion of a crop (e.g. Kennedy & Storer, 2000; Panizzi, 1997). Populations may also develop simultaneously on a number of hosts within a region, and do so continuously during suitable periods by using a succession of crop and non-crop host species. Although the successional use of host species is perhaps likely in many species of polyphagous herbivores, such systems have been subject to relatively little critical test of the underlying assumptions. Even quantitative and spatial analyses of the population processes involved are scarce.

   A quantified field survey, conducted in southern Queensland, demonstrated that H. punctigera is not uniformly distributed across regions with a diverse native 'weed' flora, which had been intimated for the species because of its polyphagous nature. Intensive surveys of three selected study sites demonstrated that within an area, H. punctigera may well be restricted to a particular host species and that other plant species, even if previously recorded on the H. punctigera 'host list', may host relatively few individuals and then only irregularly. Knowing that H. punctigera is endemic to Australia implies it must have evolved there on native host plants. The plant species that are used regularly and by relatively large numbers of individuals can be considered 'primary host species', a term already used in connection with various other insect herbivores (Wiklund, 1981; Wint, 1983). With regard to H. punctigera, such primary host species are presumed to be the ones on which the species evolved, although preadaptation could account for a proportion of them. By contrast, many recorded 'host species' are used irregularly and by relatively few individuals of H. punctigera, and such species are best seen as 'incidental hosts'. Such incidental hosts are likely to be largely irrelevant to understanding the ecology and pest status of H. punctigera (Walter & Benfield, 1994).

Other research on *H. punctigera*, in Central Australia, shows that the species can be collected regularly and in relatively high numbers from a number of native wildflower species, most of which are adapted to the arid conditions of the interior (Zalucki *et al.*, 1994). In principle, the populations on these species are the ones that should be tested for the existence of host-associated cryptic species. These recent insights into the ecology of *H. punctigera* provide a more rational basis from which to investigate the primary mechanism by which females locate and accept hosts for oviposition.

The host relationships of *H. armigera* provide a more complex problem than do those of *H. punctigera*, principally because its area of endemicity is not as readily pinpointed as it is for *H. punctigera*. Presumably because of its wide distribution, universal pest status, and the view that it is a generalist, the primary adaptations of the species have not been considered to any extent. The situation is complicated by the existence of two subspecies (Matthews, 1999), whose behavioural and genetic relationship to one another remains undefined. In broad terms, one is Eurasian and the other Australasian (Matthews, 1999). Recent results from the field in Australia and from the laboratory suggest that the species is primarily adapted to particular hosts, in much the same way as claimed for *H. punctigera*. In Australia, the Mediterranean weed *Sonchus oleraceus* attracts a great deal of oviposition by *H. armigera*, compared with other potential hosts. Further, host 'choice' experiments indicate strong 'preference' for sowthistle over other host plants, even if the female moths had been collected from the latter, as larvae, in the field (Gu & Walter, 1999). Just-hatched larvae, regardless of host plant origin of their mothers, also preferred sowthistle for feeding. Sowthistle may thus be a primary host plant of *H. armigera*, in an evolutionary sense. As such, it would have the full suite of characters that attract oviposition from *H. armigera* moths; these characters are known to have a genetic basis and are heritable (Gu *et al.*, 2001). For various reasons, a full functional analysis of those characters would be revealing and would have practical benefit. For example, an understanding of *H. armigera* oviposition behaviour in relation to sowthistle could pave the way for developing attractants to keep moths from crop plants.

In summary, *H. armigera* has major differences in its host relationships from *H. punctigera*, and this is reflected in other aspects of its ecology. These differences have ramifications for pest management. *Helicoverpa armigera* is mainly a coastal and subcoastal species in Australia, and populations are apparently maintained locally in cropping areas, on various crop hosts and on such weeds as sowthistle, but this is not well understood. Major larval populations do not seem to be common outside cropping areas and possibly contribute little to infestations of crops, but reliable quantification is not available. Exposure of H. armigera to insecticides is therefore considered to be largely uniform across the entire distribution of the species in Australia, and exposure is quite consistent through the season, although practice has been modified recently (Forrester et al., 1993). Not surprisingly, resistance to various insecticides has developed in H. armigera, to DDT, synthetic pyrethroids and endosulfan. Chemical options against H. armigera are therefore bleak. Although alternatives have been sought, most hope lies with Bt transgenic cotton.

Helicoverpa punctigera is primarily adapted to certain native species that occur across Central Australia. Suitable conditions for the primary host plants of H. punctigera occur only sporadically, usually outside of the hot summer months when any plant growth is rapidly desiccated. The sporadic appearance of suitable conditions and their unpredictable occurrence in space encourages migration of moths. However, not all moths in a population will migrate, as indicated by the continuous habitation of particular localities (Walter & Benfield, 1994), so some moths remain and oviposit near their natal site. The migrants are the ones that presumably invade cotton and other crops, from outside of the major cropping areas. They even oviposit on newly emerged cotton plants, well before flowering. They presumably do so because they have no other option for ovipositing. In other words, they may oviposit on cotton because they are physiologically stressed and their threshold for host acceptance has decreased. Such individuals do not generate local populations; they have entered a 'sink'. Helicoverpa punctigera remains susceptible to most insecticides, perhaps because only those individuals that enter the 'sink' outside their usual habitat are exposed to the toxins. Growers therefore need to identify eggs and young larvae prior to deciding about insecticide applications; synthetic pyrethroids and endosulfan are no good against H. armigera, but they do work against H. punctigera. Therefore, a molecular probe has been developed to differentiate the two species from one another as early as the egg stage.

Different species, even if morphologically very similar, are likely to have decidedly different ecologies, as illustrated by the Helicoverpa example in Australia. These differences stem from the differential adaptations of these species to particular host plant species (or suites of host plant species), the different environmental conditions in which the hosts persist, and other requirements of the life cycle. Should additional cryptic species be present in either of these moth taxa, subtleties in ecological interpretation that may be directly relevant to improving IPM of these pests are likely to be missed.

Testing for polyphagy and its functional aspects

The idea of testing alternative interpretations to derive the most reliable scientific interpretations was canvassed in Chapter 2. This method seems seldom to have been applied systematically to claimed examples of polyphagy, and this section outlines the various aspects that may warrant testing in any specific case. Without such tests of specific cases, ecologists are likely to fall ready prey to what Chamberlin termed 'precipitate explanation' (Chapter 2) when they deal with these so-called generalist species. The tests outlined below summarise and draw together the points already raised about polyphagy. Critical application of such method is likely to result in a better understanding of multiple host use by particular species, and research of this nature will undoubtedly shorten some host lists considerably; it should stop the practice of the uncritical addition of hosts regardless of their functional relationship with the herbivore in question. Under the view of species, and adaptation canvassed in Chapter 6, the incidental use of particular plant species is not seen as equivalent to the regular use of primary host species, and ovipositional 'mistakes' are unlikely to provide insight into future patterns of host use, as some have intimated (Larsson & Ekbom, 1995). The inclusion of such 'host species' in a single list is thus entirely misleading in terms of functional or evolutionary interpretation and therefore for ecological understanding (Walter & Benfield, 1994). Indeed, the idea of capturing, in a single term, the ecological circumstances demanded of the life cycle of organisms is likely to prove illusory. Ecological and evolutionary generalisations based on such categorisations will inevitably be rather weak.

The two tests that should, perhaps, take priority are those of the quantification of relative host use, in terms of regularity of use and intensity of use, which implies that negative records need to be taken into account (Novotny, 1994; Walter & Benfield, 1994; Zalucki *et al.*, 1994), and scrutiny for cryptic species (Chapter 6). Although the investigation of species status should perhaps take overall precedence, the choice of populations or samples for testing may best be indicated by appropriate interpretation of relative host use in the field. Although such procedures are seldom

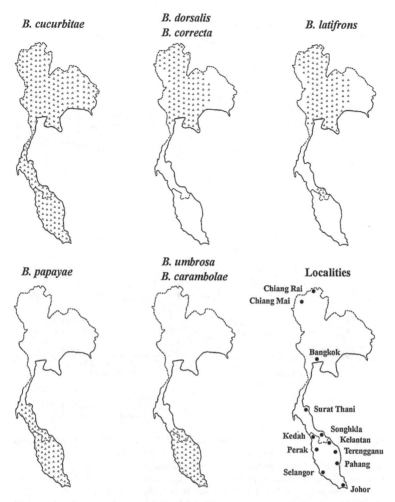

B. cucurbitae

B. dorsalis
B. correcta

B. latifrons

B. papayae

B. umbrosa
B. carambolae

Localities

Chiang Rai
Chiang Mai

Bangkok

Surat Thani

Kedah
Perak
Selangor

Songhkla
Kelantan
Terengganu
Pahang

Johor

Figure 7.3. Diagrammatic representation of the localised distribution of seven *Bactrocera* fruit flies (Tephritidae) in Thailand and Malaysia. Localities sampled are also shown. Malaysian data derive from adult traps, those for Thailand from host rearing records (Clarke *et al.*, 2001). Reproduced with permission from *Raffles Bulletin of Zoology*.

carried out, recent intensive research on Australasian fruit flies of the genus *Bactrocera* (previously included in the genus *Dacus*) does illustrate the benefits of such an approach.

For many years the most notorious *Bactrocera* has been the Oriental fruit fly (*B. dorsalis*). A systematic investigation of *B. dorsalis* revealed that the taxon comprised at least three additional cryptic species, including

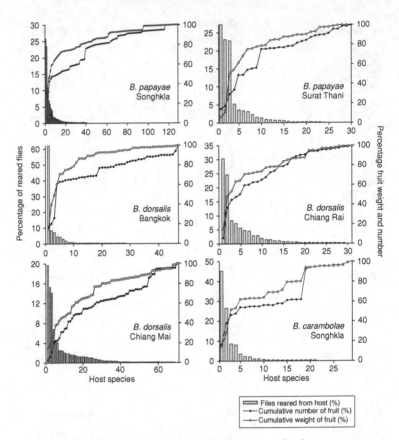

Figure 7.4. Ranked plots of the percentage of *Bactrocera dorsalis, B. papayae* and *B. carambolae* reared from fruits in Thailand (Fig. 7.3). Each column represents the number of flies reared from a particular host fruit species, presented as a percentage of the total number of flies of that species reared for the region specified in each graph. The number and weight of host fruits sampled is plotted as a cumulative percentage of the total number and weight of fruit sampled for that region. Only fruit samples that yielded at least one specimen of the particular fly species are included in these totals. Sample sizes for each plot (sequence is Number of flies reared, **Number of fruits** and Total weight of fruit (kg)): *B. papayae* (Songkhla) [118 169; **116 111**; 2815]; *B. papayae* (Surat Thani) [8479; **2708**; 85]; *B. dorsalis* (Bangkok) [24 833; **53 352**; 764]; *B. dorsalis* (Chiang Rai) [8164; **6675**; 195]; *B. dorsalis* (Chiang Mai) [20 129; **76 480**; 1059]; *B. carambolae* (Songkhla) [5893; **48 272**; 1977]. From Clarke *et al.* (2001). Reproduced with permission from *Raffles Bulletin of Zoology*.

B. dorsalis sensu stricto, B. papayae, B. carombolae and *B. philippinensis*, all of which are major pests in their own rights and are biosecurity risks (Drew & Hancock, 1994). These species have been separated on traditional taxonomic methods. The various components of their specific-mate recognition system have yet to be scrutinised and have not been investigated extensively by any of the other methods outlined in Chapter 6 that might be appropriate for these organisms. Furthermore, different populations within each of these four taxa have not been scrutinised either, so the potential for additional cryptic species should not be dismissed. Sampling data suggest that further such investigation is warranted, and also demonstrate how an understanding of species limits within the genus *Bactrocera* starts to draw attention away from the vague descriptor of polyphage or generalist and helps to refocus it on the life cycle of the organism in its entirety. At least three general points are relevant.

First, each of the species mentioned above has a relatively restricted distribution within southeast Asia (Fig. 7.3), which suggests that each relies upon or is adapted to particular aspects of these tropical areas. For a long time, tropical rainforest was treated as a relatively uniform environment with an even climate, but this view has had to be rejected (see Wolda, 1983, for example). That insect species still considered to represent archetypal generalists have relatively restricted distributions within areas of tropical forest demonstrates further the subtle diversity in these environments. Such subtlety clearly has substantial ecological impact through its influence on particular aspects of the life cycle of the organisms in question. In autecological terms, the entire life cycle must 'fit into' the seasonal changes within the area and must cope with the stochastic variation as well (Walter & Hengeveld, 2000).

Second, extensive sampling of the potential hosts of various *Bactrocera* species has been conducted in various parts of Thailand (Allwood *et al.*, 1999; Clarke *et al.*, 2001). Some of the species are reared from only a few congeneric hosts, for example *B. umbrosa* (Table 7.2). The others are reared from a wide range of host species across several plant families (Clarke *et al.*, 2001), as is generally expected. The significance of quantifying host species use is well illustrated with reference to *B. latifrons* (Table 7.2). The vast majority of these flies (90–95%) were reared from species of *Solanum* (= oligophagy in traditional terms). Ignoring that, one could emphasise *B. latifrons'* having been reared from 14 species across 10 plant families (= polyphagy by virtually all definitions) (Clarke *et al.*, 2001). As far as the behavioural and physiological mechanisms of the species are concerned, as well as its ecological requirements and consequences, the former is far more relevant. The polyphagous tag is, however, more likely to haunt the species and those who might be charged with its management. With regard to the *B. dorsalis* complex, 'as few as 5 hosts accounted for 70–90%

Table 7.2. *Host rearing records for seven species of* Bactrocera *fruit flies across five localities in Thailand (CR, Chiang Rai; CM, Chiang Mai; BK, Bangkok; ST, Surat Thani; SK, Songhkla: Fig. 7.3)*

Data have been summarised from the comprehensive table of Clarke *et al.* (2001), and additional distributional data derive from there as well (see Fig. 7.3). The table illustrates that for these fruit flies, which are notorious for their broad range of host species, most individuals within a locality (usually well over 55%; see Fig. 7.4) derive from two or three host plant species. In each locality, 600 or more plant species or varieties were sampled. The cutoff point for naming a host species below was that at least 8% of all flies from at least one locality should have been reared from that species. Data from other hosts were summed under 'Named alternatives', with the range of plant species numbers involved also given, unless a host yielded less than 2% of flies for a locality, in which case the data were summed into the 'Minor alternatives' host category. A 'O' thus indicates the range from no flies to <2% of flies in that locality, and a '–' indicates the plant was not sampled in a locality and may not occur there. The numbers after each fly species' name indicate the total number of flies recorded (with range across localities in parentheses) and, after the colon, the equivalent statistics for the amount of fruit sampled (in kg). 'x' indicates the fly species is not present in the locality.

Fly species	Thai locality				
Host plant species	CR	CM	BK	ST	SK
Bactrocera dorsalis: 53126 (8164–24833) flies: 2018 (195–1059) kg fruit					
Anacardium occidentale	24	–	–	x	x
Psidium guajava	30	19	8	x	x
Terminalia catappa	3	15	62	x	x
Syzigium samarangense	5	14	7	x	x
Prunus persica	8	–	–	x	x
Named alternatives (3–6 spp.)	11	22	12	x	x
Minor alternatives	10	30	11	x	x
Bactrocera papayae 126648 (8479–118169) flies: 2900 (85–2815) kg fruit					
Terminalia catappa	x	x	x	27	23
Psidium guajava	x	x	x	23	26
Musa paradisiaca	x	x	x	23	–
Averrhoa carambola	x	x	x	5	9
Named alternatives (4 spp.)	x	x	x	12	22
Minor alternatives	x	x	x	10	21
Bactrocera carambolae 5893 flies: 1977 kg fruit					
Psidium guajava	x	x	x	x	8
Syzigium samarangense	x	x	x	x	26
Syzigium malaccensis	x	x	x	x	8
Averrhoa carambola	x	x	x	x	45
Named and Minor alternatives	x	x	x	x	10
Bactrocera correcta 32956 (13723–19233) flies: 1572 (763–809) kg fruit					
Mangifera indica	o	9	o	x	x
Terminalia catappa	o	o	31	x	x
Psidium guajava	o	18	24	x	x
Syzigium samarangense	o	39	18	x	x
Zizyphus jujuba	o	7	15	x	x
Named alternatives (2–5 spp.)	o	17	6	x	x
Minor alternatives	o	9	6	x	x

Table 7.2. (*cont.*)

Fly species	Thai locality				
Host plant species	CR	CM	BK	ST	SK
Bactrocera umbrosa 1042 flies: 186 kg fruit					
Artocarpus altilis	x	x	x	x	4
Artocarpus heterophyllus	x	x	x	x	85
Artocarpus integer	x	x	x	x	12
Bactrocera cucurbitae 38 327 (1060–16 531) flies: 891 (14–360) kg fruit					
Coccinia grandis	34	39	33	34	42
Cucumis melo	8	0	3	–	0
Cucumis sativus	17	4	8	6	21
Cucurbita moscchata	10	0	0	0	0
Luffa acutangula	6	0	9	0	7
Momordica charantia	7	45	6	57	23
Trichosanthes anguina	–	–	22	–	–
Named alternatives (0–3 spp.)	14	7	2	0	3
Minor alternatives	3	5	13	3	5
Bactrocera latifrons 16 407 (1579–7802) flies: 221 (16–121) kg fruit					
Capsicum annuum	9	7	5	x	7
Solanum aculeatissimum	–	–	22	x	59
Solanum incanum	–	–	–	x	16
Solanum indicum	13	–	6	x	–
Solanum melongena	28	31	5	x	2
Solanum sanitwongsei	5	–	28	x	–
Solanum torvum	7	11	34	x	13
Solanum trilobatum	30	40	0	x	0
Named alternatives (1 sp.)	8	4	0	x	0
Minor alternatives	1	8	1	x	3

of all flies of a species reared', and certain host species yielded dispropor-tionately large numbers of flies (Clarke *et al.*, 2001; Table 7.2, Figs. 7.3 and 7.4). For *B. dorsalis* and *B. papayae* two hosts were particularly significant, *Terminalia catappa* and *Psidium guajava*. The former appears particularly significant relative to the small number and weight of fruits of this species collected, and is perhaps a primary host species of these flies (Clarke *et al.*, 2001).

Third, emphasis on the identification of primary host species of polyphagous herbivores, parasitoids and predators carries additional significant implications. First, it suggests that the function of secondary hosts and incidental hosts is to provide a 'safety net' for when the pri-mary host species fails or when individuals cannot otherwise locate a pri-mary host (Velasco & Walter, 1992; Walter & Benfield, 1994; Wint, 1983). For example, the unpredictability of bud burst in oaks makes this an imperative for winter moth (Wint, 1983), the spatio-temporal patchiness of suitable flowers makes it so for blossom-feeding thrips (Milne & Walter, 2000), and aridity of the environment demands it in species that specialise on reproductive parts of their host plants, as in *Nezara viridula* (Velasco & Walter, 1992) and the Australian native budworm (Walter & Benfield, 1994). Second, generalist predators such as carabids, coccinellids and syrphids also tend to specialise on particular prey species (Hodek, 1996; Sadeghi & Gilbert, 2000a,b). Predators are likely to need a trophic 'safety net' at least at some stage of their life cycle, as they tend to be motile (as are their prey) and they require many small prey items to complete development. Therefore it is easier for them to starve than for a herbivore or parasitoid that can complete most or all of its pre-imaginal life on a single individual host. Hodek (1996) refers, therefore, to principal food as opposed to substitute food, and sees mixed feeding in coccinellids as an emergency feature compelled by shortage of the 'right' food. Third, the physico-chemical features common to the primary host species are the ones most likely to provide a rational entry into an investigation of the signals that are most significant in the host-searching behaviour of insect herbivores, predators and parasitoids.

The ecology of some *Bactrocera* species has been investigated in some detail, and the species dealt with above (Table 7.2) might be considered to fall into this category. Only cursory inspection of the summarised data is sufficient to suggest that an understanding of the host relationships of even these organisms is not close at hand. And when the tolerances and requirements of the entire life cycle are included in the overall picture, ecological understanding recedes further. Even the primary hosts of the Australian species *B. tryoni* are still not known (see later) and, like the *Helicoverpa* species in Australia, it is investigated on cultivated hosts when insights on host relationships relevant to pest management are sought. Inspection of the data on the Thai species (Table 7.2) also points to differences across localities, and the basis for this variation warrants testing if the primary host relationships and the host-searching behaviour of these organisms are to be understood sufficiently well for IPM purposes.

Some insect species clearly do feed from a wide diversity of host species. At least some orthopterans fall into this category, and undoubtedly others do too. To class them as polyphagous or generalist may be realistic as a descriptor, but should not be used typologically in attempts to find correlates of this way of life. Besides testing the role of different hosts in the field, the underlying nutritional advantages to 'dietary self-selection', which is not universal across insects that are considered polyphagous (Bernays & Minkenberg, 1997), warrant testing. The nature of the host-searching mechanism in such species is bound to provide crucial information for interpreting their ecology. These insects may have relatively low thresholds across cues from a range of host types, whereas species that habitually attack only one host type may have high thresholds to all but stimuli from their usual host. This may represent the mechanistic basis for the differential pattern in host use across *Bactrocera* species that differ from one another in the numbers of host plants they attack (Fitt, 1986a), which requires consideration relative to the availability of primary host species in the field.

In summary, each generalist species, even if it has hosts or prey in common with a closely related species, will inevitably have differences in primary hosts and other aspects of its ecology. The term 'generalist' tends to hide these crucial specific aspects of the organisms' life cycle and ecology.

Understanding polyphagy and IPM application

Many insects have a reputation for polyphagy, including some of the most serious agricultural pests, most predators and many parasitoids (especially egg parasitoids). This chapter has emphasised the significance of understanding the functional aspects of polyphagy, because the correct identification of function is crucial to successful manipulation of populations in the field. This is readily illustrated by contrasting two extreme perspectives that could be offered to interpret the ecology of a particular natural enemy species. If polyphagy is interpreted as a generalist strategy to increase fitness by exploiting unused resources, and the implication is that those resources are available locally, then generalist natural enemies released on an abundant pest are expected to remain there, exploit the resources and contribute to control. If polyphagy is interpreted as an escape for when the primary resources of the released species are not available, the liberation of that species on anything but the primary host or prey should be expected to contribute little, if anything, to control. Correct interpretation of the function of polyphagy thus becomes critical for the most effective IPM application involving polyphagous species.

Several aspects of IPM are likely to be influenced differentially by alternative interpretations of polyphagy. Some of these have already been covered, including those instances in which the presence of two or more relatively host-specific cryptic species have been mistakenly included under a single species name. Also mentioned earlier was the issue of pest populations building up locally on a sequence of alternative hosts before invading a crop. Although such successive use of hosts is not unlikely, the view that this occurs in specific instances is frequently accepted simply on the basis of pattern analysis and without alternative possible interpretations being tested critically. If, instead, the pests migrated into the crop from an unrecognised distant source, then a different basis for prediction and alternative management options would have to be developed.

The prediction of population trends and forecasting of invasions will undoubtedly be improved by a functional understanding of host relationships of 'generalists'. Again, an understanding of where 'refuge' populations persist and their migration propensity and pathways will undoubtedly improve invasion forecasts. The prediction of how an invasive population will perform will again depend on an accurate functional understanding of its host relationships as well as on an appreciation of host availability within the area of invasion. Are primary host species available and in what physiological condition are they, and so on? If primary host species are not available or suitable for oviposition, then prediction may be less certain, although experience of previous invasions is likely to be helpful, as in the pest bollworms in Australia (see above). Furthermore, an 'oversupply' of individuals within an area may lead to unusual behaviour in relation to alternative hosts. Experience in weed biocontrol shows that massive herbivore populations relative to target host availability may lead to unexpected, but ephemeral, behaviour. Individuals may use a plant species that is virtually never used otherwise: this has been termed 'host substitution' (Marohasy, 1996). Such behaviour has not led to local adaptation, despite persistence of the herbivore in that area (Marohasy, 1996).

Quarantine across geographical barriers or political borders would undoubtedly be enhanced if an understanding of the host relationships of species with multiple hosts could be used to predict their ecology after invasion of a new area. Even for pest species this has proved problematic, because preadaptation to hosts never before encountered in the field is so difficult to predict in organisms that are polyphagous. For example, would the depredations of European corn borer have been predicted before maize was imported into Europe from the New World? Also, the use of host species on which the insects do not survive or reproduce well may be sufficient to sustain populations in the absence of primary hosts, but the extent to which this may occur again seems unpredictable (Milne & Walter, 2000). The way in which plant breeding, mass cultivation and irrigation influence the behaviour, host use and pest status of herbivores is still little understood. The ecological responses of insects that use multiple hosts to agricultural systems relative to natural ones cannot, therefore, be readily predicted. Consider the contrast between Queensland fruit fly (*Bactrocera tryoni*) and fruit-spotting bugs (*Amblypelta*), all of which are indigenous in eastern Australia. The fruit fly is said to be polyphagous on rainforest fruit, but is not readily collected there and breeds readily and in vast numbers in orchards (Fitt, 1986b; May, 1953). The bugs have been recorded from numerous rainforest fruit trees (Waite & Huwer, 1998) and seem to invade orchards from there (Ryan, 1994). Although both are said to be polyphagous, one seems to be a persistent resident in orchards, the other a sporadic and perhaps evanescent visitor.

Husbandry practice and plant breeding provide untested alternative explanations even for the most heralded of sympatric speciation examples, the shift of *Rhagoletis pomonella* to apples (see p. 136). Indeed, interpretation of the shift depends substantially on early records of the 'shift'. It is said to have taken place in the 1860s with the records of Walsh. How accurately this pattern of host use has been recorded is questionable, considering that economic entomology developed only after about 1850 in the USA, in an uncoordinated way and without institutional support. Government funding was made available only after 1870 (Whorton, 1974). Was the appropriate sampling conducted, or could earlier incidence have been overlooked? Furthermore, the spatial pattern of this shift remains unclear despite being crucial to understanding its cause.

One option for pest management purposes is diversification of crops through intercropping. Investigations tend to return somewhat contradictory results. Considering that the experimental design generally tests the final consequences of the complex behavioural mechanism that leads insects to plants, this is not surprising. Any number of extraneous variables could interfere with sign stimuli (visual or olfactory), disrupt 'attentiveness' of the insects (Bernays, 1999) or alter their movement (Coll & Bottrell, 1994) and thus influence outcomes. Such variation is likely to be systematised only if the primary mechanism for host location is first understood, as this reduces reliance on pattern analysis and ad hoc adjustment to speculative interpretations.

An understanding of the relative use of primary host species and other more incidental hosts is likely to impact considerably on choice of trap plants. If the crop to be protected is not a primary host but is used because little else is available, as seems to be the case for Australian bollworms on cotton, then the use of specially cultivated primary host species would be the best option. In any case, the characteristics of the trap species must match the host-searching mechanism of the pest at least as well as the crop does, and preferably better than the crop. If chemical attractants for ovipositing females are to be developed then working from primary host plants provides the logical starting point (Milne & Walter, 2000; Walter & Benfield, 1994), but preference studies frequently ignore this need. The choice of trap species is likely to be compromised, even altogether, if unrecognised cryptic species are covered by the pest species' name.

The attraction of generalist natural enemies to crops has shown considerable promise for use in IPM (Symondson et al., 2002). At least some thrips species are omnivorous, and in Australian cotton can suppress spider mites early in the season if not disrupted by insecticides (Wilson & Bauer, 1993). Several thrips species are involved, each of which uses multiple hosts but still differs from the others ecologically (Milne & Walter, 1997, 1998a,b, 2000). The encouragement of these species into cotton could well enhance IPM, but source populations need to be identified and the reason for their entering cotton needs to be established if such manipulations are to prove effective. These thrips do not persist on cotton, so their presence on the crop is likely to be as a refuge against otherwise poor circumstances. Also, the role of their 'economically useful' predation should be understood in relation to the primary requirements of the species. Primary host species may well enhance local populations but might also keep them from entering the crop in good enough numbers. Similar reasoning applies to generalist predators such as Carabidae, spiders, syrphids and coccinellids, some of which are far more prey specific than initially predicted (see above).

Discussion

The outlines of the ecology of various species presented in this chapter illustrate several points. Some are to do with method, while others deal with the utility of understanding the ecology of pests and beneficial organisms from the basis of sound principles in species theory (Chapter 6) and autecology (Chapter 5).

An investigation of the ecological processes that influence organisms, including ones referred to as generalists, is likely to be strongly misled if the pattern that first suggested the investigation has not been described accurately. Methodologically, that pattern demands testing. To be misled at such a fundamental level is surely going to misdirect the questions one asks and the interpretations one offers. This is not such an uncommon mistake in ecology. Two widely cited case histories in ecology illustrate the point, that of presumptive competitive displacement of red squirrels in Britain (Walter, 1988b) and those of character displacement in nuthatches (Grant, 1975) and aquatic snails (Saloniemi, 1993). Each one of these examples misled many, not only because they fitted preconception, but also because the underlying pattern to be explained was distorted early and remained that way for a long time. Indeed, red squirrel conservation is usually still considered primarily from the perspective of competition from grey squirrels rather than from the perspective of the primary habitat requirements of the species (e.g. Rushton et al., 2000). Of particular note in this regard is that the pattern underlying some hotly debated events in evolutionary ecology is now beyond further investigation, because of changes to the environment. Early errors could still be distorting understanding, which further supports the contention that species ecology is best understood from the basis of sound principles.

The primary step in investigating polyphagous species should be a justification of the plant species to be included in a host list, on the basis of quantified sampling within specified ecological contexts. Further, it requires criteria for inclusion of plant species as hosts. Species should, for example, be excluded if they are used only rarely and by relatively few individuals, as illustrated with reference to the native bollworm in Australia. The use of more than one primary host species should then be subject to further scientific test, because alternative explanations could hold: (i) host-associated cryptic species; (ii) the herbivore could be adapted, simultaneously, to two species that are very different in their characteristics; (iii) the alternative hosts may, incidentally, have similar characteristics in terms of the herbivore's host location and acceptance; and (iv) the alternative hosts may serve different functions. Other realistic alternatives may also be available.

Different fundamental principles drive research in different directions. This would, perhaps, provide an argument against the reliance on principle, except that there is no other way in science (Chapter 2). The use of 'facts' provides only an illusory foundation. Although the research pathway advocated in this chapter is difficult and time consuming, it is the only way to provide a sound basis from which the management of populations can be planned (e.g. Paterson, 1991; Sabrosky, 1955). Even in a political era in which short-term economic gain is pre-eminent, that approach is justifiable. Note, in addition, that the direction offered also questions seriously the flurry of activity aimed at documenting all sorts of intraspecific variation, which seems to be consequent more on technological development than on anything else. In this case, technology will be most effectively used in conjunction with sound principle; the reliance on serendipity is a sad indictment against late twentieth-century science.

In conclusion, understanding evolutionary origins is vital to an appropriate interpretation of current ecology, even in an applied context. Unfortunately, empiricist views (probably tainted with economic rationalism) have inculcated the impression that evolutionary understanding is an unnecessary luxury, even a diversion, when practical outcomes are desired. This point is considered further in the following chapter, in which the use of predatory and parasitic organisms for biological control purposes is examined. Here a lack of understanding of the organism–environment interaction seriously affects the application of knowledge.

8

Pre-release evaluation and selection of natural enemies: population and community criteria

Biological pest control . . . is considered to be an art by many scientists, although several efforts have been made to transfer it to the realm of science. A number of researchers (the scientists) defend a more scientific basis of pest control, others (the artists) do not mind too much about theoretical considerations, because they think that this scientific basis is still too small and they develop biological pest control mainly by trial-and-error methods.

J. C. VAN LENTEREN (1980, p. 369)

Introduction

Biological control has had many remarkable successes since a 'classical' introduction saved the Californian citrus industry from the ravages of cottony cushion scale at the end of the nineteenth century (Caltagirone & Doutt, 1989; Doutt, 1958; van Driesche & Bellows, 1996). The technique has since been developed and extensively deployed, often with good economic return (e.g. DeBach & Rosen, 1991; Greathead, 1986, 1994). The rationale that underpins biocontrol is the impression that many species in their native environment are limited in numbers primarily by consumers, whether predators, parasitoids, pathogens or herbivores, and which are called 'natural enemies' or, within the biocontrol context, 'beneficial species' (van Driesche & Bellows, 1996). Biocontrol is therefore aimed at restoring the original dynamic, usually in situations in which the pest species has invaded a new area without its natural enemies. The aim is not to exterminate pests, but to maintain them at lower non-pestiferous densities.

Several early biocontrol introductions were ill considered in various ways, and impacted so badly on the environment that protocol and practice were modified to prevent similar disasters (see Ehler, 1991; Greathead, 1994; Simberloff, 1992). Most of the obvious disasters involved vertebrate predators, but more subtle problems have since been detected with insects introduced for biocontrol purposes (see below). Herbivorous biocontrol agents proposed for introduction against weeds are now subject to statutory host-testing in many countries. They must be tested experimentally against desirable plants that may be of economic or conservation value. Permission to release is given if the test results indicate strongly enough that the agent species has no pestiferous potential (Greathead, 1994; van Driesche & Bellows, 1996; Wapshere, 1989). Insect predators and parasitoids have long been screened to ensure they have no undesirable traits that might disrupt biocontrol. The main aim is to intercept species with life styles that could disrupt the primary controlling agents, such as hyperparasitism. These tests are not statutory, except in Australia (Simberloff & Stiling, 1996).

The narrow host requirements of the insects that are introduced for biocontrol have for long been considered sufficient to preclude negative environmental effects. This attitude has been increasingly questioned, mainly because of growing concern for the environment. The demand for safeguards against all native organisms, however insignificant they may seem, is likely to impose additional strictures on biocontrol (Cowie, 2001; Greathead, 1986; Howarth, 1991; Louda, 1998; Simberloff, 1992; Simberloff & Stiling, 1996). Ecologists now suggest that all introductions should be screened to prevent a wider array of environmental damage, some of it exceedingly difficult to measure (Ehler, 1991; Louda, 1998; Simberloff & Stiling, 1996). The modern trend is to contrast the permanence of a biocontrol establishment with the ease of withdrawing a toxic chemical from use. Also, the toxins currently applied in developed economies are not as persistent in the environment as the early synthetic organics.

Entomologists therefore suffer the onus of having to judge the future ecological impact of natural enemies before they are released into the field. Economic efficiency also demands accurate predictions about natural enemies, as does the desire among entomologists to make biocontrol more effective (Ehler, 1990, 1991; Greathead, 1986; Waage & Mills, 1992). The pool of species with at least some potential to control a pest species is often quite large. Many of these species may be excluded by practical considerations, which are often arbitrary (Waage, 1990), but a choice is likely to remain and this is where theory is seen to have a critical role (see Miller & Ehler, 1990; Waage & Mills, 1992). It should help to decide whether the number of introduced species should be restricted to prevent inferior species from interfering with control, and whether species should be introduced in a particular sequence. Choice of the most suitable or promising species for release is also encouraged because all species with biocontrol potential cannot be given the same treatment (Godfray & Waage, 1991; Waage & Mills, 1992), especially in screening for host specificity and in mass rearing exercises for inundative or augmentative release. Also, target species should be selected for their likely susceptibility to biocontrol (Greathead, 1986).

A desirable natural enemy is one that is going to be effective in biocontrol and is not going to have adverse environmental effects. This chapter scrutinises the pre-release evaluation procedures that are currently used in selection of the most promising species for release. Because ecological theory is so diverse (see Chapter 5), several lines of reasoning influence pre-release evaluation goals, criteria and procedures, and these have to be teased apart if they are to be considered in an effective way. Further, the premises of what features make a biocontrol agent successful have been challenged (Murdoch et al., 1985; Strong, 1990), which adds yet another dimension to the discussion because the challenge is based more on empirical evidence than on novel ecological principles.

The overall message of this chapter is that although biocontrol is a successful and desirable technique, practice could be improved. The weakest part of biocontrol seems to be the lack of a direct and realistic connection between theory and practice, an aspect that is quite generally acknowledged (e.g. Ehler, 1991; Godfray & Waage, 1991; Greathead, 1994; Mills, 1994a, p. 214; Waage, 1990; Waage & Hassell, 1982). This disjunction has been attributed to practice demanding immediate outcomes whereas population ecology theory is concerned with long-term dynamics (Kareiva, 1990, p. 70). The disjunction may be better explained, though, by current ecological theory relying on principles (or underlying premises) that were developed without cognisance of the critical ecological influence of the physiological and behavioural mechanisms of individual organisms (see Chapters 5–7).

This chapter therefore scrutinises current efforts to amalgamate theory and practice in biocontrol, particularly in relation to the criteria that are recommended for selection of the species most likely to reduce populations of the target pest. What is said applies also to other aspects of biocontrol. The chapter outlines the direction in which current theory is headed in relation to biocontrol practice, and provides a critique of that approach. Specific arguments are developed at some length because the theoretical developments currently espoused in the literature are multidimensional. A criticism of a particular aspect of biocontrol theory is therefore readily shored up by an advocate citing another aspect or recent development that is reputed to show promise. The problem is that the flaw in the newer statements may not be so readily detected, despite them being based on premises as unrealistic as those originally criticised. In this way the 'frontiers of science' seem to be moving positively, but the flawed foundations remain unadjusted. The assessment of biocontrol presented in this chapter, and the suggestions for improving practice in the following chapter, differ from most previous challenges in being based on an application of alternative fundamental principles, which have been developed and justified earlier (mainly in Chapters 2, 5 and 6).

Pre-release evaluation of natural enemies

Biocontrol practitioners are not unanimous that pre-release evaluation is worthwhile, and those who believe it has promise do not agree on what criteria are relevant or to what extent the impact of biocontrol can be predicted. The division in opinion is not clearly demarcated and has to be dealt with in somewhat general terms. On the one hand, pragmatic biocontrol practitioners keep pre-release work to a minimum and do not try to evaluate future performance (e.g. van Lenteren, 1980). This 'empirical approach' (Ehler, 1990) tends to concentrate on basic aspects of taxonomy and host associations, with an important aim being to screen out species that would be detrimental if they were released. Note that 'empirical' does not imply a lack of care, for the recommended guidelines are extensive and far-reaching (van Driesche & Bellows, 1996, pp. 147ff.). The alternative school of thought is the 'predictive approach' (Ehler, 1990), which accepts that current ecological theory should provide grounds, as well as further promise, for assessing the pressure that natural enemy species will exert on pest populations after their future release into the field.

The pragmatic approach undoubtedly persists because the predictive approach has not yet demonstrated much utility at all, despite a fairly extensive history. For instance, experience shows that little can be predicted about the future performance of any species (see Chapter 9). Some would add that the complexity and heterogeneity of the environment and the subtlety of a species' requirements and tolerances render accurate forecasting impossible (also see Chapter 9). The resistance to pre-release evaluation seems, in addition, to be aimed at particular aspects of the theory. Some react against calls for 'long-term basic research' (e.g. van Lenteren, 1980), which presumably means the type of research that is driven by population dynamics theory. Included would be such aspects as k-factor and life table analyses and life system studies. Also, practitioners have seldom responded positively to the mathematical models that purport to mimic the most significant processes in successful biocontrol. Such models have never influenced action in a classical biocontrol programme, although they are still given textbook treatment as representative of the general features of biocontrol systems (e.g. van Driesche & Bellows, 1996, p. 397). Reasons for the link between biocontrol practice and mathematics being so tenuous are also considered in this chapter.

The predictive approach has an entirely different outlook from that of the empirical school. It accepts that theory should provide a sound basis for selecting the most appropriate species, or suite of species, for mass rearing and release, and should also set priorities for research effort (Ehler, 1991; Godfray & Waage, 1991; Hochberg & Hawkins, 1993; Murdoch et al., 1996a; Waage & Mills, 1992). The predictive approach draws input from two areas, population dynamics and community theory (see Chapter 5). Each circumscribes particular ecological criteria that are considered essential in a biocontrol agent if it is to be successful. Those criteria specified by population theory are referred to as reductionist, whereas the community specifications are seen as holistic. These two sets of criteria are perceived by many to be complementary (Waage, 1990; Waage & Mills, 1992) through their common underlying principles (see Hassell & Godfray, 1992, pp. 287–288).

Each of the various demarcation lines drawn so far in this section is somewhat arbitrary in practice. Individual researchers tend to synthesise an eclectic approach, and 'even the most fanatic opponents of basic research' may develop fairly extensive lists of criteria that an effective natural enemy should possess, although they may never screen for them (van Lenteren, 1980). Two implications follow. First, views and practice are open to influence from arbitrary conjunctions of ideas and premises, a problem that apparently does not usually affect the application of principles in such sciences as physics and chemistry, and which surely signals problems with the fundamentals of ecology. Second, if any current statement about pre-release evaluation is to be scrutinised effectively for its validity and utility, it will first have to be broken down into its component parts for analysis. One has to analyse specifics whilst maintaining the logic of the broad issues. The perspectives offered in this chapter are intended to be applied in this way and they concentrate on the predictive approach. The empirical approach is considered in the following chapter.

Prediction, demographic principles, and sacrifice of realism

Practising entomologists and ecologists generally agree that natural enemies control pests by reducing their density to a low stable equilibrium level (Murdoch, 1992, p. 198). The predictive approach incorporates this view into a means for developing a general concept of the features that characterise successful natural enemies. The principles that underpin this approach derive from demographic ecology, in which biotic processes are assumed to dominate in ecological settings, and are believed to function deterministically. Density-dependent processes such as competition, predation and parasitism (Chapter 5) should, therefore, stabilise or regulate populations. Understanding the source of the density dependence needed to stabilise the biocontrol target at a low equilibrium density is the basis for assessing the future potential of a natural enemy species (see Murdoch & Briggs, 1996).

Factors other than density-dependent ones are frequently seen simply as 'noise', and population trends are frequently acknowledged to be an interplay between 'precise' and 'noisy' factors (e.g. Hassell, 1986). Nevertheless, demographers still believe it 'remains valuable to set the stochastic elements to one side and focus on the deterministic processes . . . to understand how some key features of an interaction can promote population regulation' (Hassell, 1986, p. 202). When environmental influences are included in demographic models, they are usually simplified into a single survival probability (e.g. Godfray & Waage, 1991). This is unrealistic and potentially misleading in that the mortality factor with the major influence on population change (the key factor) is virtually always density independent in nature, and driven by the environment (e.g. Varley et al., 1973), and not amenable to such simplification (as described by Walter & Hengeveld, 2000).

Density dependence is still invoked to specify fundamental cause in biological control (e.g. Waage, 1990, pp. 147–148), and the methods used in the pre-release evaluation of candidate natural enemies are generally influenced by the demand that successful biocontrol agents confer stability on the pest population. Both of these approaches to biocontrol are questionable, for two significant reasons. First, the concentration on density dependence and population regulation ignores the fact that the ecological importance attributed to density dependence is highly questionable (Chapter 5). Second, the interaction between environmental influences and natural enemy autecology is more likely to influence population trends in the pest species. These points are raised again in this chapter to criticise the current approach. In Chapter 9 they are used to help to develop suggestions for an alternative approach to the demographic one.

The ecological importance accorded the attributes of organisms under current pre-release evaluation programmes (i.e. reductionist and holistic) has been questioned fairly extensively in the literature. Aspects of the reductionist approach have been criticised in the biocontrol literature itself. For example, van Lenteren (1980) and Waage (1990) both point out the limitations of the evaluation criteria usually isolated for consideration, a long summary list of which has been extracted from the *Trichogramma* literature by Smith (1996). Nevertheless, most suggestions for improving the situation advocate persistence with the same basic direction (e.g. Godfray & Waage, 1991; Waage & Mills, 1992). By contrast, the holistic approach has not yet been subject to much published scrutiny from within the discipline. However, criticism of the premises and interpretations provide fairly extensive reading in the ecological literature (see Chapter 5). Furthermore, biocontrol is frequently, even if only implicitly, seen as applied population dynamics (e.g. Luck, 1990; Murdoch & Briggs, 1996) rather than applied community ecology, which suggests that the holistic approach is not widely accepted as a theoretical basis for biocontrol.

To avoid repeating the published criticisms of these approaches to developing biocontrol theory, especially those related to analyses of the historical frequency of success (see van Driesche & Bellows, 1996, pp. 134ff.), attention is drawn to shortcomings in the ways in which evolutionary and ecological theory is applied in the quest for a useful theoretical background to biocontrol practice. Although the various aspects are considered under separate headings they are, in reality, interrelated. A way around the impasse imposed by a lack of theoretical underpinning is tackled in the following chapter, where the application of autecological principles to biocontrol practice is considered.

Emphasis on population stability

Successful biocontrol is interpreted as a two-step process. An initial reduction in overall pest density is followed by the maintenance of pest

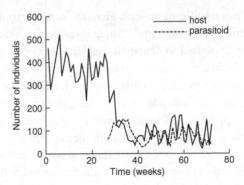

Figure 8.1. Example of population trends aimed for in biocontrol. This is a laboratory example involving a seed-eating bruchid beetle and its pteromalid parasitoid. Note the two phases in the pest's population dynamics: the initial lowering of density and the subsequent maintenance of a lowered density. Reprinted from Hassell & Godfray (1992) with permission from Blackwell Publishing.

populations at a new lower level (or 'equilibrium') (Waage & Greathead, 1988, Waage & Hassell, 1982) (Fig. 8.1). The process that reduces the population initially (population depression) may operate independently of the process that imposes stability (population regulation). The initial 'transient dynamics' are, however, usually neglected in biocontrol theory and in pre-release evaluation procedures (Kareiva, 1990), presumably because (i) population reduction is seen as a temporary part of the biocontrol process and (ii) natural enemies that act in density-dependent fashion, by aggregating, are assumed to achieve both goals. Field quantification reveals this second assumption does not inevitably hold (Roland, 1994), and population models suggest that density dependence and stability are achieved at a cost to the level at which the equilibrium density is set (see Luck, 1990). The most direct statements about the influence of natural enemies in depressing pest populations (= lowering the equilibrium density) claim that natural enemy species should have a high successful search rate per individual (Murdoch & Briggs, 1996) or a reproductive rate commensurate with the rate at which hosts are renewed (Barlow *et al.*, 1996; Price, 1972).

Practical assessments of a natural enemy species' potential to impose stability on the pest population fall into two categories. On the one hand, behavioural and reproductive features believed to dictate the response of organisms to host density are quantified, usually in the laboratory. Various parameters are thus specified and the future impact of the species may be assessed with the aid of mathematical models. Biocontrol potential can also be compared among species in this way, with a view to selecting the one that will be most effective as a control agent. The second method of assessment is carried out in the field, and is less common than the first. The impact of a natural enemy species on its host's population dynamics is quantified, preferably from life table data and in the host's home country (Waage & Mills, 1992). The aim is to judge how effective the species will be in imposing density-dependent mortality when introduced elsewhere. Neither of these general methods has contributed to the pre-release evaluation of any species, but both have been consulted retrospectively in efforts to identify the critical ecological mechanisms in successful biocontrol. They are still widely seen as the way forward in endeavours to improve biocontrol practice. Each approach is now considered further.

1   Laboratory-based assessment criteria

Population dynamics conceptualises the key elements in the ecology of a natural enemy species as its response to the distribution of its hosts or prey in the environment. The population-level response to the density of resources is called the total response, and has two components. The functional response deals with the rate at which individual natural enemies kill hosts within a patch. The numerical response quantifies the way in which the population as a whole gathers and grows in response to patch size.

The functional response (Fig. 8.2) is considered to offer a good conceptual framework for understanding natural enemy searching behaviour within a patch of hosts. It is reputedly relevant mainly to inundative

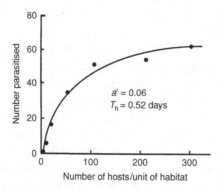

Figure 8.2. A typical functional response of an insect parasitoid, with the attack rate (a') and handling time (T_h) given. Reprinted from Hassell (1976) with permission from Hodder Arnold.

release of biocontrol agents rather than to inoculative release (Waage & Greathead, 1988, p. 116), but is used in far more general ways, as indicated below. The functional response model quantifies the behavioural and physiological aspects of the response of natural enemy individuals to different densities of hosts or prey (see O'Neil, 1990). The upper limit on the graph (Fig. 8.2) is imposed by behavioural and physiological attributes of the natural enemies, namely their *per capita* searching efficiency and handling time. These are estimated indirectly by substitution of the derived values for parasitisation rates at different host densities into an equation, modified from Holling's disc equation:

$$N_a = [Ta'N/1 + a'T_hN]P,$$

where N_a is the number attacked, T is the total search time (duration of experiment), a' is the searching efficiency or coefficient of attack, N is the initial number of prey available, T_h is the handling time, and P is the number of predators or parasitoids (Varley *et al.*, 1973). Attributes such as searching efficiency and handling time are considered to be among the key attributes of natural enemies that enable them to impose stability on the pest population (e.g. Waage & Mills, 1992). Others include stage of host attacked, larval survival, life span and aggregation ability. These features are estimated and assembled into somewhat general analytical models of intermediate complexity (e.g. Godfray & Waage, 1991). Such models are favoured over situation-specific simulation models and more simple general models because (i) of their expected potential for providing insights about other systems, which specific simulations cannot do, (ii) they do not require information on a large number of parameters, as do more specific and detailed simulations of particular systems (see Godfray & Waage, 1991), and (iii) the various features anticipated to characterise an effective natural enemy are said to be assembled more realistically than in more simple and generalised models (Murdoch & Briggs, 1996; Waage & Mills, 1992).

The rate at which parasitoid or predator individuals collect on patches of hosts or prey of different densities is quantified into the numerical response (Fig. 8.3). Note that the origin of the natural enemies is not specified; they may switch prey locally or be recruits from reproduction or migration. The functional response is investigated more frequently than the numerical response, presumably because it is more amenable to quantification. The numerical response may, however, be more important in the suppression of pest populations (Pedigo, 1999, p. 299), but defining a 'patch' of hosts in a non-arbitrary way is problematic, even if one accepts

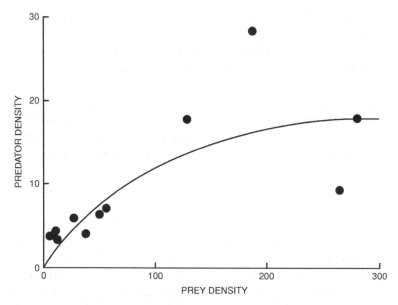

Figure 8.3. A typical numerical response of a generalist insect predator. Reprinted from Hassell (1986). Copyright 1986 with permission from Elsevier Science.

that 'patch' is the most appropriate abstraction around which to interpret natural enemy ecology.

Host or prey specialists are generally advocated for biocontrol purposes, and have frequently been very successful. Generalists are, nevertheless, considered to be far superior in their numerical response, because individuals can switch immediately from one host species to another, so their population size remains relatively constant compared with the fluctuations in abundance of a particular host species (Hassell & Godfray, 1992). Specialists have first to reproduce, and so a time delay in their numerical response is likely (Hassell, 1986).

Generalisations such as those outlined above would be useful if they could provide reliable practical guidance. Unfortunately they have not done so, and because the principles upon which they are built are inappropriate (Chapter 5), they are unlikely to do so. Also, the evidence about such generalisations is not encouraging. For example, the functional response, from which attack rate (or searching coefficient, searching capacity or area of discovery) is measured, is not as constant as originally thought (O'Neil, 1990; Rosen, 1985). It is affected by changing prey densities, different densities of conspecifics, change in the substrate (through plant growth altering prey density), and so on (O'Neil, 1990). One can thus picture how variable the functional response would be in the field because of all the factors (whether biological or physical) that influence presence and abundance of natural enemies (of all species relevant to the prey or host species of interest) and all the prey or host species relevant to the ecology of each natural enemy species. Each species involved would be independently influenced by the prevailing ecological influences, and these influences would, in turn, vary spatially, even on a local scale. Factors that influence natural enemy behaviour, such as plant density, would also alter the situation spatially. Furthermore, field evidence even contradicts the expectations that natural enemies should preferentially aggregate in areas of high host density, or that percentage parasitism correlates with host density (e.g. Jones *et al.*, 1996; Reeve, 1987; Smith & Maelzer, 1986).

The variability outlined above does not provide a reliable basis upon which to develop generalisations or accurate predictions, which implies the underlying concept of 'total response' is an unrealistic abstraction. That is demonstrated by the difficulty of differentiating, in practice, elements of the 'functional response' from the 'numerical response', because of migration (Wratten, 1987) for instance. Migration is one of the primary behavioural and ecological processes conducted by individual organisms and falls under the definition of 'numerical response'. Yet the movement of organisms obscures the clarity between numerical and functional response. This confusion indicates an inappropriate emphasis on population processes as opposed to the processes that are primary to the existence and reproduction of individual organisms, and thus to their ecology. That the functional response is an inappropriate abstraction can be justified further. Experimental estimates of 'searching efficiency' decline with increasing prey density (Fig. 8.4), which seems counterintuitive. But they do so because estimates of 'searching efficiency' are influenced by estimates of 'handling time' (O'Neil, 1989), and both are estimated indirectly from the population consequences of behaviour, not directly from the behaviour itself.

2   Life table estimates of parasitoid impact

Life table analyses of biocontrol successes have all been retrospective. The understanding generated by the analyses would never have predicted the outcomes actually achieved, to date at least, and some introductions even achieved control contrary to the predictions derived from life table studies (see Greathead, 1994). Undoubtedly 'this reflects in part the different situations in which natural enemies act in their areas of origin and

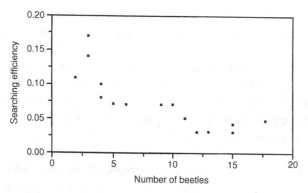

Figure 8.4. Decline in 'searching efficiency' of a predator, despite prey density (beetle larvae) being increased and exposure time remaining constant. The decline is caused by increased 'handling time' of the natural enemies as they eat more individuals. The contradiction is a consequence of working from a population pattern down to the behaviour of individual natural enemies, instead of working from individual behaviour outwards. Reprinted from O'Neil (1989) with permission from the *Journal of the Kansas Entomological Society*.

introduction' (Waage & Mills, 1992), which is supported by numerous ecological studies of organisms in different parts of their distribution (e.g. Hengeveld, 1990). In other words, species do not carry with them their perceived role in conferring density dependence or their postulated role in a community when they are transported elsewhere. Whereas the individuals carry equivalent adaptations in different areas, their abundance and seasonality is, more often than not, quite different. This is a fundamental point, both to pragmatic biocontrol practitioners and to autecology (Chapter 9).

Concentrating on population stability in the way outlined in this subsection has not proved useful in practice, and has been criticised from several perspectives, not least for its having a rather abstract mathematical quality (Strong, 1990). Other flaws include the following: (i) The primary and variable influence of environment is ignored. (ii) The behaviour of organisms is interpreted in relation to population-level interactions rather than in relation to the primary determinants of the behaviour of individuals. (iii) Any density effects are likely to be influenced by context, are unlikely to be linear, and are likely to be significant only at unusually high densities, and so do not provide a suitable basis for effective ecological generalisation.

To relate the properties of organisms to a peripheral abstraction, such as population stability, effectively eliminates organisms and their properties from the concept of the system that is to be manipulated.

Furthermore, stability is not necessarily correlated positively with pest mortality and is obviously irrelevant to some of the best-known biological control successes, in glasshouses for example (Strong, 1988, 1990). More telling, perhaps, is that the persistence and dynamics of natural enemy and prey can be explained without reference to stability (Strong, 1990). Some ecologists have therefore sought more meaningful criteria for the pre-release judgement of biocontrol potential, and we consider these in the following subsection.

Non-equilibrium, regional or metapopulation dynamics

Many of the problems identified in the previous subsection have already been recognised in the ecological literature as problematic or unrealistic. The solutions proposed so far in the literature are superior to the original interpretations. They include non-equilibrium perspectives of populations, regional influences on local ecology, and the concept of metapopulation. All, however, are still supported by the original principles that underpin population dynamics theory. Consequently, the solutions they propose eventually run into similar problems to those they were designed to overcome. Although the relevance of invoking a 'population equilibrium' is increasingly questioned, for example, the usual alternative offered is that the world should be seen as being in a state of 'non-equilibrium' (e.g. Kareiva, 1990; Murdoch, 1992). This proposal is ineffective because it centres on the existence of some sort of 'non state' (Chapter 5) that can be defined only in relation to a non-existent 'equilibrium', so the proposed solution has too much in common with the earlier interpretations to solve the problem effectively. Its underlying assumptions remain in the rejected dimension, which probably explains why some recent interpretations include statements that do not differentiate non-equilibrium from equilibrium ones for practical purposes. Even non-equilibrium populations are considered to be regulated, through 'stochastic bounding' (e.g. Murdoch & Briggs, 1996). The approach thus retains the emphasis on population-level processes and works down to the ecology of individuals, rather than the other way round, as in autecology (see Chapter 9).

Because there has been no deep-rooted shift in perspective, the pre-release criteria developed from the alternative perspectives of non-equilibrium and metapopulation dynamics do not differ significantly from the earlier criteria, even though the spatial element of the environment and migration of natural enemies may have been incorporated. To illustrate: the characteristics envisaged for an effective natural enemy have been related to the speed with which they aggregate at isolated, high density patches of prey, which implies that one would need to measure the rate at which a predator travels through a field in the absence of prey (e.g. Kareiva, 1990), among other things. What is being assessed here is the ability of the natural enemy to extinguish incipient outbreaks before the pest is able to 'swamp out the regulatory capacity of the control agent' (Kareiva, 1990). But the ability to accomplish control in this way may not require the sort of aggregation characteristics predicted under demographic ecology, and rate of movement between 'patches' may not be a primary influence on the dynamic. Until the primary causes of population change and persistence are included in biocontrol analyses, the role of natural enemies will not be well understood. In any case, the speed of movement in the absence of prey tells one little about the ecology of the organism in question, and especially not in relation to its ability to find pests and kill them. How much difference would it make if a parasitoid moved at 10 km an hour or 20 km an hour? Probably very little when one considers that egg production rates are so low in many parasitoids (Donaldson & Walter, 1988; Fernando & Walter, 1999; Walter, 1988a). Again, unhelpful abstractions seem to be driving interpretation and expectation, which warns that ecological understanding should be sought elsewhere.

When the behaviour of individual natural enemies has been generalised in relation to population processes, it has usually been cast in terms of optimisation (e.g. Godfray, 1994; Luck, 1990). The behaviour of searching natural enemies is therefore seen, from the optimisation perspective, to be a process of making decisions in the interests of maximising fitness. Because oviposition rate changes under such influences as egg load, experience and learning (Collier et al., 1994; Jervis & Copland, 1996; Murdoch et al., 1997; Rosenheim & Rosen, 1991), these factors are said to influence the 'decisions' made by the organisms. The behaviour of the insects is therefore simulated with dynamic variable models. This approach to generalisation brings new difficulties, most fundamental of which is that optimisation cannot be practised by natural enemies (Pierce & Ollason, 1987; Rapport, 1991). Whether the aim of maximising fitness can drive behaviour and evolution is therefore also questionable, because of the nature of complex adaptations and the structure of the genetic and developmental system (Hengeveld, 1997; Walter, 1993c). Two additional complexities undermine the optimisation approach. For example, the generalisations developed in this way transcend species-specific differences in behaviour and ecology, which are primary influences on eco-logical processes (Walter, 1995b). Also, the influences of experience and learning are not yet tested in the field, where conditions are complicated to the extent that these aspects may have relatively little role in the ecology of insect natural enemies. This has yet to be investigated.

Ecological dichotomies, continua and correlation

Numerous efforts have been made to characterise natural systems accord-ing to features that are considered central in the ecology of organisms. Most such generalisations take the form of ecological dichotomies, the end points of which delimit a continuum. Criteria upon which they are based include aspects of the environment (e.g. habitat stability, habitat predictability, stage of succession), the nature of selection pressures as-sumed to be acting on the organisms (e.g. r-selection vs. K-selection), or the ecological strategies that organisms are considered to have evolved (e.g. degree of competitiveness of natural enemy species (Myers *et al.*, 1989; Zwolfer, 1971)) or amount of immunity to parasitoids evolved in the host species (Myers *et al.*, 1989). Parasitoids, for example, have been charac-terised according to their perceived status as being intrinsically superior competitors, in the larval stage, or as being extrinsically superior as adults searching for hosts (see Hassell, 1978). Also, generalist or polyphagous species have been contrasted with specialists, as they are considered, for example, to spread a web of interaction as opposed to comprising a more discrete ecological entity (Hassell & Godfray, 1992, Lawton, 1986; Wratten, 1987, but see Chapter 7 for an alternative view).

Dichotomies that are even more arbitrary have also been developed. They involve judgement of whether the ecology of the natural enemies is influenced more by 'intrinsic' factors than by 'extrinsic' ones (e.g. Ehler, 1990, p. 123). Such abstractions are too restricted to represent or encap-sulate accurately the vast number of factors that influence the ecology of conspecific individuals, even before one takes account of the spatial het-erogeneity of the environment (see Chapter 5). In any case, the distinction between extremes of proposed ecological dichotomies tends to fall away as one introduces more factors (and thus more realism) into the analysis (Horn & Dowell, 1979). The lack of realism of ecological dichotomies is fur-ther apparent when one considers that it is the interaction between organ-ism (intrinsic) and environment (extrinsic) that is important, as stressed in the autecological approach (Chapter 9), and that this interaction is species specific and context specific. Ecological generalisations need to be designed to cope with the idiosyncrasy of species if they are to be robust and workable.

Once species have been assigned to a branch of a dichotomy, or to a position along a continuum, the degree of biological control achieved by species on that branch (or position) is contrasted with that achieved by species on the other branch (or at another position). In this way life history and population parameters are explored for correlations with known outcomes (Waage & Greathead, 1988). The probability of establishment after release may be dealt with in similar fashion. Other such unidimensional correlations have been investigated. They involve such series of discrete character states as life stage of host attacked (Mills, 1994a), species richness (Hawkins, 1993), whether parasitism is internal or external (Mills, 1994a, p. 219), degree of concealment of the host (Hawkins, 1993), whether the association of host and natural enemy is old or new (Ehler, 1990, p. 120; Hokkanen & Pimentel, 1984), whether the natural enemy is a 'high density' or 'low density' one (Mills, 1990; Myers *et al.*, 1989; Waage & Mills, 1992), and the degree of resistance reputedly evolved by the host in the postulated 'arms race' between host and parasitoid (Hokkanen & Pimentel, 1984; Myers *et al.*, 1989). Various demographic parameters have also been analysed in this way, including fecundity, degree of egg aggregation, voltinism, size, life span, and so on (e.g. Crawley, 1986; Lane *et al.*, 1999).

Again, in selecting a particular dichotomy or continuum all the other ecological factors known to influence the local abundance of species significantly are ignored, as pointed out by Andrewartha & Birch (1954, 1984) and Krebs (1995). Is this a reasonable degree of omission or is the 'Bronowski cut' removing too much that is relevant (Chapter 2)? Local abundance varies so much with changes in environmental conditions that even when the results of analyses do suggest that such measured features as species richness, host mortality and biocontrol success rates do co-vary, we should still be cautious. Correlation does not imply causation, and even indirect causation in such cases is unlikely if one is dealing with by-products such as species diversity, population size, and so on (see Chapter 5).

Several continua or sets of character states are sometimes combined into some sort of synthesis, in efforts to make the abstraction more realistic, and to cope with more than two natural enemy species simultaneously (e.g. Mills, 1994a; Price, 1991). The theoretical structures developed by Price (1991) and Mills (1994a) are appealing, not least for their neatness. Price (1991) postulated a functional relationship between the increasing 'stability' of different stages along successional gradients (from herbs, to shrubs and then trees), the feeding habits of herbivores (external feeding to concealed feeding), and the relative host specificity of parasitoids. Too many exceptions exist and the treatment is too frugal in biological detail. And even when the dichotomies and continua are related logically to one another their product has never been sufficiently robust to explain the ecology of organisms or to predict what they will achieve within a new setting. Specifically, organisms comprise such a complex of ecologically significant features that the variation in the combination of features is immeasurable. Although we might expect an association between particular characteristics, that does not occur with sufficient regularity for any such generalisation to be useful.

Emphasis on resources, and species considered typologically

An emphasis on resource quantity as the primary ecological driver is common to both population and community ecology (Walter, 1995b; Walter & Hengeveld, 2000). Those pest individuals that do not suffer attack from natural enemies, for example because they live within the canopy of their host plant rather than on the outer leaves of that plant, are seen by population ecologists as living within a 'refuge' and thus contributing to the stabilisation of the population as a whole. Community ecologists treat host or prey individuals as resources, and all 'unused resources' need to be eliminated for effective control, even those in refuges (Hochberg & Hawkins, 1992). Each approach is considered in turn.

The population ecology approach sees refuges in two ways. First, if a low proportion of the pest population is in a refuge, the population as a whole will be 'highly exploitable' by natural enemies (Hawkins et al., 1993). In other words, a large proportion of pest individuals would be exposed to attack. This makes intuitive sense, but how general a phenomenon this could be is difficult to say, as sometimes 'refuges' themselves provide a source of natural enemies, for example those that invade certain orchards (Caltagirone & Doutt, 1989), so the value of refuges may be context specific. Furthermore, various problems exist in the method used by Hawkins et al. (1993) to estimate 'refuge size' (Myers et al., 1994), and Williams & Hails (1994) recognised that 'refuge size' cannot be treated typologically, for it varies in relation to host density, parasitoid density, or both simultaneously. 'Refuges' are therefore poor candidates for developing robust general theories for biocontrol. The approach is weakened even more when all the other ecological factors that impinge on pest and natural enemies are taken into account (see previous subsection).

Refuges are considered important in a second way by population ecologists. They are seen as being useful in stabilising the pest population, an insight derived from population modelling (Luck, 1990). The main point about such a general contention is the question of whether population stability is as important a goal as frequently accepted. If not, as argued above, then the only consideration about refuges relates to the points made in the preceding paragraph.

Biocontrol theorists who favour the community approach analyse suites of species, within guilds or communities, for the biocontrol potential of the entire collective (e.g. Miller & Ehler, 1990, pp. 162–163). This approach focuses on the interplay between resource availability and resource use by the particular combination of species, and this defines the primary determinant of ecological structure. Attention is thus generally directed at patterns of resource use among natural enemy species. For instance, a relationship may be sought between the 'structure' of the guild, or pattern of resource use, and the collective impact of the guild on the host population (Miller & Ehler, 1990, p. 166). Some of the unstated assumptions and implications of this approach need to be considered, to show how realism is sacrificed when interspecific patterns based on resource use are given prominence over the species-specific properties of organisms and their role in determining ecological outcomes.

The species included in each guild are, in certain respects, dealt with as equivalents because they are seen to have a 'role' relative to the other member species. Such typology is unacceptable (see Chapters 5 and 6), in part because each member species is assigned its 'role' as a resource user in the guild without reference to the range of features known to impact on the ecology of the individuals that comprise each of those species. For an empirical example of this, see van Klinken & Walter (1996), who demonstrated with reference to drosophilid flies that the *coexistence* or community approach to explaining local ecology pushes aside those features of each species that allow them to *exist* locally. Local existence is the problem, coexistence is not an issue (Sale, 1988; Walter, 1988b). Several adaptive complexes are likely to be involved in the existence of organisms in any given area, as in their relationship to particular environmental conditions. Any changes in local environmental conditions will change the equation again. It is these interactions that dictate the locality-specific ecology of each species. The local ecology of each species is played out mainly independently of the other species in the guild, except for any incidental competitive interactions (see Chapter 5). In other words, the prevailing environmental conditions, which influence the local abundance of species significantly and differentially, are excluded from consideration in the community approach to natural enemies. So are the species-specific properties of the organisms involved (Chapter 5). Of most relevance, perhaps, is the host-searching behaviour of the natural enemies, which influences directly which host species will be attacked by individuals of each natural enemy species, and, to some extent, the intensity of the attack on each host species. This begs the question of what signifies that a particular natural enemy species is a 'member' of the guild of interest? Some parasitoids use the species of interest only as a minor host, others use it incidentally, and so on. Clearly, they cannot be accorded equivalent status in relation to the host population as a resource. The other criticisms of the community approach outlined in Chapter 5 are also relevant here, and need to be appreciated in any judgement of the validity and utility of this approach to biocontrol.

The 'individualistic' nature of the ecology of species reveals that guild structure is a by-product of the primary interaction between individuals of the various species represented and the environment, whether physical or biological (see Chapter 5). That this primary role of the 'autecological' interactions is denied in the community and guild approaches is patent in conclusions that claim, for example, that 'parasitoid guilds are determined by three factors: the mode of parasitism, the host stage attacked, and the form of parasitoid development' (Mills, 1992). Because this statement also ignores the fortuitous nature of speciation, it must accept implicitly that speciation is an adaptive device, with natural selection acting directly to produce a suite of species that will use the environment efficiently (see Chapter 6). Insurmountable difficulties arise when working from a complex pattern rather than from a fundamental process. An almost universal experience in ecology is that identified community patterns are evanescent. The idiosyncrasy of species simply overrides them sooner or later (see Miller & Ehler, 1990, p. 166, and Chapter 5). These problems are best circumvented by working up from the fundamental processes associated with a focus on individual organisms and their species-related adaptations (Chapters 5 and 6) towards the complex pattern apparent at the time and place of interest.

The community approach sees pest populations as unused resources. Partial biocontrol is thus taken as a problem of why an obvious abundance of resources (hosts) is not more fully utilised by the natural enemy species present. That explains why refuges from parasitism are emphasised in explanations of why pest populations are not lowered by the natural enemies present, and why other species in the natural enemy guild are considered to complete the complement in the natural enemy community. Resources are seen to be unused because a proportion of the pest population occupies a 'refuge', and that refuge represents the niche of another natural enemy species (see, e.g., Hawkins, 1993; Miller & Ehler, 1990). A change in fundamental principles offers a more realistic perspective on such refuges (Chapter 9).

The topic of the role of species has another dimension that has been developed to some extent in the biocontrol literature. This involves the postulate that if a species is introduced for biocontrol and it has the same 'role' (or niche) as a species already present, one will affect the other negatively through interspecific competition. Biocontrol data sets have therefore been analysed in efforts to clarify such issues. The numbers of natural enemy species introduced are examined in relation to success in establishment or control, because the competitive effects from one are expected to disrupt the biocontrol efficiency of others (Ehler & Hall, 1982). Exceptional cases, realistic alternatives (e.g. Keller, 1984) and theoretical objections (see Chapter 5) argue against the validity of such an approach. Despite the published objections to the conclusions drawn from such analyses, the original competition-based interpretations are still used in developing practical guidelines (e.g. Waage & Mills, 1992). The incidental nature of interspecific competition and its variable effects (Chapter 5) suggest that this approach is unlikely to lead to realistic theory or robust practical measures.

Modelling to evaluate proposed determinants of biocontrol success

What actually determines biocontrol success? Is it a feature of the control agents alone, a combination of features of the pest as well as the biocontrol agents, or are aspects of the environment relevant as well? Mathematical population models are seen as crucial to evaluating the determinants of biocontrol success and are frequently portrayed, intentionally or unintentionally, as *the* theoretical underpinning to ecology. The parameters making up the structure of population models tend to be treated as the only valid currency in ecological theorising, as part of the ecological laws identified by Turchin (2001). This seems to lead readily to the view that population models can test interpretation and even verify it. Claims that an author has demonstrated or proved a point need to be assessed critically when, in fact, they have developed a model only that is consistent with the original verbal interpretation. This is not such a convincing feat when the outcome of the equation is predetermined (Wangersky, 1978). In other words, models are essentially mathematical expressions that represent a particular body of theory, with the parameters merely representing conceptualisations of a limited number of ecological features. Alternative theories will support different models (see Walter & Hengeveld, 2000).

Several consequences stem from the demographic population modelling approach. These are outlined below, and in general show how easy it is to mould a biocontrol success to fit the predicted cause(s) or determinant(s) of that success. This is an important detractant because almost invariably it is the successful cases that are modelled in the verification process, even though they are likely to represent a biased sample. They are 'self-selected' because of their conventional history and rapid success. These cases tend to be ones that have been relatively straightforward in that a host-specific natural enemy (or one that is nearly so) has reduced population levels soon after introduction. However, most programmes are quite different from this, and the following points warrant consideration in this regard.

1   The general biocontrol models borrowed from population ecology assume that a character of the natural enemy species, or a particular suite of characters, was responsible for conferring the recorded biocontrol success. That character, or suite, is also assumed to be 'demographic' in nature (e.g. 'r', the intrinsic rate of natural increase), because that is the way in which the population models are set up. These general models are used to search for characters that relate to biocontrol success, implying that a single specifiable cause is responsible for success and is common to other biocontrol successes, perhaps even all of them. However, the reduction in pest densities may not be related at all to the potential demographic 'causes' being modelled. Could the lowered abundance imposed on the target species, which is an end product of all ecological influences that act on the target individuals, ever be traced to the measured demographic properties of the biocontrol agents? This seems unlikely, since the organism–environment interaction that is so crucial to the existence, rate of reproduction, species-wide distribution and local abundance of

the organisms is ignored (Hengeveld & Walter, 1999; Walter & Hengeveld, 2000). The population ecology approach is likely to bolster unjustifiably the image that the measured demographic parameters have been the most significant influences in achieving biocontrol, rather than the behavioural and physiological adaptations of organisms, which undermines the realism claimed for most models (e.g. Godfray & Waage, 1991; Murdoch & Briggs, 1996). Perhaps the demographic approach to identifying key parameters would not be so readily dismissed if the measured parameters stayed reasonably constant, but this is not even true of the more tangible features, such as 'r' and the functional response (see 'Emphasis on population stability', above), so they are not likely to allow the development of general insights about other systems.

2   Natural enemies with host-searching behaviour appropriate for locating the target species are assumed to be easily recognised, whether from field samples or laboratory tests. The inference that they attack the target species habitually is also readily made, as reference to almost all models of successful projects shows. This may not be such an easy prediction to make accurately, for two reasons. First, field data need to be comprehensive and extensive before incidental or sporadic attack can be ruled out (see Chapter 7). If such 'sporadic attackers' are introduced, they will not inevitably tackle the target species, even though, to hopeful human observers, the target may represent an abundant unused resource. If such organisms were to adapt to these conditions, as is frequently assumed (Chapter 6), we would surely see significantly more successes than we see now. Second, laboratory data can be seriously misleading, especially when predators, egg parasitoids and ectoparasitoids are involved. Many of these species are considered to be polyphagous generalists, a view based mainly on records of 'sporadic attacks' in the field and on laboratory attacks (Chapter 7). Laboratory attacks may not be a good indication that a species is a habitual natural enemy of the pest in question.

3   No obvious relationship has yet been spelled out between theory in ecology and any measure of success in biocontrol. 'Success' is not an ecological phenomenon but a subjective economic judgement (Mills, 1994a, p. 214). Thus the only conclusion we can make about a successful case is that the minimum requirements for success have been met (see the two points above). Alternative avenues to generalisation are available from this point and warrant investigation different from the search for demographic correlates of success.

4   Modelling only successful cases ignores those situations (or aspects) that are obviously contradictory, as pointed out by Caltagirone & Doutt

(1989). These awkward observations stand firmly in the way of generalisation, to the extent that Caltagirone & Doutt (1989) even label the 'undisputed prime example of a successful biological control agent' as exceptional in several ways.

5   Simulation of successful cases ignores the fact that each biocontrol programme usually represents a series of specific problems, which need both identification and solution. This seems to be the only way in which to deal with any subject that provides such diverse challenges. Although lessons can invariably be derived from successes, the research behind the success may reveal more about the requirements for success than about any correlates with success. Any such correlates may well be much more general than the demographic correlates investigated thus far, a point pursued further in Chapter 9.

The points outlined above suggest that biocontrol might have to be viewed differently from the focus offered by demographic ecology, an issue pursued in Chapter 9. Why projects have failed in the past also needs consideration, to address the issues of how to fulfil the minimum requirements for a successful outcome, and whether we should do more than concern ourselves with the minimum requirements in our efforts to ensure successful control. Finally, it may be necessary to consider two dynamics in biocontrol, namely the initial stage when the pest organism is abundant and later stages when the density has been lowered.

Adaptationist approach ignores primary processes

Biologists frequently treat the concept of adaptation in a rather trivial way. Adaptation is inferred without justification, evidence, or recourse to principle. This attitude or approach, labelled the 'adaptationist program', has been criticised (e.g. Gould & Lewontin, 1979; Lewontin, 1978; Williams, 1966), but the practice remains. In population ecology, adaptation is generally treated as an ongoing process of optimisation, because individuals are considered inevitably to compete with one another to the extent that they have to allocate their foraging time optimally to patches of hosts of different quality (e.g. Godfray, 1994; Hassell, 1986).

This approach to adaptation is based on the inappropriate dissociation of ultimate (evolutionary) factors from proximate (behavioural) ones (see Chapter 5). Optimisation is seen to drive, by natural selection, change to the host-searching mechanism. How this is achieved remains unspecified, because ultimate explanations deal with the evolution of searching efficiency through its influence on decisions by the natural enemies about which patches of hosts to visit and how long to spend in each (e.g. Waage & Hassell, 1982). The proximate behavioural aspects themselves are seen simply as descriptive, and not really useful to theory in population ecology and biocontrol.

The failure of natural enemy behaviour to conform to expectations generated by optimisation theory is seldom accepted as a refutation of the optimising principles, despite the questionable validity of those principles. Rather, more subtlety in the means of optimising is expected, and sought. The real ecological consequences of behaviour are thus readily missed. To illustrate: females of the egg parasitoid *Anagrus delicatus* Dozier usually leave a patch after parasitising less than 8% of available hosts (Cronin & Strong, 1993). Bouskila *et al.* (1995) downplayed the 'spreading of risk' explanation that was offered originally, and advocated thinking of the wasps as being egg limited; they can detect variation in host quality and to optimise their reproductive output oviposit only in selected high quality hosts. Several points caution against invoking such abilities on the part of the wasps: (i) The wasps cannot discern whether a host is already parasitised (Cronin & Strong, 1993), and superparasitism is not uncommon. If optimisation was so important, would it not be expected that mechanisms to detect earlier parasitism would have evolved before any optimising abilities? (ii) Greater numbers of female wasps on a patch of hosts result in higher parasitism rates, so host eggs rejected earlier must be 'judged' suitable by other individual parasitoids. They are not taking these hosts as 'second best', because the wasps are unaware of previous parasitism. (iii) For parasitoids to adjust their behaviour according to whether they will ultimately be egg limited or time limited implies they have perfect knowledge of future host availability, which is impossible (Rapport, 1991; Walter & Donaldson, 1994). Not surprisingly, populations of wasps are likely to contain both egg-limited and time-limited individuals (Heimpel & Rosenheim, 1998), so generalisations around this proposed determinant of behaviour are unlikely to be robust (Walter & Donaldson, 1994).

The acceptance of optimisation principles bolsters the view that an 'evolutionary arms race' persists between host and parasitoid (e.g. Lively, 1993; Myers *et al.*, 1989). Although such views are questionable (see Chapter 7), they are used to infer which host associations are likely to be the most successful in biocontrol. For example, 'natural enemies that are relatively rare in the native environment may be the best biological control agents because they will readily attack the host . . . but the hosts will not have evolved resistance or tolerance to them', and '[t]he species that respond to artificially dense populations of hosts should be good biological control agents' (both from Myers *et al.*, 1989).

If host-searching behaviour is a complex mechanism that has as its end (or ultimate goal) the location and parasitisation of hosts with which a particular suite of chemical and physical characteristics can be associated (see Chapters 5 and 9), then a 'rare' natural enemy may be only apparently rare because it is a 'sporadic attacker' with a vast number of additional representatives that are simultaneously attacking another species. If so, increasing host density may simply increase the chance of incidental parasitism, through accidental encounters or because of a coincidence of unusual circumstances. The probability of establishment of such sporadic attackers may be low compared with a species that is specifically adapted to attack the target species and does so habitually rather than sporadically (see Chapter 7).

Adaptation within local populations of species is expected under the demographic paradigm (see Chapters 5 and 6). This has had two practical consequences in biocontrol, neither of which has proved beneficial in practice. First, released natural enemies are expected to adapt to the local area of release, at least to some extent (e.g. Caltagirone & Doutt, 1989; Roderick, 1992). If the released sample is representative of the genetic variability in the species as a whole, it may adapt to some extent through changes in frequencies of polymorphic alleles, but that represents only limited adaptive potential (see Clarke & Walter, 1995; Marohasy, 1996). Second, the importation of additional local 'strains' or 'biotypes' is unlikely to improve biocontrol prospects. No unequivocal example of success by this means has been published; the only successes have been a consequence of the fortuitous inclusion of a cryptic species that controlled the pest (Clarke & Walter, 1995; see also Chapter 6). Fortuitous successes are acceptable and welcome, but if the biocontrol success rate is to be improved, reliable principle is more likely to help than is serendipity.

In summary, too strong a belief in the powers of local adaptation may discourage sound release practice, especially as regards ensuring that the organisms for release will (i) attack the target habitually under field conditions in the target area, (ii) persist and reproduce under the new suite of climatic conditions, and (iii) become established through choice of appropriate release sites and times. Coincidental matching of these three sets of criteria is sufficient to explain how limited introductions of certain species not only establish, but also achieve successful control (see Chapter 9).

Discussion

If the demographic approach to developing ecological generalisations for biocontrol worked effectively, ecology would gain considerable kudos. Besides the difficulties outlined above, three additional lines of reasoning warn why this popular approach to generalising about biocontrol needs reorientation, and indicate that our expectations of ecological theory need to be tempered (these expectations are reassessed in Chapter 9). First, the demographic principles from which the approach has been developed (and which provide the only justification for the approach) do not relate directly to individual organisms (see Chapter 5). Second, the evidence available from ecological studies in general is not encouraging. For instance, the role postulated for density dependence is highly questionable, as indicated by the retreat of protagonists to the point where they have even claimed it is important but not readily measurable as it may be swamped by random environmental variation (Hassell, 1985, 1986). Also, despite intense effort over almost a century, no robust patterns have been detected in ecological 'communities' (e.g. Greig-Smith, 1986; see also Chapter 5). Third, even the advocates of a greater role for ecological theory in biological control admit that theory has not improved practice, although they are optimistic that the demographic approach will yield the desired improvements (e.g. Murdoch & Briggs, 1996).

Demographic ecology (population and community theory) has the potential to mislead biological control, for it is based on premises that inappropriately direct understanding away from organisms and their species-specific adaptations, generate unreasonable expectations and maintain the focus on the same general approaches to ecological research. Ecological theory must be adapted to accommodate individual organisms in their natural setting as well as the idiosyncrasy of species. It must desist from imposing the expectations and desires of humans on natural systems. The determinism and idealism of demographic ecology impose on natural systems too much that is clearly not there. Demographic ecology is consequently far too optimistic about the prospects for predicting what will happen in circumstances that are all too obviously influenced by the vagaries of unpredictably changing climatic and other factors.

Persistence with demographic principles is, in itself, not a problem, but perhaps more progress could be made in applied ecology if that approach were not portrayed as the only 'legitimate' one to the general problem. This is a consequence of the hegemony of demographic ecology (Chapter 5). Finally, the autecological approach to ecology, which provides an alternative to the demographic one (Hengeveld & Walter, 1999), suggests that the idealism inherent in demographic ecology cannot support the desired practical benefits.

Chapter 5 contrasted demographic principles with autecological ones. If autecology is more realistic than demographic ecology, it should provide practical guidance for biocontrol. This would be a significant advance, as the 'pragmatic school' of biocontrollers lacks theoretical direction from realistic principles. Chapter 9 shows that autecology does provide the required guidelines, but also warns that ecologists' expectations of theory are overextended. We need to temper our expected abilities in applied ecology, but we can enhance our performance considerably by noting the principles of autecology, applying them with care, and by not allowing our expectations to exceed the limitations of our current abilities in ecology.

9

Autecological research on pests and natural enemies

[T]here should be applicable ecological theory to help choose natural
enemies, plan their release, and perhaps alter the agroecosystem.

W. W. MURDOCH & C. J. BRIGGS (1996, p. 2001)

Introduction

The point made by Murdoch and Briggs in the text above is of central
importance to improving pest management practice in general and biocontrol in particular. The difficulty lies in deciding upon the appropriate
direction to take in seeking that theory (Chapters 4–6). Demographic theory in ecology currently holds a monopoly in attempts to understand the
ecology of pests and natural enemies (Chapters 5–7) and to apply theory to
biocontrol practice (Chapter 8). Autecology has seldom been mentioned
as an alternative source of theory and insight. A problem for autecology
is its frequent misrepresentation as simple natural history observation.
Even when natural history observation has been mentioned as useful, it
has been portrayed as intuitive observation whose utility is to fill in gaps
of fact.

This chapter therefore examines how a changed perception of autecology can provide a scientifically sound and helpful theoretical basis for
enhancing biocontrol theory and practice. Autecology is based on the integration of physiological, behavioural and ecological principles into an
evolutionary framework that is consistent with such principles and which
has its basis in the recognition concept of species (see Chapter 6). Natural
selection provides a common basis for these principles, but is not seen as
an agent of optimisation. Natural selection acts directly on the mechanisms, not their postulated efficiency within a competitively driven world.

Autecology is thus built explicitly on the notion that ecology represents the expression of behavioural and physiological processes in nature, in similar spirit to Kennedy's (1967) view that behaviour is the expression of physiological process. This approach cannot promise to fulfil the type of predictive role currently envisaged for ecological theory by demographic theoreticians, a point elaborated further in the body of the chapter. Suffice to say the more realistic approach provided by autecology deals explicitly with individual organisms in nature and thereby defines different limits to prediction in ecological systems. Although demographic ecology promises greater predictive ability than does autecology (Hengeveld & Walter, 1999; Walter & Hengeveld, 2000), it has not delivered in this regard. It is the subject matter that specifies the limits to predictive ability within a discipline and these limits exist independently of hope or desirability.

The issue of what makes a successful biocontrol agent has yet to be tackled in any detail from the autecological perspective, except in the most general terms. Aspects of autecology are evident in the pragmatic approach to biocontrol as it is currently practised, because species are dealt with on a case-by-case basis. In other words, species are accepted as being most realistically treated individualistically (see Chapters 5 and 6). However, 'pragmatic biocontrol' has not been formally linked with the principles of autecology, presumably because autecological theory has not been well developed and, until recently, had no evolutionary underpinning (Chapters 5 and 6). The rest of this chapter introduces a general approach to investigating natural enemies from the autecological perspective.

Because the autecology of parasitoids and predatory insects is not well known, the development and justification of general ideas in autecology cannot be achieved with reference to biocontrol agents alone. I have therefore drawn examples from other organisms, mainly pest species, to illustrate the potential that autecology holds for pest management in general and biocontrol in particular. This chapter thus also links back to Chapter 7, in which polyphagy was discussed in some detail.

The chapter is divided into two principal sections. The first introduces the autecological principles relevant to understanding the ecology of parasitic and predatory (together comprising 'entomophagous') insects. The emphasis here is on ecological relationships deemed relevant through their definition in terms of (i) the individual and the species, (ii) organisms and their environment, (iii) behaviour and the environment, and (iv) the movement of individuals related to ecological scale. The second section introduces a practical approach to the pre-release evaluation of biocontrol agents, by going through a three-step procedure that is designed to ensure the natural enemy released has the appropriate characteristics to attack the pest species habitually in the field. The discussion section draws together the points made and considers the nature of ecological generalisations and what is likely to be predictable in biological control.

Entomophagous insects and autecological principles

Behaviour, physiology, environment and successful projects

The significance of behaviour, physiology and the environment to developing general ecological principles has been downplayed. They are widely considered to be somewhat peripheral to the development of demographic generalisations (Chapter 5). More specifically, behavioural and physiological mechanisms are seen to be proximate factors rather than ultimate or evolutionary ones (Chapter 5). However, recent developments in evolutionary theory provide a basis for redressing this misconception (Chapter 6). Although demographic ecologists do recognise that behaviour, physiology and the environment are relevant at the pre-release evaluation stage, these aspects are seen as 'practical criteria' (e.g. Godfray & Waage, 1991, p. 434) and are thus portrayed as aspects that are straightforward to understand, by simply gathering the appropriate facts.

Admittedly, behavioural and environmental 'matching' has been readily achieved in successful biocontrol programmes, but that may be encouraging the illusion that such aspects of an organism's ecology are readily understood. The understanding of just how organisms match their ecological circumstances presents special problems. As yet only rudimentary theory is available (Hengeveld & Walter, 1999; Walter & Hengeveld, 2000) to deal with this process, although it underlies the idiosyncratic species distributions, local ecologies and responses to climatic influences that represent the most consistent patterns detectable in ecology (Hengeveld, 1990; Walter & Paterson, 1994).

Failure to develop autecology may go part way to explaining the high proportion of unsuccessful biocontrol programmes, because demographic theory removes the focus from the organism and organism–environment interaction and directs it at hypothetical equilibria (Chapter 5). The minimum requirements of a successful biocontrol programme have not been met in so many instances because the reality in understanding autecology is different from its portrayal as a simple fact-gathering exercise in natural history observation. This implies that biocontrol failure has not necessarily resulted from a failure to identify the demographically defined characteristics that best correlate with measures of biocontrol success. Failure to consider autecological principles provides a more likely explanation (although not all failures will be explicable in this way).

Emphasis on individual organisms

Autecological theory suggests that if natural enemies are abundant enough in the environment and have a mechanism that enables them to locate individuals of the target species, even when it is at low densities, they are likely to suppress the pest population. The phrase 'abundant enough' draws attention to such biological features of the natural enemy species as survival of all life stages within the area of interest, as well as reproductive success and movement into or away from the area. The continuous movement of the natural enemies, which is considered in a subsequent subsection, is fundamental to understanding their presence and abundance as well as their impact on the pest species within the area of interest. This emphasis on individuals deflects attention from such abstractions as population, equilibrium, stability, and so on.

The autecological perspective advocates concentrating on the fundamental point that the natural enemy must match the environment into which it is to be introduced. Biotic factors comprise aspects of that environment and include such features as the host or prey species, the host plants on which they are found, vegetation structure, and so on. The difficulty is the identification of the primary components of the match between organism and environment, for they are diverse, frequently subtle and sometimes inconvenient to investigate. For example, the mechanisms that require study may be behaviours that involve distance attraction of natural enemies to their host or prey, or to the host plant upon which the target organism feeds. In addition, they may be subtle impositions of the physical environment on individuals of a particular life stage. The ultimate test is the natural enemy in the field, a situation that is impossible to emulate in the laboratory or to predict with general models, for parameter estimation is technically close to impossible for conditions still alien to a species (Hengeveld, 1999). Empirical investigation of environmental matching is considered in the following major section.

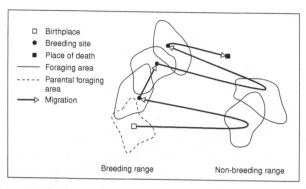

Figure 9.1. Lifetime track of an individual organism (Drake *et al.*, 1995). Reproduced with permission.

Individuals, species and ecological scale

Chapter 6 details how physiological and behavioural mechanisms of individual organisms, within their usual environmental context, dictate the limits of species. A duality thus exists between the conceptualisation of the species and the individuals that make up each species. The most important practical consequence of this duality is for our understanding of the appropriate scales of observation and interpretation in ecology. The geographical distribution of the species dictates one scale of ecological resolution, which is the geographical extent to which the activities of the individuals are generally confined. This scale reveals the extremes with which the individuals can cope and the range of environmental settings that are inhabitable. The species-specific adaptations, as manifest in the behavioural and physiological properties of individuals, dictate a second scale of observation, the local region within which the individuals of interest to the ecologist will carry out their life and reproductive processes. An important part of the lifeline of individuals (Rose, 1997) is their lifetime track (Drake *et al.*, 1995; Howard, 1960) (Fig. 9.1). The lifetime track will reflect species-specific traits. Interpretation of local ecology thus entails an understanding of the reproductive limits of species, the geographical scope of the activities of the individuals that comprise the species, their adaptations as they relate to host organisms and other features of the environment, and their lifetime track.

The individual-based view of ecology just outlined is underpinned by the recognition concept of species, and justifies the emphasis on species-specificity in ecology (Chapter 6), as opposed to the suppression of species' unique features in the search for general patterns among emergent phenomena such as postulated population equilibria, population stabilising mechanisms and guild structure. Emphasis on individuals and the lifetime track they follow as they move within the environmental setting introduces complexities around which ecological generalisations have yet to be developed. Movement has generally been treated in its own right, as migration, and demographic ecology has incorporated movement mainly with respect to such aspects as life history strategies (e.g. Southwood, 1988), patch quality and optimisation (Hassell & Southwood, 1978), and population stability (Jones et al., 1996). Demographic ecology sees a disjunction between the behavioural processes that relate to movement, on the one hand, and the ecological processes that are considered important, on the other (e.g. Jones et al., 1996, p. 318). Autecology works in the opposite direction, from behavioural processes to ecological consequences, after the style of Lima & Zollner (1996) for instance, but without being built upon ideas relating to decision-making by organisms, maximisation of fitness, and metapopulation dynamics (see Chapters 5 and 6).

Movement of individuals

Organisms that undertake, in large numbers, more than just local movements are treated as migrants per se. Migration is generally associated with particular physiological conditions of the participating individuals and with specific behaviours, sometimes in relation to currents of wind, as occurs with the invasion of brown planthoppers into Far Eastern rice paddies (Fig. 9.2) and bollworms into Australian cotton (Fig. 9.3). Such movements are 'undistracted' in that the organisms do not respond to their usual resources or home ranges (Dingle, 1996). Other movements, in contrast, are local and appetitive (Kennedy, 1992), and include foraging and territorial behaviours. Because appetitive movements tend to be governed by resources and home range, they generally do not result in significant displacement of individuals, and have been classified as 'station keeping' (Dingle, 1996).

Between these two extremes, another type of movement has been characterised. This has usually been thought of as 'dispersal', however, strictly dispersal is a population consequence, so Dingle (1996) prefers the term 'ranging', to indicate exploratory movement by individuals. Ranging differs from migration in that it ends, by definition, when suitable habitat or territory is located. In terms of ecological consequences, the most significant aspect of this behaviour is that the movement is undirected, but away from the habitat of origin (den Boer, 1990b). The general

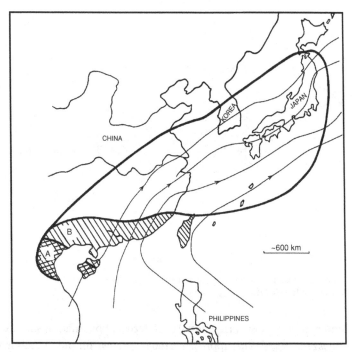

Figure 9.2. Wind-assisted summer migration pathways of the brown planthopper (*Nilaparvata lugens*), a pest of rice, from China to Korea and Japan. 'A' is a suspected source area, 'B' is an area of sporadic overwintering, and the solid line demarcates the main rice-growing area of temperate east Asia, into which migration takes place regularly. The winds represent those recorded during a mass migration episode. After Kisimoto & Sogawa (1995) and Kisimoto & Rosenberg (1994). Reproduced with permission from Drake *et al.* (1995).

view is that individuals undertake such movement when conditions become unfavourable. That may be true of migration per se, but despite the safe assumption that 'all animals will try to flee from sites where conditions are becoming adverse', the ranging of terrestrial arthropods cannot be considered merely an 'escape from adverse conditions' (den Boer, 1990b). Many organisms respond positively to environmental conditions that favour their movement, despite their immediate surroundings being favourable for their survival and reproduction. Thus, 'aerial dispersal occurs each year and everywhere' (den Boer, 1990b).

How do the movements of natural enemies match these categorisations? The study of parasitoid or predator movement usually concentrates on foraging among localised groups of hosts within a small area, and is frequently interpreted functionally in terms of optimisation (e.g. van

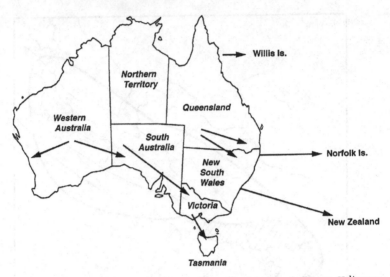

Figure 9.3. Recorded migration pathways of the native Australian bollworm, *Helicoverpa punctigera*. Modified from Gregg (1994).

Alphen & Vet, 1986; Waage, 1979) or in terms of population stabilisation (Jones *et al.*, 1996). This approach is too restrictive, for individuals may be displaced considerable distances, as in the mymarid egg parasitoids that seasonally re-invade islets up to a kilometre from source sites (Antolin & Strong, 1987; Strong, 1988) (Fig. 9.4). Consequently, the distinction between foraging and ranging is not an easy one to make. Furthermore, egg-laden female wasps may move away from an area containing suitable hosts (Cronin & Strong, 1993; Fernando, 1993), which is a behaviour more closely aligned with migration, so again distinctions are blurred. Many parasitoids have been found to move over distances more usually associated with migration. For instance, a parasitoid of locust eggs, *Scelio fulgidus* Crawford, moves with wind systems in the upper air (100–300 m) (Farrow, 1981). This tendency seems not to be solely related to the distribution and nomadic life style of the host, for many other microhymenopteran parasitoids were recorded in the aerial samples taken by Farrow and the other authors he cited. That individuals ignore suitable resources (or habitat) in this way is not restricted to parasitoids or insects, but is evident also in vertebrates (Howard, 1960).

Rather than dwelling on general categorisations of movement, let us bring into focus again the behaviour of individuals and examine it

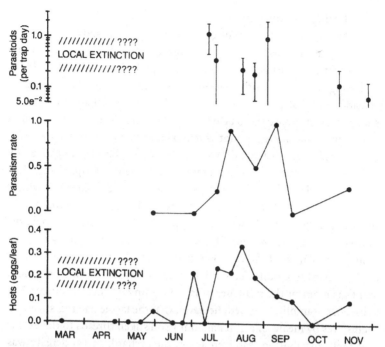

Figure 9.4. Evidence of the migratory behaviour of the mymarid egg parasitoid *Anagrus delicatus*, which seasonally recolonises islets off the Florida coast after winter extinction of its planthopper host *Prokelisia marginata*. After Strong (1988). Copyright 1988, reprinted with permission from Elsevier Science.

for its ecological consequences. Information about movement of natural enemies is still scant, so other types of insects will also be considered. The contention developed is that in insect natural enemies, as well as pests, there is an ongoing movement of individuals that results in their being continuously redistributed within the area of ecological interest. Furthermore, ongoing movement into the area of interest, as well as out of it, also adds to this redistribution. The extent of such redistribution (or 'shuffling'; Howard, 1960) has the consequence that levels of parasitism or predation, and their spatial distributions, will be best understood from an integration of information on (i) the movements of individual natural enemies, which is a species-specific aspect of their behaviour (e.g. den Boer, 1979; Jones *et al.*, 1996; Meijer, 1974; van Huizen, 1979), (ii) reproduction and survival rates of the natural enemies under the conditions in the locality of interest, and (iii) their host or prey relationships.

Ecological consequences

The ongoing redistribution of organisms, described above, takes place within generations as well as over a geographical scale considerably more extensive than is usually included in interpretations of 'local ecology', including studies of optimisation relative to patches of resources or considerations of population stability. The ecological implications of such movements have yet to be worked into general theory. They occur at a scale and intensity that is not as readily detectable as mass movements over considerable distances (Davis, 1984). Also, they are not as easily followed as are the small-scale foraging movements of organisms. Nevertheless, their ecological impact is fundamental. Consider the winter moth, for example. Female winter moths are flightless and have to walk up tree trunks to reach oviposition sites in the vicinity of leaf buds. Moths can easily be prevented from ascending trees, by placing barriers around the trunk. Holliday (1977) banded a subset of trees within an apple orchard to modify the spatial pattern of winter moth oviposition, and thus monitor the influence of larval feeding on their resources. His subsequent sampling showed, however, that the larvae had redistributed themselves by means of aerial ballooning on silken threads to the extent that their distribution was even across the orchard. Larval density was thus almost completely dependent on larval dispersal.

Such within-generation redistributions take place in other species that had been considered largely sedentary until investigated in an appropriate way, including codling moth (Schumacher et al., 1997), tent caterpillars (Wellington, 1977), chrysomelid forestry pests (Clarke et al., 1997), newly emerged Trichogramma evanescens wasps (Smits, 1982, cited by Smith, 1996) and drosophilid flies (Kimura et al., 1978). Although individuals of some species do leave their natal site only at relatively low frequencies (e.g. some cerambycid beetles (Davis, 1981, 1984, 1986) and tussock moth larvae (Harrison, 1997)), the hypothetical species whose movement can be ignored in efforts to understand its ecology has been derided as that 'ill-conceived Autochthone, the mythical animal that never migrates' (Taylor, 1986). Even in carabids, a large proportion of the population (up to 60%) walks away from suitable habitat, despite flight being the major process in post-diapause redistributional movements (den Boer, 1970).

The movements considered above are unexpected in relation to current ecological theory, because individuals leave resources that are ostensibly suitable, as judged by other individuals staying there (e.g. Davis, 1981, 1984, 1986; Vorley & Wratten, 1987). The physiological 'trigger' for individual insects leaving such situations is not known, but hormonal influences have been implicated in vertebrates (Howard, 1960). We should not interpret this behaviour, or its ecological consequences, in terms of the maximisation of fitness, for mistakes are frequent. Significant numbers of carabids land in habitats unsuitable for reproduction, but cannot leave because of flight muscle autolysis (van Huizen, 1977). Parasitoids oviposit into hosts from which successful emergence is precluded (Carroll & Hoyt, 1986), and herbivorous insects oviposit on to unsuitable host plants (Thompson, 1988). Further, population density, and thus density-dependent influences, are not likely to be significant, for 'the number [of carabids] taking part in dispersal by flight are similar for both sparse and numerous species' (den Boer, 1990b). The same would be true of many other species (e.g. Davis, 1981, 1984, 1986).

Individuals, movement and ecological models

Models of population change, in time and across space, that are developed from a basis in individual behaviour and physiology will look different from those developed around the feedback influences of density, but have been attempted only for invading species (Lensink, 1997; Van den Bosch et al., 1992). Invaders are more readily studied from this perspective, because the consequences of their 'redistributional' movements are fairly easily measured while the population is expanding into new areas. The post-release movement of natural enemies has been measured directly in several cases, as well as indirectly (e.g. Follett & Roderick, 1996).

Rates of movement are species specific and take place regardless of suitable hosts (or habitat) being abundant (or available) in the area which the individuals had left (den Boer, 1990b; Howard, 1960). Rates of invasion of such species are indicative of the scale over which redistribution is likely to occur in non-expanding populations of the species. This point needs to be checked for insects generally, and natural enemies specifically, but it holds for at least some vertebrates (Van den Bosch et al., 1992). Lensink (1997) did record locality-specific differences in invasion rates, but whether they reflect differences in scale of movement or differences in reproductive output is not yet clear. The distribution of some natural enemy species expands relatively slowly, at least in terms of the aims of biocontrol deadlines. An example is the coccinellid Curinus coeruleus Mulsant on Indonesian (Soehardjan, 1989) and Hawaiian (Follett & Roderick, 1996) islands. Data from Indonesia have apparently not been published, and the Hawaiian measurements were taken after the species had already distributed itself widely across the archipelago. Since rates of distribution change were estimated indirectly by means of genetic analysis, alternative explanations of the population genetics structure cannot yet be ruled out entirely (Follett & Roderick, 1996).

Note also that individuals 'must contend with a shifting set of transitory habitats which can suddenly change from suitable to unsuitable between consecutive generations' (Wellington, 1977), and that these conditions are species specific and may change even within a generation. We therefore need to consider the ecology of all organisms from the viewpoint of the geographical setting of suitable environments for the species in question.

For long-distance migrants the geographical setting may well be the continental species distribution. In any case, the ecology of a diversity of species, from mammals and birds (e.g. Lensink, 1997; Van den Bosch et al., 1992) to insects of all sizes (e.g. Kisimota & Sogawa, 1995; Parmesan et al., 1999; Rainey, 1989), is known to be understandable only if this geographical perspective is added to the suite of properties considered to be fundamental to ecological interpretation. Furthermore, the geographical distribution of a species will inevitably shift with time as a result of the colonisation of new areas and the local extinction of other populations (e.g. Hengeveld, 1990).

Adaptation and ecological interpretation

The outline of ecological scale, above, suggests that an appropriate understanding of the ecological consequences we term presence and abundance demands the integration of an area-wide consideration with the perspective from a finer focus that considers the ecological consequences of the adaptations of organisms in particular localities. Here we find differential abundance in relation to spatially and temporally changing environmental conditions (including biological influences, as detailed in Chapter 5). A common oversight in ecology is to treat the environment of a locality as providing equivalent physical conditions for all the species that live in that area, as is done for instance in interpretations of 'community structure' (Chapter 5), and in biocontrol models that seek demographic correlates of success (Chapter 8). That this is inappropriate is readily demonstrated (Fig. 9.5; see also Chapters 5 and 8).

To deal with the problem of species carrying adaptations that are idiosyncratic and individualistically adapted to a particular suite of

WITHIN A SINGLE LOCALITY

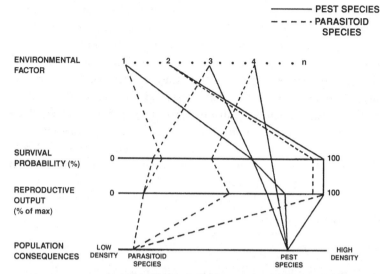

Figure 9.5. Various environmental factors impinge on organisms. Several such factors are represented here, as they impinge on a pest species of interest as well as a natural enemy. Should factors 1–4 reach values that decrease the survival and reproduction of the prey, but simultaneously increase the survival and reproduction of the natural enemy, then relatively higher rates of parasitism or predation can be expected. One would therefore expect rates of natural enemy-induced loss to change with time and across space. This view is acknowledged by practitioners through their development of augmentative release programmes to overcome the negative effects on natural enemies of environmental influences.

environmental conditions (which includes also biological components; Chapter 5), two developments in ecology have been necessary. First came the logical reorientation in the conceptually intricate area of species and individuals, as detailed above. The second development necessitated a move away from the emphasis on resource use as a point for developing generalisations. Rather, the specific adaptations shared by individuals of the same species dictate the relationship between organisms and environment, which is independent of those other species postulated to 'share' a common pool of resources and thus 'community membership' (Chapter 5).

With theory shifting focus from resources as a common currency for which species compete, the abstraction of the environment into an abiotic sector and a biotic one is not useful, particularly when the biotic is almost always equated with demographic influences related to resource use (Walter & Hengeveld, 2000). The important aspect is to understand the species-specific relationships between individuals and the suite of physico-chemical and biotic aspects of the environment with which their survival and reproductive mechanisms interact. Frequently, physical factors relate to physiological tolerance limits and biological ones to behavioural responses to environmental sense data (e.g. Bernays & Wcislo, 1994; Dusenbery, 1992), and that dictates which parts of the environment those individuals will inhabit. Harrison (1997) noticed that such influences of adaptation on population levels are neither 'top-down' nor 'bottom-up', which are the avenues usually pursued by ecologists. Again, this view of abundance being primarily a consequence of the interaction between the adaptations carried by organisms and their local environment is consistent with the principles of the recognition concept of species (Walter & Zalucki, 1999).

Pre-release evaluation of biocontrol agents

The release of biocontrol agents into a new environment and their successful colonisation are events crucial to success in biocontrol. The initial small size and limited distribution of the newly liberated population place it at risk from various influences that would not affect a large population in the same way. For example, environmental variability or even demographic stochasticity may cause extinction (Kareiva, 1990). To increase chances of success at this point, we need to characterise the environmental relationship of the organisms. This will often be subtle, especially in environments that are not strongly seasonal or with organisms that survive adverse conditions in ways that are not obvious.

Autecology provides at least the beginnings of theory for biological control to accommodate these problems, as expanded below. In brief, it suggests a three-step research protocol for the pre-release investigation of natural enemies. The research is not aimed at selecting the species that shows most potential, the goal at which demographic ecology aims. Autecology intends to make sure that only natural enemies that are functionally equipped for the target situation are considered. Other species are unlikely to perform adequately. The three steps are (i) the establishment of species status, (ii) confirmation that the host relationship with the target pest is functional, and (iii) definition of the environmental relationships of the natural enemy species. These are dealt with in turn.

Behavioural systems that delimit species gene pools

The importance of defining the limits of species unambiguously before release has already been emphasised (Chapter 6). Such procedures are also important for providing a clear record of what was done in the programme and allow accurate interpretation of any problems that may develop subsequently. For example, if successful biocontrol suddenly fails or if levels of success have been incorrectly portrayed, meaningful corrective measures can be taken only on the basis of the appropriate scientific understanding (Clarke, 1990; Sabrosky, 1955). In the case of multiple introductions, one should be confident of whether the same species or different species are being introduced when 'biotypes' or 'strains' are considered. Such situations, if not resolved at the outset, are likely to confound interpretation of the outcome, and will also leave unnecessary complications for any biocontrol efforts that might be needed subsequently (Clarke & Walter, 1995).

Behaviour of individuals: habitat and host range

A minimum requirement for successful biocontrol is a natural enemy species that will, upon release into the environment, habitually attack individuals of the target species. This requirement is frequently acknowledged, but only infrequently investigated in an appropriate way prior to release. An additional complication lies in there being alternative models of the behavioural basis that underlies patterns of host relationships. The different practical consequences of these alternatives are not widely discussed, which impedes the development of appropriate theory.

Natural enemies will habitually attack individuals of the target species only if the multiple steps of their host-searching behaviour correspond with the cues that are associated with the 'target' organism and its surroundings. Correspondence will occur if the natural enemy is primarily adapted to encounter individuals of the 'target species', or is preadapted to attack the pest species through its adaptation to another species, perhaps a close relative. Successful control with organisms that attack the pest incidentally will inevitably be far less likely, even if (i) the pest does represent a readily available unused resource, and (ii) the natural enemy occasionally does attack the pest in large numbers.

The ecological correspondence between natural enemy and pest is not simply a question of natural enemy behaviour. Their coincidence in the field may be influenced by additional factors, such as the synchrony between the natural enemy and the target. That could derive through diapause (Quicke, 1997), although diapause may in some cases have to be broken to maintain synchrony (e.g. Gilkeson & Hill, 1986; Hoy, 1990). An alternative host may be necessary to bridge periods when the target host is not available (e.g. Corbett & Rosenheim, 1996; Doutt & Nakata, 1973; Vorley & Wratten, 1987), but here timing is essential and the alternative hosts must be primary ones or ones to which the natural enemy species is preadapted. Incidental hosts may not serve as well, but this has not been investigated. 'Timing' may also be achieved by natural enemies immigrating from areas or situations well outside of the biocontrol system (Corbett & Rosenheim, 1996). Such migration introduces a perspective on persistence that differs from that usually espoused in biocontrol. Instead of hinging interpretation around the persistence or coexistence of natural enemy and pest populations, emphasis is switched to the persistence and reproduction of individuals within the context of the ecological scale dictated by the species-specific properties of the individuals concerned, as well as the 'provision' of resources within that context.

The scale of operation of individuals of the natural enemy species of interest may well be different from that of the target species. In an open system this may result in problems when the natural enemies are introduced, for the number of introduction points and their distribution would affect the time needed for complete spatial coverage to be achieved (e.g. Follett & Roderick, 1996). The scale of operation of individuals may be locality specific (e.g. Lensink, 1997), so localities that differ in climate or vegetation, for instance, may warrant different release strategies (or landscape management; Corbett & Rosenheim, 1996). Local extinctions and the need for re-invasion may cause problems subsequently. Release strategies in closed systems (or systems that are effectively closed to the natural enemies being released, through time constraints for example) will need an appropriate spatial plan for success. Practitioners may also have to consider the scale dictated by the geographical distribution of the species concerned, when the distribution of the introduced natural enemy does not coincide with the pest's distribution, for example. Additional natural enemy species that are 'climatically complementary' (e.g. DeBach, 1974, pp. 70, 171; Huffaker & Kennett, 1966; Rochat & Gutierrez, 2001) would be needed to achieve complete coverage of the pest's distribution. Climatic conditions in the areas not covered by the introduced natural enemy will indicate the climatic regions that may repay additional exploration. Climatic matching is considered in the following subsection.

A good start has been made in integrating biocontrol aims with an understanding of the behavioural and physiological aspects of natural

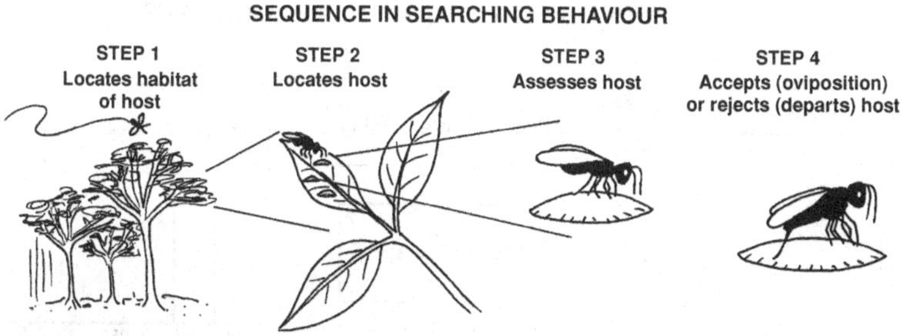

SEQUENCE IN SEARCHING BEHAVIOUR

STEP 1	STEP 2	STEP 3	STEP 4
Locates habitat of host	Locates host	Assesses host	Accepts (oviposition) or rejects (departs) host

Figure 9.6. Diagrammatic representation of Salt's (1935) hierarchical model of host-searching behaviour by parasitoids.

enemy biology that relate to their host use and habitat requirements. These efforts (e.g. Lewis *et al.*, 1990; Vinson, 1998) are essentially autecological in that they work from the level of the individual organism (and its properties) towards the consequences for population size. Approaches that embrace optimality in explaining host relationships do not fall into this category, for they ignore physiological and behavioural mechanisms, and their interpretation is reliant on a competitive view of natural selection that is ongoing and directional (Chapters 5 and 6).

Two general models for host-searching behaviour have been proposed.

1   The original model was pioneered by Salt (1935) and was essentially hierarchical in that each step focused the activities of the searching wasp within the environment (Fig. 9.6), such that successive steps eliminated non-host species (Flanders, 1953). Subsequent addition of the behavioural components, and the environmental sense data that influenced them, made the model more realistic when it was formalised as the 'find and attack' cycle by Lewis *et al.* (1976) and Vinson (1977).

2   The 'reliability–detectability of information' model (Fig. 9.7) was developed specifically in relation to semiochemicals used by searching natural enemies (Tumlinson *et al.*, 1992; Vet & Dicke, 1992). Its acknowledged basic premise is that the great variability and dynamic nature of these cues requires a highly sophisticated and flexible response system. The generalisation is developed around the differential properties of the information used by the insects, as well as the ability of natural enemies to learn and the variability inherent in their behaviour (Lewis *et al.*, 1990; Vet *et al.*, 1990).

These two models have now been united, by Vinson (1991, 1998) into a single functional model (Fig. 9.8) along the lines of the concept of

INFOCHEMICAL USE BY NATURAL ENEMIES

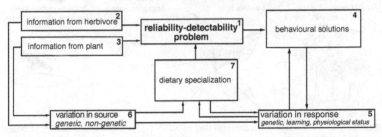

Figure 9.7. The reliability–detectability of semiochemicals model depicted by Vet & Dicke (1992). Reprinted with permission from the *Annual Review of Entomology* © 1992.

fertilisation mechanisms (Fig. 6.3, Chapter 6). A few deletions, modifications and additions to this scheme make a more accurate general model of what might be termed 'parasitisation mechanisms' and 'predation mechanisms' (Fig. 9.9). This proposed amalgamation is consistent with the more recent behavioural models developed by weed biocontrol practitioners for their specificity testing of herbivores intended for introduction against weeds (Marohasy, 1998; Wapshere, 1989). It pays attention to the complexity of the mechanism and its sequential nature, the diversity of modes of information that are used (Vinson, 1977), the environmental context of the information used by the natural enemy individuals, and their behavioural responses. This model thus extends the more simple 'long-distance/close range' dichotomy that tends to be used in discussions of semiochemicals and host finding by parasitoids.

The host-finding mechanism of parasitoids, and probably even 'generalist' predators (e.g. Obata, 1986), is complex in that it has many steps that follow a distinct sequence, starting with detection of a long-distance component (Lewis & Norlund, 1984), the localisation of the host, and its acceptance or rejection. Each of these steps is in turn complex, being comprised of subcomponents. The mechanism thus 'ensures' a particular end result. As such it is a catenary sequence in which each component fulfills a particular requirement that is triggered by reaction to cues from the environment and host (see Wapshere, 1989). Together these steps lead sequentially and ultimately to deposition of an egg into or on to a site with particular characteristics, and that would usually support development.

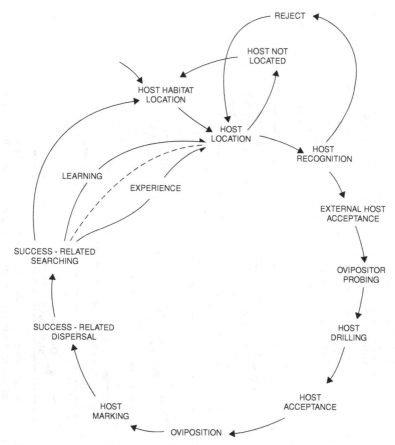

Figure 9.8. A composite model of the 'parasitisation mechanism' of parasitoids. This incorporates aspects of the models depicted in Figs. 9.6 and 9.7. Reprinted from Vinson (1998), copyright 1998, with permission from Elsevier Science.

The general reconstruction of Fig. 9.9 carries several consequences for understanding parasitism and predation of pest organisms in the field (Walter, 1993c; Walter & Donaldson, 1994). First, the system is unlikely to undergo change locally. Reputed cases of local adaptive change therefore need re-inspection, as does the perception of ready evolution of host-adapted 'strains'. Second, the parasitisation mechanism is likely to be species wide, as evolutionary change is expected only in small populations put under severe stress, such as when the usual hosts are absent (Mayr, 1963; Walter, 1993c). Third, the mechanism should be seen against

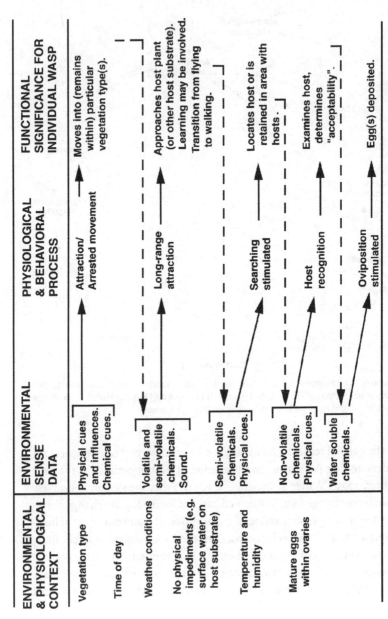

Figure 9.9. General model of the sequential physiological and behavioural processes that result in parasitism of the usual hosts of a parasitoid species. An act of parasitism involves the detection, localisation and recognition of an appropriate host within the usual environmental context of the organisms concerned. The model has been developed from schemes proposed by Vinson (1991, 1998).

the background of a dynamic and heterogeneous environment, within which the movement of individual organisms must be taken into account. Fourth, natural enemy individuals follow 'paths' or 'trajectories' within the environment, under the influence of components of the parasitisation (or predation) mechanism and the surrounding environment. 'Patches' of hosts (and 'patch quality') may not influence their behaviour as much as predicted by the premises of demographic ecology. Rather, the information available to the organisms as they move through the environment may influence rates of parasitism more than the traditional density-related processes are thought to. Fifth, the 'reliability vs. detectability' distinction is important in relation to location of hosts, but genetic change in response to reliability or detectability of signal should perhaps not be expected, as the signals change in relation to such dynamic aspects as host population density. Finally, learning is unlikely to take precedence over genetically fixed behaviours (Vinson, 1998), because variable conditions in nature are likely to limit the utility of learning as much as they are supposed to limit the utility of fixed responses.

In short, we are dealing with a functional interaction rather than an incidental interaction, like competition (Walter, 1993a), and it is expected to be under stabilising selection (for reasons see Chapter 6). The 'parasitisation mechanism' model therefore removes those influences that intrude from the demographic perspective, including the view that significant local adaptive change is ongoing and that hosts and parasitoids are engaged in 'evolutionary arms races' (Chapter 6). Concepts like optimality, often expressed as 'maximisation of reproductive success', and characterisations of behaviour derived from population models, such as searching efficiency and handling time, are therefore also of questionable utility.

The widely accepted notion that natural enemies can recognise the habitat of their host also warrants re-examination. Rather, the organisms recognise information stimuli that may derive from a diversity of sources and which may be physical or biological in origin. They respond accordingly and may thus move into the habitat of the host. The most frequently documented stimuli are plant-associated chemicals. Response to these takes the parasitoid closer to the vicinity of potential hosts, but that does not imply that the insects can recognise or respond to the 'habitat' of their host. The 'parasitisation mechanism' approach is more consistent with behavioural observation, and explains why hosts on some plants are attacked less frequently, or not at all (Lewis & Norlund, 1984), and why even unsuitable hosts are persistently attacked (Carroll & Hoyt, 1986), as seen also in herbivorous species (Thompson, 1988). That is, the species may be found in different biotopes across its distribution, but this does not imply local adaptation to the different biotopes (Walter & Hengeveld, 2000).

The 'parasitisation mechanism' concept provides leads for investigating and interpreting the observed host relationships and predation patterns of natural enemies. Although this is frequently portrayed as a mundane matter, it is where mistakes frequently intrude, especially in quests for general ecological patterns across species or for evidence supportive of general models (Chapter 8). This implies that the variability in host relationships that is frequently claimed for species in the literature needs scrutiny, especially since it has been accorded an important place in developing models of host-searching behaviour and the interpretation of predation. Several categories of observations on parasitoid–host relationships and predator–prey relationships are relevant in this context. Misinterpretation at this level impacts negatively on biocontrol actions, as expanded in the treatment of each such category in the points below.

1   Behavioural variation

Variation in behaviour and consequently in host relationships does occur at times to a measurable extent. This may be internally dictated, by physiological condition for example. Physiological stress may, in turn, be externally imposed, by environmental or laboratory conditions. Although the observed outcome does indicate a degree of behavioural variation, placing emphasis on that variation at the expense of the usual regularity of the parasitisation mechanism may mislead.

2   Incidental parasitism and predation

The results of field sampling that is too limited also lead to inappropriate extrapolations about the usual host or prey species of the natural enemy of interest. For example, if a few parasitoid individuals are reared from the 'pest' in its country of origin, the assumption is frequently made that parasitoids of that species have host-finding behaviour that will inevitably lead them to that host species. Such an assumption is not necessarily valid. Incidental parasitism and predation occurs in the field for various reasons, and may even happen at a relatively high level on occasion, for whatever reason. Mass release of such organisms is unlikely to lead to successful control; the released insects may seldom actually locate and destroy individuals of the target pest species, or they may not reproduce well on them, as may have happened with the leucaena psyllid predator described in the following point. And the nature of the host (or prey)-finding mechanism is such that the released organisms are unlikely to adapt gradually to the pest species.

3   Factitious laboratory hosts

Laboratory evidence is frequently exposed to unacceptable extrapolation, to the detriment of biocontrol practice. A natural enemy species may be assumed to have potential as a biological control agent after successful laboratory rearing on the pest. This seems to be common with egg parasitoids, notably *Trichogramma* species, and ectoparasitoids (Sayaboc, 1994), since the eggs and larvae of these organisms do not actually face the immune response of the host. The immune system is not well developed in the egg stage (Salt, 1968), and ectoparasitoids feed from the outside of their host. Consequently, both types of parasitoids can be reared successfully from many factitious hosts in the laboratory. They do so because in the laboratory they will oviposit quite readily into insects that they do not normally attack in the field, presumably because confinement removes the need for them to use the long-distance components from their searching behaviour, a point recognised by weed biocontrol practitioners in their pre-release work (Marohasy, 1998; Wapshere, 1989). *Trichogramma* species are considered to be more (micro-)habitat specific than host specific (Pinto & Stouthamer, 1994; Smith, 1996), but this view needs critical appraisal. Factitious hosts and the unresolved status of strains undoubtedly cloud the issue considerably.

Once such host-specific natural enemies are released into the field the tacit assumption is that they, having been judged to be polyphagous or generalist, will locate and attack the target pest. Perhaps this perception is aided by the additional beliefs that insects as minute as *Trichogramma* do not habitually undertake extended movement, and that the pest is so abundant as to represent an unused resource that is bound to be encountered by the parasitoids. Experience frequently shows that these assumptions do not necessarily hold (Chapter 8). 'Generalist' predators may also be subject to such misjudgements. The leucaena psyllid predator *Curinus coeruleus* (Coccinellidae) remained uncommon in Hawaii for over 60 years, following its 1922 introduction from Cuba against the coconut mealybug. It persisted at low levels, on various scale insects and aphids, until the Cuban leucaena psyllid *Heteropsylla cubana* Crawford became established in the early 1980s, when the coccinellid became abundant (Follett & Roderick, 1996). The common failure of the many 'generalist' predator species introduced in many parts of the world after the success of the vedalia beetle (Lounsbury's 'ladybird fantasy'; Caltagirone & Doutt, 1989) may be explained, at least in part, by an inappropriate matching of predator to prey species.

Artificially selected lines of natural enemies are also advocated, and some have been used in control (Hoy, 1990). The limits to what can be achieved by this approach are going to be dictated by the way in which the overall searching mechanism of the laboratory-selected organisms operates under field conditions. Distortion of the mechanism may not inhibit host or prey finding under laboratory circumstances, but under field conditions the mechanism may work ineffectively, or not at all.

4   New associations

The interpretation offered in the previous point may shed light on another debatable issue in the biocontrol literature, that surrounding the use of so-called 'new associations' as opposed to 'old associations'. New associations are those combinations of natural enemy and host species that have been brought together by humans and therefore have not shared an evolutionary history. The host species involved in old associations are expected to have evolved more resistance to the parasitoid (Ehler, 1990; Myers et al., 1989). The 'age of association' issue is unlikely to lead to effective or useful generalisation, because adaptive change to host relationships is more likely to take place under special circumstances than according to the 'arms race' model (Chapter 6). However, a natural enemy species may, through preadaptation, have the appropriate mechanism for successful attack of the pest, and no reason should preclude its use. So one cannot generalise about 'new associations', except to say that the location of such preadapted natural enemies is likely to be almost entirely a chance affair and with no certainty that such preadapted natural enemies will exist for each pest species.

5   Partial success

Another practical issue is that partial or incomplete control by natural enemies is sometimes seen as a problem of why an obvious abundance of resources (hosts or prey) is not more fully utilised by the natural enemies that are present. The community ecology approach suggests that a species (or more than one) with a different niche should be used to boost the community or guild (e.g. Hochberg & Hawkins, 1992). But additional species will work only if they fulfil the autecological environmental match, not because their 'niche' is empty (Chapter 5). In some biocontrol projects a part of the pest population is considered invulnerable to parasitism, through its location on a particular plant part not searched by the established parasitoids or because it is of an age class that is not parasitised (Chapter 8). Again, this is not a problem of guilds, although it may be a problem requiring additional species that match environmental conditions and search the 'refuge'. For example, only one of several trichogrammatid species tested by Browning & Melton (1987) actually searched where the eggs of the relevant pest species were hidden. Also, different species may predominate at different vertical levels of vegetation, as *Trichogramma pretiosum* Riley is usually recovered close to the ground whereas *T. minutum* Riley is mainly higher in the vegetation (Thorpe, 1985). However, results from field sampling (e.g. Newton, 1988) frequently reveal just how inherently variable such vertical patterns may be.

Conclusions for this section

The implications of the points raised in this section are clear. The appropriate evidence to support the choice of parasitoid or predator species for biocontrol should be gathered prior to selecting agents. The species must have a parasitisation (or predation) mechanism that ensures the pest is going to be a habitual host (or prey species) of that parasitoid (or predator) in the environment of release. Further, the development of general theories in biocontrol will only be as good as the evidence about the organisms that is cited in support or contradiction of hypotheses and interpretations that relate to biocontrol. In particular, good evidence about their species status, host relationships and other adaptations that relate them to the environment is required.

Considering parasitic and predatory interactions from the perspective of individuals directs attention at the identification of the adaptive aspects related to their interactions and whether that behaviour will lead the natural enemies to the target organisms in the field. It also focuses attention on the survival of the natural enemy individuals in a locality and on how well they will perform reproductively. Whether successful biocontrol agents have characteristics additional to those just mentioned will have to be assessed in terms of their ovarian physiology, and their consequent patterns of ovipositional activity and movement (Antolin & Strong, 1987; Cronin & Strong, 1994; Donaldson & Walter, 1988; Fernando, 1993; Reeve, 1988; Strong, 1988; Walter, 1988a). The shift of attention to the behaviour, in the field, of individual natural enemies should help to introduce a more accurate spatial dimension to biocontrol theory and thus to interpretations of what characteristics of individual behaviour may lead to desired patterns of parasitism in the field. These aspects are mediated by adaptive features that are primarily physiological and behavioural, and which are species specific (Donaldson & Walter, 1988; Fernando & Walter, 1999; Walter, 1988a).

Environmental relationship: selecting the time and locality for release

The approach to the problem of matching organisms to environmental conditions in another locality is influenced by whether one is making inoculative or inundative releases of natural enemies. Inoculative release requires a climate match that will ensure continuous survival across seasons, whereas inundation does not have this requisite. The latter demands, instead, synchrony of release with the availability of suitable hosts, and the coordination of release with physical conditions that are appropriate for the organisms. Neither of these is easily achieved, as shown by experience with *Trichogramma* wasps (Smith, 1996).

Prevailing climate clearly affects the impact of natural enemies, for one or more of a variety of reasons. Climate matching has been approximated through selecting source areas for particular natural enemies that have a climate similar to that in the target area, and is seen as a 'first rule of thumb' in biocontrol (Myers *et al.*, 1989). Nevertheless, negative impacts of climate resulting in the poor performance of released agents is frequently discounted, presumably because no obvious connection between climatic conditions and performance could be discerned. Practitioners tend to be convinced only in those cases in which extreme climatic conditions have had an obvious impact, such as occurs with drought or very low temperatures (Myers *et al.*, 1989).

1   Climate matching

The original approach to climate matching was based principally on intuition. More recently, climate diagrams have been used for comparative, explanatory and predictive purposes. For instance, Samways' (1989) comparison of the match between (i) the expanded distribution of the coccinellid *Chilocorus nigritus* (Fabricius), (ii) the climatic types covered by its distribution, and (iii) the climatic types within localities where it was introduced intensively but did not establish (Fig. 9.10) explains the limits to this species' expanded distribution and why further efforts at establishing the species outside that area would not be worthwhile. The original distribution of the species was within the Indian subcontinent, within

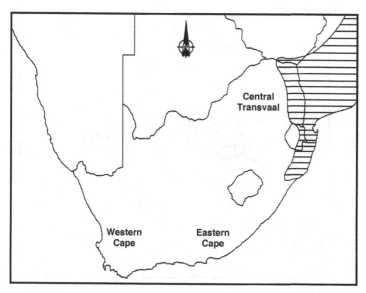

Figure 9.10. Limit to the distribution (hatched area) of the coccinellid *Chilocorus nigritus* in southern Africa. Despite intensive introductions from rearing facilities the species did not establish in the central Transvaal, eastern Cape or western Cape. Reprinted from Samways (1989) with permission from Blackwell Science and The Royal Society.

which it had been confined by geographical barriers. This example is valuable in demonstrating the scope of the autecological approach in applied insect ecology. Other examples may not prove so straightforward (e.g. McGeoch & Wossler, 2000). That autecological approaches, methods and models can be improved through better understanding of the various subtleties involved in the way in which organisms interact with climatic variables can be demonstrated as follows.

1   Although the predacious carabid beetle *Bembidion fumigatum* has a fragmented distribution in southern Europe, the fragmentation is somehow related to prevailing temperatures, for beetles within the various localities occur under similar temperature conditions (Fig. 9.11).

2   Figure 9.5 illustrates in general terms why different species within an area are differentially affected by local climatic conditions. That is, in part, a consequence of each species being sensitive in particular ways to ambient physical conditions. Various parts of the life cycle may be involved, as may different physiological processes. Plants have been better investigated from this perspective, and work on the bluebell in

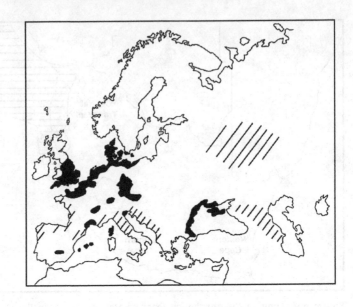

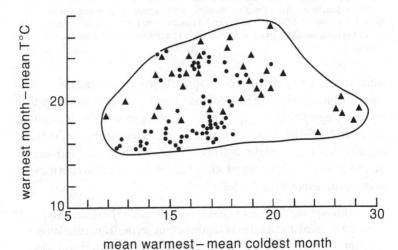

mean warmest – mean coldest month

Figure 9.11. Distribution and temperature regime of the carabid *Bembidion fumigatum*. Despite a fragmented distribution of the species, it can be seen that individuals in different areas are exposed to similar thermal conditions. Solid areas on the map and dots on the graph represent precise locality data. There is less certainty about hatched areas and triangles. Reprinted from Coope (1986) with permission from the Royal Society.

Table 9.1. *The ecological requirements of bluebells (* Hyacinthoides non-scripta*) dictate which areas will be suitable for their growth and reproduction. Here, only two biological processes are considered, and then in relation to only two physical factors. That is enough to indicate just how much 'autecological factors' influence local ecology (or presence and abundance of a species in a locality). Other factors that would need consideration include edaphic factors, as well as mycorrhizal and light requirements and the time of their availability relative to relevant stages of the life cycle*

Biological process	Season	Moisture level	Temperature
Seed ripening	Spring	Low	>26 °C
Seed germination	Autumn	High	<11 °C

western Europe (Thompson & Cox, 1978) shows how subtle the physical influences may be. Table 9.1 shows the temperature and moisture demands of bluebells. Note that only two physical features are included and two physiological processes, but the consequences are clear. The species has restricted physical requirements, and these limit its distribution to particular parts of western Europe. Many other factors influence the survival and reproduction of bluebells, including light availability, mycorrhizal composition of the soil, and so on (Gonzalez Sierra *et al.*, 1996; Merryweather & Fitter, 1995; Pigott, 1990), so the species specificity of this organism's requirements is assured, as is the circumscribed circumstances within which it can survive.

3   The evidence that physical influences affect local ecology in insects is starting to build again after decades of disinterest. In particular, Ford (1982) drew together the evidence to show that even small changes in average temperature in a locality were sufficient to affect the presence and abundance of species. How subtle those influences may be has not really been explored, but characterising the physical environment of natural enemies may not be as straightforward as commonly assumed. Preliminary evidence indicates, for example, that minimum temperature may not be the most significant influence on survival of overwintering *Aphytis lingnanensis* pupae, but that the level of winter maxima may be critical, as sufficient warming may be needed after exposure to the lowest temperatures (DeBach *et al.*, 1955).

4   To invoke physical influences, as above, is not to deny that biological influences may be just as significant. Herbivorous and parasitic organisms cannot, for example, exist in areas where their host

requirements are not met, even if the physical environment is suitable (e.g. Hengeveld, 1990). The important aspect to note from this example is that although a biotic influence is considered important, it is not a demographic or density-related influence (Chapter 5).

The issues and examples covered above indicate that the accurate matching of an organism to a new locality is not as straightforward as it is frequently portrayed. Some species, upon introduction to a new area, expand readily across fairly broad geographical fronts, but most species do not establish readily or do not expand rapidly and widely (Carey, 1996; Lodge, 1993), and presumably such situations are well represented among the introduced natural enemies that have failed to establish. Specialised requirements of this nature need not preclude species from being used in biocontrol; they mean only that a more informed approach to deploying natural enemies needs to be developed.

New techniques in climate matching are available, such as the software package CLIMEX (Sutherst *et al.*, 1995), but their use seems generally to have been restricted to well-known pests with wide distributions, which presumably have wide tolerances (e.g. Worner, 1988). However, a good match seems to be achieved more by adjusting the model's parameters to a known distribution than by refining the basic data to generate the fit independently. The effective use of such models will undoubtedly depend a great deal on understanding the organisms' ecology more precisely. Perhaps the question we need to consider in this respect is whether we know enough about the process of climate matching to be confident not to persist with a species that is predicted to be unsuitable because of a mismatch between its characteristics and the climate in the area of release. Some agents have performed well in areas where this was not expected at all (e.g. McGeoch & Wossler, 2000), so reliance on models in this way is doubtful, but it is a goal at which to aim.

2   Biocontrol and the climatic dynamic
Biological control takes place within the context of the climatic dynamic. Environmental conditions change, and such changes are likely to affect parasitoid (predator) and host (prey) differentially (Fig. 9.4). The complicating influence of this dynamic is generally omitted from biocontrol models, but it does suggest alternative explanations for various biocontrol-related phenomena, as outlined below. These suggestions have not been tested, but if the basic premises upon which they are built are sound, their validity or otherwise does warrant empirical scrutiny.

High population densities are generated when survival and reproduction are enhanced under particularly favourable conditions (see Chapter 5). Nevertheless, the influence of climate in such cases is frequently downplayed, even if the appropriate research and analysis has not been conducted. Population densities, originally elevated during favourable environmental conditions, may be maintained at high densities even when the more suitable environmental conditions pass and survival rates and reproductive success drop somewhat. Replacement-level reproduction should hold populations to the high levels, attained during the good period, until environmental conditions lower survival and reproductive output. Alternatively, high levels may be forced down by natural enemies, as in successful biocontrol and drastic overfishing. Even if the predation pressure is reduced (as in the cessation of fishing), populations may not automatically increase again, because reproductive success would remain at replacement level. This would explain post-biocontrol densities of pests remaining relatively low even in the face of reduced predation or parasitism rates. This interpretation also accounts for the continued need of natural enemy input, by inoculation, under circumstances where environmental conditions are artificially, and incidentally, maintained beneficial for pests through agronomic and husbandry practices, as in citrus orchards.

3   Lessons from invasion research

A commonly perceived problem in biocontrol is a lack of precision and predictability: 'unless biological control is both more reliable and predictable, it will not be generally adopted and integrated into IPM systems' (Tauber *et al.*, 1984). To this end, Kareiva (1990) advocates the study of invasions into communities and the influence of competition after release, so that we can predict disasters in advance, and presumably also predict success in biocontrol introductions, for they are human-assisted invasions designed for agricultural benefit.

As usual in ecology (e.g. see Chapter 5), invasion research is practised under different systems of thought. Demographic interpretations are available (Williamson, 1996), as are autecological ones (Hengeveld, 1990). Consequently, we have different perceptions of what is achievable in dealing with invasions. Demographic ecology supports the view that Kareiva's (1990) aim for prediction can be achieved, even though it is acknowledged that the ability to predict from this basis is still elusive (Lodge, 1993). By contrast, autecology considers, from experience to date, that such predictions are not likely to be possible, because of the species-specificity of organisms' characteristics and the context-dependence of the expression of those characteristics. The essential problem is that although we might know what parameters to measure (e.g. Samways, 1989; Van den Bosch *et al.*, 1992), we cannot predict the values they will assume in a new area. The experience of the introduction itself is likely to remain the only way in which to achieve any degree of certainty (Hengeveld, 1992).

Discussion

The autecological principles and approach outlined in this chapter argue against the common perception that a robust general model for biocontrol will provide a basis for selecting among species for release and will predict outcomes of natural enemy releases (e.g. Greathead, 1994). Instead, autecology bases its generalisations around the species-specific adaptations of organisms, and how they function in the interaction between organism and environment. This emphasis provides a basis for understanding how individuals of a species will match the environment in which they are intended for release, and thus provides the most appropriate basis for assessing the likelihood that they will establish and survive in the area of release.

Four basic points made in this chapter further legitimise the autecological approach to biocontrol.

1   The autecological approach provides an alternative, realistic starting point for conducting pre-release investigative procedures, for developing generalisations about natural enemies in relation to biocontrol, and for interpreting what it is that makes a natural enemy successful in biocontrol.
2   Understanding the autecology of pests and biocontrol agents is not simply an exercise in fact-gathering natural history. Autecology provides a valid scientific alternative to demographic theory. It is based on alternative premises to those underpinning demographic theory. The autecological premises relate to individual organisms, their species-specific properties and their functional interaction with their surroundings, and provide a strong contrast to those underlying demographic interpretations. Demographic premises relate less to individual organisms and species than to postulated population and community equilibria and non-equilibria.
3   The complex physiological and behavioural adaptations of natural enemies dictate how individual organisms of each species 'operate' in

nature and thus influence the abundance of pest organisms within a
particular locality, or under a particular set of circumstances.

4   An autecological interpretation of organisms in nature is amenable to
modelling, as indicated by the availability of successful
individual-based models (Lensink, 1997; Van den Bosch *et al.*, 1992).

The autecological solution proposed here is predicated on the view
that the search for workable biocontrol theory from a demographic perspective has been driven by an idealistically influenced notion of nature
as an economy. Natural systems are not economies, and they cannot be
altered or organised after the fashion of an economy. The orientation
in ecology needs to be shifted from what we wish we could achieve to
what the nature of ecological systems will allow us to achieve. Externally
imposed limitations to prediction are a reality to biologists, as evidenced
by the lack of such predictive abilities despite the considerable research
effort so far expended. A significant consideration at this point is that the
approach one takes to the issue of how to improve biocontrol practice also
relates very much to one's view of scientific methodology (see Chapter 2).
For instance, Waage (1990, p. 149) has claimed: 'The main purpose of a
scientific approach to classical biological control is . . . to improve success;
the improvement of understanding is a desirable but secondary aim'.
The approach taken in the present book suggests that an inversion of the
statement above would put us in a better position to break the influence
of some of the traditional approaches, which show few signs of improving
practice. To some extent, we need to start afresh, from more realistic
premises about organisms in nature.

The complex functional interactions that dictate the behaviour of individuals in the field are likely to constitute the strongest basis for reliable generalisations. A typical such generalisation is the model of the parasitisation mechanism (Fig. 9.9), which was developed from the work of
Vinson (1998) and others. Nevertheless, the best chance of successfully extending our biocontrol capabilities will be if we begin with realistic principles about individual organisms, and then work outwards towards an
understanding of the system as a whole. For example, the parasitisation
mechanism and climatic requirements (see Hengeveld, 1990, for example) of species deemed to be serious biocontrol candidates can be tested,
and can be used to predict the final distribution of species with expanding
ranges (Chapter 9), including introduced natural enemies. Species should
therefore not be seen to have failed to establish or spread because they are

'not adaptable' or 'not very good colonizers' (e.g. Caltagirone & Doutt, 1989); they should be judged on appropriate evidence about their properties and the local environment.

The practical implications of all this are that species need to be approached somewhat differently than from the currently popular demographic perspective. Although the individuals that make up each species are variable in their features, whether in their morphology, molecular constitution or behaviour, this should not be overemphasised, for in the ecological sphere their similarities far outweigh most such differences. The species status of populations being considered for biocontrol needs rigorous testing, and full documentation; this should be seen as a minimum requirement, to put the science on a firm base and to render biocontrol as efficient as possible (Paterson, 1991). Institution of such an approach would require funding bodies and practitioners to forego speculative projects in favour of well-researched ones, and to forego the urge to commit smaller sums to certain natural enemies only because they are readily available in culture. The short-term costs of a rational, scientific approach to biocontrol will pay off in the long term.

The most basic prediction we can derive from autecology is that the ecological properties of individuals that are potentially part of the same gene pool in nature should be stable across space and through time, which provides a basis for extrapolating information and making predictions about aspects of the ecology of these organisms (i) from one place of derivation to another, (ii) from place of derivation to place of release, (iii) from one place of introduction to another, or (iv) from one time to another. However, the population consequences of these properties and the interactions they mediate will vary with locality and time, as environmental conditions vary (Walter, 1995b). In other words, the population consequences of a natural enemy's activity in its area of endemicity cannot necessarily be expected to be mirrored in areas where it is introduced (e.g. Waage, 1990).

The points just made imply that the ecology of natural enemies cannot be characterised as being limited only by intrinsic factors or by extrinsic ones, because it is the interaction between both types of factors that is important. Natural enemies that are perceived to be limited by intrinsic factors should not 'be expected to display the same limitation when introduced into a new region', as Ehler (1990, p. 123) claims they should. Since one is dealing here with complex adaptations with several components, including distance components, close-up recognition, and so on, it is a complex sequence expressed within the environmental context that needs to be understood. That context and the adaptations impose the particular scale at which each functional aspect of behaviour will take place.

In summary, autecology stresses the primary importance of the adaptive features carried by the individuals that make up a species. This, in turn, directs research emphasis and interpretation at the organism–environment interaction and emphasises the species-specificity of the complex adaptations that mediate those interactions, as well as the functional interactions that take place between individuals of different species. Considering the interaction from the perspective of individuals directs attention at the identification of the adaptive aspects of that behaviour and, in the case of natural enemies, at whether that will lead to parasitism or predation of the target organisms in the field, and whether those individuals will survive in the locality of interest and how well they will perform there reproductively.

Part 3

Synopsis: Ecological research for IPM today

10

Synopsis, practical implications, and modern society

The twentieth century's only claim to have benefited humanity rested on the enormous triumphs of a material progress based on science and technology. Paradoxically, this era ended in a rejection of these by substantial bodies of public opinion and people claiming to be thinkers in the West.

E. HOBSBAWM (1994, REDRAFT OF ORIGINAL STATEMENT, p. 11)

Introduction

In this chapter the major conclusions from the preceding chapters are linked, to illustrate the advantages of a generalist approach to the interaction between insect ecology and pest management. In general, a much narrower view tends to be prosecuted today, often indirectly through research programmes being artificially constrained by short-term productivity demands and bureaucratic impositions on the one hand, and technocratic leanings in researchers on the other. Two themes especially relevant to the future of insect ecology research for pest management purposes are developed from this synopsis.

The first theme is developed to support the proposition that effective pest management is going to rely on the appropriate use of strong scientific principles. Unfortunately we live in times when the advantages of logic and scientific principles are being sacrificed, with the 'shift from the modernist culture that reified the liberating power of science to a culture of so-called post-modernism – a post 1960's culture that has shrugged off the emancipatory certainties of science and erected commercial marketplace values and pluralistic images in its stead' (Hill & Turpin, 1994). Consequently, the value of research is increasingly judged by 'intellectually facile considerations of marketability' as science is put to national and economic utility (Hill & Turpin, 1994). Similarly, education systems suffer and opportunities open for expediency in its various guises. The comments on this decay and its negative implications for society are made to help to keep the strong points of our scientific heritage in service. The public suspicion that increasingly surrounds science provides an added reason to present this argument. Perceptions of science failing in many areas are widespread (Allaby, 1995; Hobsbawm, 1994; Holton, 1993; Perutz, 1991; Rees, 1993); examples are the problems of disposing of nuclear waste, the emergence of uncontrollable pests resistant to the very pesticides that were supposed to eliminate the problem, malaria and other diseases re-emerging as an everyday threat, and the regular forecasting of 'doomsday' scenarios consequent upon the activities of society. Blaming science for these problems is unfortunate for various reasons, and the nature of science and its utility and limitations need to be seen in a much broader context. Science seldom produces a solution that has no ancillary effects, some of which may be undesirable. Frequently, but not always, these by-products are known before technologies are implemented, but decisions by this stage have usually moved from the realm of science to the sociopolitical context (Allaby, 1995, p. 117). In any case, scientific evidence informs decisions, it does not take them. Medawar (1984, p. 24) exemplified the point, thus: 'If the termination of a pregnancy is now in question, scientific evidence may tell us that the chances of a defective birth are 100 per cent, 50 per cent, 25 per cent, or perhaps unascertainable. The evidence is highly relevant to the decision, but the decision itself is not a scientific one, and I see no reason why scientists as such should be specially well qualified to make it'. Romesburg (1981) is succinct on this: '[s]cience uses fact as its standard for selection, whereas planning uses values'.

The second theme is developed from the groundwork laid by the first. It extracts the practical implications of what has been said in all the preceding chapters. How should we set off, in general terms, to solve serious pest management problems? These views are offered at the risk of being presumptuous, for there have been many exciting pest control successes and there are many highly competent people in pest management. A general framework is nevertheless needed to capture the essence of good pest management practice, and the derivation of such an 'organising' tool needs some sort of provocative initiation.

Synopsis of early chapters

Chapters 2–9 cover a diverse range of topics for a book on pest management. Each aspect was deliberately selected to illustrate the general argument that is developed from chapter to chapter. This highlights the practical relevance of fundamental theoretical principles, drawn from several relevant sources, for the practice of pest management. Two general points emerge. First, if an idea is changed at a basic level of understanding, derivative interpretations will inevitably be affected. In turn, the pest management actions that rely on that understanding will also be affected. Second, improvements in underlying interpretations are already available in some areas, for example in species theory and ecology, and are likely in others too. One can ignore such fundamental understanding and logical dependence, but few succeed in producing lasting understanding or efficient practice with such a laissez-faire attitude to knowledge. The three penultimate chapters (Chapters 7–9) concentrate on specialist topics, including the ecology of polyphagous pests and theoretical underpinnings for biological control, to illustrate how the application of the principles developed earlier (Chapters 5 and 6) can assist pest management in practical ways. These principles are, however, also applicable to other areas of pest management, such as interpreting and predicting abundance, development of pheromone-based technologies and use of trap crops, but the specific consequences have yet to be worked out in detail for them.

Chapter 2 summarises the idea that one's very view of method in science is likely to influence the way in which one approaches theory, interpretation and derivative practice. Although scientific method traditionally falls within the realm of philosophy, that should not discourage participants in pest management from considering the practical lessons this aspect of philosophy has to offer. Errors that arise from faulty logic or conflation, for example, remain errors even if one dismisses theory lightly. Such errors will ultimately be recognised for what they are, even if that does not happen immediately. The main practical lesson from this chapter is in the deconstruction of interpretation for scientific testing. There are many approaches to developing and extending knowledge, but the scrutiny of underlying principles and the subsequent development of risky empirical tests are particularly strong weapons in advancing understanding.

The history of insect pest management, covered in Chapter 3, reveals periodic change in general attitudes to pests and in the consequent approaches to pest management. Such general attitudes, or intellectual frameworks, are generally referred to as paradigms, a concept introduced into the pest management literature by Perkins (1982). Inspection of the progression of paradigms in pest management, from the time of the earliest available records to the changes we have seen over the past few decades, reveals several points relevant to developing a framework for pest management. First, more recent shifts in attitudes have frequently been subtle, not as obvious to us as some of the early changes. Detection of current difficulties is thus likely to require dedication of thought and active pursuit of the problems if their consequences are going to be appreciated and widely understood. Second, difficulties with any particular approach to pest management, as seen with the chemical paradigm, cannot be remedied by a simple return to previous approaches and methods. Changes to society and the expectations of society are reflected in changes to agricultural production. New solutions have to be found to satisfy the changed expectations. The interaction between science and society in the development of social expectations is likely to prove critical in this respect. For this, we need an education system that works effectively to open minds to the general strengths of logic and the scientific approach to analysis and synthesis. Current trends in many parts of the world work against such an aim, as academics are forced to compete for students and thus undermine their own conception of what is educationally worthwhile (Hill & Turpin, 1994).

Integrated pest management is the current pest management paradigm (e.g. Perkins, 1982), but definitions of IPM are generally quite vague. They do not, for instance, say much about the intricate structure of IPM, and they give no idea of the overall social and scientific context of IPM. Chapter 4 therefore develops a detailed statement about the scientific structure of one aspect of IPM, that of understanding pest ecology and developing applications with reference to the 'IPM chain' (Table 4.1). Although Table 4.1 provides considerable detail, only the entomological influences in agricultural systems are considered. For a more complete picture the other major influences, socioeconomic, botanical and so on, also need to be incorporated.

Two particularly relevant points emerge from consideration of the IPM chain: (i) IPM actually represents a complex social philosophy; (ii) entomological understanding for purposes of IPM is driven in only a general way by IPM, with details of the research necessarily being influenced by biological theories. Consequently, the derivation of the entomological knowledge upon which effective IPM can be built is not discussed at any depth in IPM-related texts. Although the type of information required is often described, ecological theory is usually presented in the form of population dynamics theory, although community and ecosystems approaches are sometimes advocated. Population ecology is almost invariably dealt with as if that is the only approach to understanding ecology; indeed the premises of population ecology are all too often treated as fact. However, the fundamentals of population ecology are increasingly questioned in the scientific literature, a point elaborated and explored in Chapter 5.

A survey of approaches to insect ecology (Chapter 5) shows that there are several directions in ecological research. Each is built upon different fundamental premises, and each has different implications for aspects of the practice of pest management. Only one approach, autecology, deals specifically with individual organisms and their species-specific adaptations within their usual habitat. Autecology thus copes, at a fundamental level, with the unique character and ecology of each species, as well as the variable influences, both through time and across space, of the environment on the local ecology of a species. The other approaches to ecology–population, community and ecosystem ecology–seek principles that transcend environmental variability and the ecologically relevant differences among species. So far, such overarching principles have proved elusive, and the problems that autecology has been developed to overcome can explain why such desirable principles do not exist, or, at least, will never work effectively in practice.

Autecology is predicated upon an understanding of species. Although most biologists feel comfortable with the concept of species and use it confidently, Chapter 6 demonstrates that 'species' is a term with several meanings, and an awareness of the subtleties that separate these meanings is important for the most realistic understanding of organisms in nature. The consequences for agricultural applications are significant. Chapter 6 spells out these subtleties and their significance in relation to the behaviour and physiology of individual organisms, especially those aspects related to sexual behaviour, because that is what defines species most realistically and reliably. Reconstruction of the circumstances that force species formation is therefore possible. The crucial nature of this information lies in our being able to identify the environmental circumstances and primary driving forces providing the directional selection that ultimately guided adaptive change in the species of ecological interest. The identification of the primary adaptations of organisms is crucial to understanding the ecology of a given species within a particular environmental context. This point drives much of the interpretation that follows in Chapter 6.

The methods used in detecting the limits to species are reviewed in Chapter 6. This is a process that is not as straightforward as frequently portrayed or implied, because the procedure is not strictly technological. Instead, the role of technology is to assist in answering specific questions about species limits. Carefully phrased questions based upon the most robust principles will generate more appropriate data than will questions based on discredited premises. In other words, the mere application of technology will not inevitably provide the data needed to clarify a particular issue. The onus falls squarely on the researcher to phrase meaningful questions. Failure to do so may well yield data that are uninterpretable or, worse still, misleading. Facts do not inevitably emerge from this type of research; rather, understanding needs to be eked out through formulating meaningful specific questions and then addressing them with appropriate techniques.

Chapters 7–9 illustrate and explore the consequences of accurately interpreting species status for understanding pest and beneficial organisms. They provide illustrations of how a difference in the premises about species can alter ecological interpretation and thus approaches to pest management. The most appropriate action is easily missed or overlooked if an inappropriate theoretical lead is followed. The material covered in Chapters 7–9 is a small subset of the material that could have been included, but it should illustrate the strengths of the approach advocated. First, polyphagous or generalist species are covered (Chapter 7), as this is an aspect crucial to much further understanding in evolution and ecology, and also to the improvement of pest management.

To illustrate: the dichotomous classification of organisms into generalists and specialists is frequently used to aid ecological understanding and to help develop theory for pest management purposes. Such an approach is entirely inappropriate for understanding ecology and for generating further ecological generalisations. Even the arrangement of species along a continuum from specialists, or monophages, through to generalists, or polyphages, is neither realistic nor useful. The terms 'polyphage', 'generalist' and so on are not in themselves so bad; the trouble is that they are somewhat arbitrary classifications that remain without strict definition and so are used inconsistently. The real problem arises from ecological generalisations being sought around these categorisations, in which situations they are then likely to mislead. Similarly, the terms do not reveal enough about the ecology of the species deemed to fall within or between these typological categories. To understand the ecology of such species for predictive and manipulative purposes, their host/prey relationships need to be understood from a functional perspective, a new basis for which is outlined in Chapter 7.

The approach recommended in Chapter 6, and exemplified in Chapter 7, can be applied to pest management practice. This is done in Chapter 9 with reference to biological control practice involving insect pests. To provide a counterpoint for Chapter 9, in Chapter 8 the current approach to biocontrol theory and practice is outlined. Current approaches have devolved from demographic ecology principles, which were criticised in general terms in Chapter 5. In Chapter 8, the specific demographic developments within biocontrol are criticised from several angles, to provide added justification for the autecological developments offered in Chapter 9, and to help in the assessment of the differences between the two approaches and selection between them. Without a contrast, the distinction is easily lost. Indeed, blurring of the distinction is sometimes practised, perhaps inadvertently, and this contributes to the maintenance of the status quo in theory development.

In Chapter 9 autecological principles are used to show how best to develop generalisations for applied ecological purposes. This approach cautions with regard to the extent of prediction in applied ecology because ecological theory cannot transcend the unique nature of species and the context-dependence of their ecology. Predatory insects and parasitoids are examined from the perspective of the functional interactions that link them with their prey or hosts. Such interactions are complex and of a physiological and behavioural nature, so this approach to ecological generalisation is built from a consideration of the individual within its usual environmental context. Species are relevant, but only to the extent that individuals derive their physiological and behavioural characteristics from their parents, both members of the same species gene pool. Functional interactions are thus species specific and species wide. Climatic and other influences are then incorporated to provide a theoretical basis for the location and selection of natural enemies. Prediction of the population dynamics and biocontrol performance of released natural enemies is not possible, and is not built into the theory as it is in demographic approaches. Therefore the processing of additional candidates, coupled with effective post-release monitoring, is recommended.

In summary, the role of science in the management of insect pests is specified in the early chapters, and then expanded and exemplified in the later chapters. However, the role of science takes place within a broader context than that of science itself. Today that context is a democratic social one, although that may well change. The consequences of the current sociopolitical context are examined in the following section, mainly to show how the principles and practices advocated in this book have a place in modern society. This is warranted because current attitudes leave science in an uncertain position, as indicated by Hobsbawm's quotation at the beginning of the chapter.

Sacrifice of scientific principles in modern society

'Applied science' is often portrayed in social forums, and in parts of the IPM literature, as clinical, factual and disciplined. Conversely, and often through omission or by implication, the unruly relative of application, 'pure science', is commonly seen as a luxury, perhaps one that society carries at too great a cost because it is not put immediately to direct and obvious use (see Allaby, 1995, Chapter 25). That such views are espoused even by those whose professional activities are covered by the umbrella of 'science' demonstrates just how much individuals forget about the various advantageous roles that science has played in the development and maintenance of modern society.

Indeed, the scientific way of thought continues to provide rational direction for society in many facets, and also plays a cultural role as significant as any other cultural activity that we may favour (Baltimore, 1978). The damage to science from within is compounded by various antiscientific and anti-rationalist sentiments. These latter issues are covered in the detailed books of Holton (1993) and Allaby (1995). This section examines aspects of the denigration of science and how IPM is being negatively affected. The main point made is that a sacrifice of scientific principles is inimical to rationally developed IPM strategies, and that it is important that all interested in improving the performance of IPM should understand and appreciate the role of science and scientific principles in this endeavour.

The damage from within

Generally, both the research and development sides of science have suffered funding cuts since the 1960s. Naturally, questions of priority and participation have been discussed (e.g. Brooks, 1978). More recently, economic rationalism has been in unquestioned ascendence, and with it have come the most intense demands for immediate practical outcomes that scientists have yet seen. This pressure poses serious risks for science in all manner of ways (see Allaby, 1995). Where application is concerned, ecology (including IPM) may well suffer disproportionately, for specification of desired outcomes is frequently rather vague; and, in any case, measurement of the effectiveness of any remedial actions is not straightforward or is not even monitored (Botsford & Jain, 1992; Perkins, 1982, p. 138).

The impact on IPM of the views outlined above is potentially insidious, damaging from within the scientific basis of the IPM approach. Competitors for limited research funds can readily undercut one another's bids, promote their immediate research interests as of greater relevance, or invulnerably inflate their promises of success, frequently without fear of detection because outcomes cannot be specified accurately. Even the sympathetic expression of a need for understanding sometimes masks a biased attitude against scientific investigation, perhaps to silence opposition. The proposals and actions of individuals need to be assessed if an accurate judgement is to be made of their attitudes to science, their understanding of scientific method, and even their willingness to sacrifice scientific principle for personal gain.

Notwithstanding these issues, or even in the face of them, almost all scientists would acknowledge that science does not provide a 'quick fix' for any problem. However, the specification of what falls into the category of 'quick fix' is unlikely to be universally agreed, for different scientists see science, and the relationship between application and understanding, in different ways (see Chapter 2). Those who emphasise utility at the expense of understanding, even if only tacitly, will certainly have different attitudes to application from those who give prominence to understanding. Applied ecology may well be further undermined, for it suffers the added complexity of a diversity of viewpoints of what ecological aspects need to be understood in any given system (see Chapter 5). These are issues that warrant deep consideration, active decisions and further development in applied ecology.

The development of scientific innovation, principle and theory has, until recently, been seen as an individualistic pursuit. Elements in society and even within science now oppose this view. Individuality, even individual excellence, is increasingly pushed aside and is sometimes even denigrated as egocentric. Administrators strive to homogenise expectations across units and individuals, to unify directions and objectives (Hill & Turpin, 1994). This development is aided by perceptions of 'big science' being the only good science, together with the view that the work of research teams is the only way forward. The concentration of funds in this way tends to concentrate power within the research establishment. Many areas of research will not be served well in such a way, and ecology may be especially at risk since there is no clear best road ahead (Weiner, 1995). Research teams can clearly be effective if they are led well, by scientists who are suitably generalist across the necessary subdisciplines (Lindsay, 1986; Petrie, 1976), but such leadership lapses if a democratisation of scientific priorities is allowed to predominate. Damage through democratisation is likely to be more comprehensive if the process is extended beyond researchers, for many are happy to pronounce and vote but few are likely to understand enough about the ways and principles of science to decide wisely. And here, even in general terms, we find the potential for erroneous policy (Holton, 1993, p. 148).

The information age and the sacrifice of understanding

The 'information age' is widely heralded, but uncritical acceptance of this slogan is likely to contribute substantially to a view of science that is somewhat medieval (Chapter 2). Unfortunately, such slogans seem to pervade the development of policy and the financing of many scientific endeavours that involve application, for many feel we have long had enough facts, and that what we need to do now is more a matter of finding the required information, transmitting it and applying it (e.g. Surtees, 1977). Indeed, Holton (1993) points out that it is yet again currently fashionable to talk of the 'end of science' (e.g. Horgan, 1996), not because of public or economic attitudes killing science, but because of a feeling that the basic questions have been answered. But anyone who is satisfied, for example, with an explanation of biodiversity based solely on Darwin's formulation of natural selection (e.g. Horgan, 1996, pp. 112–114) has not only to be thinking entirely in terms of naive induction (Chapter 2), but also very superficially (see Chapters 5–9 for examples of unresolved fundamental issues in ecology and evolution). Predicting the 'end of science' has in any case a fairly long history (see, e.g., Holton, 1993).

Few who herald the 'information age' consider seriously enough the issues relating to the quality and scientific relevance of the information with which we deal, as well as to the basic need for synthesis, understanding and the extension of knowledge. Insight, wisdom and effectiveness in society do not come simply with the possession of information. Nevertheless, economic rationalism is easily convinced that information is a commodity. If one equates data with information this is perhaps not that unrealistic, especially if one wishes to mould the information to suit one's 'product'. Even so, the quality of such data and its realistic evaluation and interpretation remain an issue.

Outstanding counters to the widespread seduction of the information age are found in the small book by Lewontin (1991) and a recent article of Commoner (2002), which outline the still little-appreciated limitations of the Human Genome Project. Consider also the rush for molecular data in many phylogenetic studies, which often enough tramples accurate perceptions we previously had from the organisms themselves (e.g. Fryer, 1996). Current attitudes to technology and information hark back to naive induction and its limitations. Perhaps the greatest tragedy here is that the ascendency of technology and its associated specialisation is likely to subvert real intellectual advances in many disciplines. This is not to suggest that technological specialities have no place in science. Rather, the appropriate use of such technology in scientific advance requires careful consideration and the appropriate integration of information. Although this may sound obvious to many, it is clearly lacking in some exceptionally well-funded projects (see Commoner, 2002; Lewontin, 1991).

To emphasise information but turn one's back on understanding, especially in the education of applied scientists of the future, is to condemn others to unenlightened prosecution of worn principles. When it comes to the application of understanding, society should be doing its collective utmost to shed the 'weight of unreliable knowledge' (Romesburg, 1981). To achieve maximum efficiency in solving pest-related problems, even if that be measured in economic terms, the ultimate goal should be solutions based on rational, scientific understanding. Applied scientists must therefore be educated in a way that will show them how to assess the quality of available information, how to integrate it, how to extend it reliably and how to apply it (see Romesburg, 1981). 'Applied' ecologists, like ecologists in general, should be both theoreticians and empiricists (Weiner, 1995). The rest of this chapter expands aspects of this final point, specifically in relation to improving IPM.

The practical solution of pest management problems

Practical solutions have been achieved in many pest management projects, but rates of successful intervention could undoubtedly be enhanced considerably. The earlier chapters provide background to a suite of scientific principles that should help in the development of a general approach to improving IPM outcomes. The scope of a general model has been spelled out by Dent (1991, 2000), with particular areas being expanded in Chapter 4. The suggestions made below about pest management relate mainly to those comments about improving pest management that appear in the general IPM literature. This direction has been chosen on purpose, mainly to illustrate the subtlety of some of the alternatives that are available to the development of a pest management framework.

Statements to IPM scientists and practitioners about improving IPM practice frequently draw responses that relate to issues of economy. In short: 'A fine suggestion, but funding is unlikely'. That may be the current state of affairs, but it need not be the future of IPM. Successful IPM is desirable, both environmentally and economically. If IPM can be improved, and the endless repetition of expensive, avoidable mistakes and unnecessary failures avoided, then 'getting it right' has to be a sound investment, even if considered solely in economic terms. We cannot afford to think any differently from this, and the rest of the chapter is written in that spirit. Somehow the legitimate place of ecologically sound pest management in the overall economy has to be established if society is going to benefit fully from the implementation of IPM, and institutional and other support will be needed. But these are different issues and they have been pursued by others (e.g. Carson, 1962; Dent, 2000; Pimentel, 1997; Pimentel *et al.*, 1980; Rees, 1993; Stoner *et al.*, 1986; Zalom, 1993).

1   Emphasis on specific understanding

IPM projects are complex: 'There is never a universal recipe, nor is there a definitive conclusion as the outcome is never the same, either in time or in space. In no case is there a definitive solution' (Labeyrie, 1988, p. 23). This statement implies that the unthinking repetition of successful techniques is likely to run into the problems typically encountered when dealing with complex systems, including changes to important variables, nonlinear interactions between variables, and cascades of influences.

Appreciation of the fact that IPM solutions will differ across commodities, and across localities for a single commodity, will save one from the fallacy of reliance upon the methods of successful case histories (see Chapter 8 and Hilborn & Ludwig, 1993). Concentrating on successes and trying to extrapolate general rules from 'how they worked' is an inadequate approach based on too many groundless assumptions; for example: (i) that pest systems and pest management interventions will inevitably 'work' in the same way; (ii) that other problem situations are not subject to novel problems, flaws or difficulties; (iii) that one can develop, in purely inductive fashion, an appreciation of the existence of general principles and an understanding of how they operate, simply by examining a number of cases, each of which, in itself, may not be fully understood; and (iv) that success in applied entomology may be achieved in a certain number of cases simply because so many attempts are made across the world. Although one can extract some generalities from such an approach, they are likely to represent minimum requirements, not underlying principles (Chapters 2 and 8). What is quite clear from the history of applied ecology is that successful advances in application have come from advances in fundamental understanding, such as that detailed by Sinclair & Solemdal (1988) in relation to fisheries in the late nineteenth century. It seems tedious that this lesson has to be learned countless times in connection with applied ecology. In IPM terms, why do we have to relive Lounsbury's 'ladybird fantasy' in various incarnations? Why does society and its institutions learn so poorly from previous mistakes? Before the most appropriate application of knowledge can be made, one requires at least a basic understanding of the situation, and particularly one needs to be aware of the subtleties that might undermine successful application. To repeat, 'applied solutions' that are truly scientific are based, first and foremost, on understanding. Before the most appropriate application of knowledge can be identified, in pest management for example, one requires an adequate understanding of several facets of the situation. This statement clearly does not advocate doing nothing to alleviate pest incursions before full understanding is achieved. That would be impractical and foolhardy; rather, 'the challenge in applied ecology is often to reach the best possible decisions on the basis of present information' (Allaby, 1995, p. 60; Newman, 1993, p. 2). Such necessities should not, however, be used as an excuse for turning application in ecology to professional 'ad hoc-ery' and unquestioned acceptance of current theories, a real danger given today's sociopolitical climate and its negative influences on education.

The alternative is to build interpretation and application around fundamental understanding, that is, to build the generalised approach from first principles. This is taking place in IPM (Chapter 4), but progress is slow. The problem with IPM, as pointed out above, is that its success is not easily demonstrated. By contrast, powered flight or spanning rivers with bridges demands a given outcome that can be readily judged by all, at least if aesthetics are disregarded. Shortcuts that undermine the success of the technology are thus readily detected. Practitioners of IPM have no such widely appreciated measures. Any intervention or technology can there-fore be claimed to have been successful (Chapter 4), as in several biological control programmes deemed to be 'landmark cases', but which could not have been successful in the way claimed (e.g. Clarke, 1990). This is a prob-lem that needs to be redressed in IPM.

2   IPM needs coordination of functions

The idea that there should be a formal model to reflect the general struc-ture of IPM is relatively recent (see Chapter 4) and it illustrates the need for various and distinct roles to be fulfilled if an IPM project is to suc-ceed. Although research and implementation are usually treated indepen-dently in the literature, the distinction between them seems frequently to be lost when the practical and financial details of a programme are de-cided. The entire suite of roles portrayed in Table 4.1 consequently falls easily and seemingly tacitly to researchers or to implementers, without any clear devolution of particular tasks. Lack of a clear general procedure at the start of IPM programmes seems to reflect a lack of direction or lead-ership within relevant governmental agencies and industry bodies. Con-sider the narrowly circumscribed role of the research entomologist in the insecticide industry. Entomologists screen potential toxins, which others may develop or locate, for their killing power against different species. Yet other entomologists, with expertise of particular crops and pests, are responsible for testing application rates under field conditions, and the development of application technology is yet a further distinct field. Compare the place of the research entomologist here with the diffuse role they have inherited with the development of IPM. The need to bring to-gether multidisciplinary teams has been frequently mentioned, but a gen-eral procedure for doing so in specific situations is apparently not available (Dent, 2000).

Perhaps the most major omission in the development and deployment of pest management lies in the lack of coordination across all aspects that impinge on the acquisition of knowledge and the design and implementa-tion of strategies (Dent, 1991, p. xv). In practice, most individuals currently charged with pest management programmes may have expertise in one or two areas represented in Table 4.1, but are seemingly expected to have the knowledge and vision to cover the other areas relevant to implementation. Alternatively, they may have to claim such abilities to gain funding. How such a coordination role is set up is likely to be crucial to the success of IPM. Dent (2000) is one of the few who have explicitly tackled this issue, but he did not favour the 'team leader' approach because 'the success or failure of the approach is totally dependent on the skills of the leader ... and even a good leader would have problems with understanding, assimilating and making use' of the range of contributions. This seems to be a difficulty that needs to be addressed directly rather than avoided, however. Without leadership the conjunction of specialist research and application cannot be achieved.

Judging from practice in other applied fields, this lack of a coordinating role is unusual, and it must lead to massive inefficiencies, wasted opportunities and poor economic outcomes. In engineering, for example, planning is conducted independently of executive functions, with both of these processes each having at least three subsidiary spheres of operation. Furthermore, the executive function is closely coordinated to ensure the planned product is delivered (e.g. Calvert, 1986). Although the extension of knowledge by research is seen as fundamental to improving methods in general, the research function is considered independent of planning and execution. Researchers have their task, and those who coordinate and apply the techniques are independent of the researchers, although their future activities are likely to be influenced by research results. Just who is responsible for translating research into practice is often an uncertain issue, and it seems that such a translating or development role should be a separate task altogether (Calvert, 1986).

Although civil engineering is unlikely to provide a very close analogy for IPM developments, the comparison is worth making, if only to emphasise the point that even when a product, a new building for instance, is readily definable and can have specifications clearly set, a coordinating role is a necessity. In pest management, where the product is far more abstract and difficult to specify, the coordination role is left out of the equation. Frequently researchers are expected to perform the coordination, linkage and communication. When this is the case, the development of adequate understanding is going to suffer, or the application will suffer, so this situation needs to be rectified. One area in which IPM differs substantially from civil engineering is in the research component, since each situation or problem in IPM will be unique to some extent. Reference to the IPM chain (Table 4.1) indicates that as far as the developmental side of research is concerned, a coordinator may prove as necessary as one for implementation. The 'discovery' side of research, even if it is aimed at enhancing IPM, is likely to be left more independent (see below).

In pest management, the analysts of problems, designers of control strategies and coordinators of implementation and research all require intellectual strengths and experience in several areas of endeavour (see Table 4.1). The solution to this problem cannot simply be furnished by the demand for interdisciplinary research, which is frequently stated or implied in the general IPM literature (e.g. Burn *et al.*, 1987; Dent, 1991, 2000). Although interdisciplinary research will be needed in almost all fields of endeavour covered in Table 4.1, specialist input and specialist research is going to be essential in various areas of the overall research agenda. And some of the most specialised research is likely to be interdisciplinary (e.g. pheromone research).

Effective coordination will rely on a sensitive understanding of the diversity of research requirements and the subtleties behind those requirements. Coordinators will have to be sufficiently well placed scientifically to appreciate the assumptions upon which a proposed study of pests or beneficials is to be based, and to identify the assumptions that underpin the general approach suggested for dealing with the problem, as well as those basic to the design of a specific control measure to be implemented. Perceptive participants will therefore prove invaluable, for coordinators will be hampered if the participants in each link of the chain are not on top of the diversity of theory and techniques within their area of responsibility (Petrie, 1976).

The above interpretation of the structure and execution of an IPM project raises several significant points, besides cost. Many of these have been recognised, and many implemented. But no programme seems to have attended effectively to all of the issues that are now raised; certainly, they have not been translated into a helpful and workable generalised scheme.

3   Desired end points and the direction of science

Solving problems through an understanding of the system in question, and based upon the application of scientific principles, places constraints on the participants. These constraints are not present when 'curiosity-driven' research is conducted. Specifically, a desired application or end product maintains the direction of the investigative science (Wigglesworth, 1955), to a considerable degree at least. This is frequently misunderstood to mean that the path of the research has to be dictated by that end point. Mixing intentions and the role of science in this way is unacceptable, but is nevertheless a common failure of modern management in science. Indeed, maintenance of the originality of scientists involved in extending knowledge is critical to success in IPM. Society needs funding mechanisms that harness the strongest aspects of scientific enquiry, diversity and difference of opinion, for excitement and competition are important motors in scientific advance, whether that be between individuals or collaborating groups (Chapter 2). In corollary, what is not needed is a mechanism that directs funding selectively to powerful groups (Weiner, 1995) or to the lowest bidder, the one who is prepared to make baseless promises of an immediate outcome.

Managers, if science needs them, should heed the warning of Fox (1994): 'Science is not fixed, rather it is a vital entity, basic or applied, and its lugubrious progress is more easily crippled by attempts to regularize it than not'. Freedom and resources for exploration are vital; indeed, exploration of what may appear to be side tracks may ultimately prove the only way forward, even when application is pre-eminent. Science meanders, simply because we still lack a reliable method or guaranteed algorithm for this process, so we have to make do with the fallible capacities of human thinking (Holton, 1993, p. 137). The real issue is what understanding is relevant, and how much is relevant? These difficult questions verge on the impossible, for continued scientific probing may well turn up understanding that could lead to better IPM. Universally effective biocontrol may yet be developed against tephritids, for example. How are we to know? Failure to understand or acknowledge the validity of this element to IPM, and science in general, seems to be a central one of the technological mind and currently popular views on running economies.

4   The place of reductionist science in IPM

Disillusion with 'reductionist science' is fairly widespread (Allaby, 1995, pp. 120–122), and is evident in the general IPM literature. It is even evident in the best synthetic IPM texts. For instance, Dent (1991, p. xv) suggests: '[W]e have lost direction and the holistic perspective of the subject in the rush to carry out specialist research and develop specific control techniques'. He does explain, though, that in-depth enquiry into the complexities and subtleties of insect biology and ecology is needed, but he still feels a danger exists in that the research can get so far removed from the original objectives that the final results are inapplicable (see also Newman, 1993, p. 2). Several influences have undoubtedly led to this state of affairs, including the perception that specialist research has usurped money to the detriment of development of usable products. It may also reflect the failure of ecological theory to deliver usable outcomes despite prodigious efforts. A vibrant autecological theory may well help to bolster the scientific basis of IPM and the useful role of reductionist science.

Frequently, applied sciences such as IPM are being identified as 'holistic' and therefore good, and thus independent of reductionist science (e.g. Tait, 1987). However, such a view centres on a naive caricature. One cannot achieve in other ways what reductionist science can achieve (and here 'reductionist science' is used non-pejoratively). Workable IPM systems are dependent on understanding, much of which can be derived only through reductionist research. It is a basic principle of IPM, or at least it should be, that one understand the components of the system.

'Reductionist science' was never set up to solve problems in which multiple causation is endemic (Hilborn & Stearns, 1982), as in systems involving the testing of pest management applications. Reductionist science cannot solve such problems, it can only make predictions about outcomes. And those predictions will always be based upon assumptions, which are, unfortunately, usually unstated in practice.

Ideally, one would like to set up large-scale experiments that are more 'systems oriented' to test outcomes (see Table 4.1), but that introduces special difficulties, ones that cannot be blamed on reductionist science. Kogan (1988) suggests that the inherent difficulties of very large-scale experiments in pest management may outweigh progress made towards achieving the intended objectives. He writes that 'the flow of irrelevant data can be overwhelming'. Obtaining sufficient replication from such an approach to be confident of identifying causative factors presents difficulties that also should not be overlooked, a point that may be more readily appreciated by those who have estimated, from preliminary sampling, the sample sizes needed in ecological fieldwork to reduce error to acceptable levels (e.g. Southwood, 1978, p. 21), in sampling pest densities for example. The question of how to deal with this difficulty remains, but systems models have been developed to explore the consequences of IPM actions (Dent, 1991; Tait, 1987). Here we also enter the sphere of 'scientific uncertainty' (Shrader-Frechette, 1996) and the need for scientists to be aware of the limits of their science so as not to create unrealistic expectations (Botsford &

Jain, 1992; Hilborn & Ludwig, 1993; Ludwig, 1994; Slobodkin, 1988). Not all issues can be dealt with in a scientifically rigorous way, and when attempts are made to 'raise management targets to the level of scientific hypotheses . . . we only confuse the issue and give managers a false sense of assuredness' (Drew, 1994).

In summary, 'reductionist science' is an integral and indispensable part of the IPM chain. Its role is to extend knowledge, and should not be conflated with other roles that are also essential to achieving a working end product. It should not be seen as a competitor to other essential parts of the chain, which share complementary roles with it.

5   IPM needs scientists as well as engineers

Scientists are increasingly seen as a fundamental cause of IPM not being developed or more widely deployed, mainly because it is they who do not participate in defining what are seen to be the real problems and constraints of the end users (e.g. Surtees, 1977). 'Natural scientists continue to show a strong distaste for becoming involved in the social and policy sciences; modellers are unwilling to engage in practical field experiments; economists have a tendency to view the scientific and political aspects of a problem as of lesser importance' (Tait, 1987, p. 203). Why should scientists be forced to participate as hinted at above, as a potential solution to implementing IPM effectively? The extension of knowledge (Chapter 2) is frequently conflated with the application of knowledge when the role of the scientist is considered, when in fact research scientists extend knowledge while engineers apply it.

The implementation of IPM requires dedicated 'environmental engineers' to do the task of coordinating, directing and managing the integration and application of knowledge (Dent, 1991, p. xv). For example, one could hardly expect an IPM programme to be developed by an entomologist conducting interdisciplinary research on pheromones and pheromonal function, just as an 'IPM engineer' could not be expected to launch an effective study of a particular pheromone system. This suggests that a more vocational type of education is needed for practitioners, as is available in other areas of applied science (e.g. engineering, medicine); at least, a satisfactory understanding of the overall structure of IPM and the diversity of roles is needed, so that each individual knows their place in the entire scheme, along the lines of Browning's (1998) 'general practitioner plant doctor'. Currently, conflation of several issues, as outlined in various parts of this book, clouds roles, confuses direction and inhibits progress. Successful IPM needs an active interface, probably even more than one (Table 4.1), between science and practice in the field.

The citrus industry, in different countries, has developed various successful solutions to this problem (e.g. Papacek & Smith, 1998). A major difficulty with IPM in industries that are not organised in some such way is that research at particular levels of the IPM chain tends to be dropped. It may be seen, for instance, as duplication, or even as unnecessary. In such cases, an analysis of the situation against the backdrop of the 'IPM chain' model proposed in Table 4.1 should help to organise ideas and to relate suggested means of achieving control to specific assumptions. In reality, scientists who have no model against which they can justify their views may well be pushed into the position of making promises that would succeed only by chance or serendipity. Usually, a rational analysis would reveal such weaknesses, but such analyses are seldom sought. In one sense Tait (1987) is correct when he writes of the 'unbalanced supremacy of the scientific paradigm', for science cannot solve certain problems associated with the application of technology and, indeed, has not made claims in this regard.

6   Selection of management techniques

No theoretical development can reveal exactly which means of population suppression is most likely to work against a particular pest, or which combination of techniques should be used. The few attempts to link ecological theory to control method are not compelling. Even if the suggestions are taken in the broadest possible way, they provide little insight. The frustration of having no workable scheme is compounded by the attempts that have been made yielding conflicting conclusions (Dent, 1991, p. 445). Whether an effective system is possible is still an open question. On the one hand, population dynamics theory in particular is seen to have good potential for predicting future problems (Cherrett, 1977) and choice of appropriate control strategies (Conway, 1981). On the other hand, autecological theory (Chapter 5) suggests that such refined levels of prediction are unlikely in ecology (Hengeveld & Walter, 1999), no matter how desirable.

Control decisions are more frequently influenced by economic considerations, with this approach simply eliminating relatively expensive options. An economic analysis has discounted sterile insect release against sheep blowflies in Australia (Spradbery, 1994), and biological control is frequently favoured against weeds on extensive areas of relatively low quality land. In this connection, the point of 'empowered users' conducting research, or even deciding on research priorities for others to carry out, warrants comment. Empowering of users is increasingly seen as important in the extension of IPM technology to growers. This approach has clearly achieved a great deal (e.g. Matteson *et al.*, 1994). However, discoveries relevant to the growers' understanding of and confidence with the technology, which form an integral part of the empowerment process, should not be confused with discoveries related to original scientific understanding and the experimentation and synthesis of knowledge that place the discovery in context. Growers may indeed make significant discoveries, but exploring the consequences is usually a very different issue.

Closing comments

Integrated pest management can be improved substantially if we can develop a structure for it that incorporates all relevant aspects realistically, and if the various roles within that structure are specified unambiguously. For particular IPM projects the targeted outcome has to be clearly and unambiguously specified. Research and development designed to contribute to the desired outcome has to be coordinated responsibly by an individual with the appropriate vision and breadth, to ensure that the entire suite of underlying assumptions is appreciated, and the necessary 'diversions' in research direction are appreciated for what they are. The tension between research and application needs to be resolved satisfactorily. Integrated pest management measures are readily undermined by lack of understanding, even in relatively well-studied systems (some of the reasons for this are discussed in Chapters 5–9). Application must go ahead, but the research needed to underpin effective IPM needs support, at all levels of the IPM chain.

References

Aarssen, L. W. (1997). On the progress of ecology. *Oikos* **80**, 177–178.

Abeeluck, D. & Walter, G. H. (1997). Mating behaviour of an undescribed species of *Coccophagus*, near *C. gurneyi* (Hymenoptera: Aphelinidae). *Journal of Hymenoptera Research* **6**, 92–98.

Abele, L. G., Simberloff, D. S., Strong, D. R. & Thistle, A. B. (1984). Preface. In *Ecological Communities: Conceptual Issues and the Evidence* (eds. D. R. Strong, D. Simberloff, L. G. Abele & A. B. Thistle), pp. vii–vix. Princeton University Press, Princeton.

Adams, M., Baverstock, P. R., Watts, C. H. S. & Reardon, T. (1987). Electrophoretic resolution of species boundaries in Australian Microchiroptera. I. *Eptesicus* (Chiroptera: Vespertilionidae). *Australian Journal of Biological Science* **40**, 143–162.

Adkisson, P. L. (1973). The principles, strategies and tactics of pest control in cotton. In *Insects: Studies in Population Management*, Memoir 1 (eds. P. W. Geier, L. R. Clark, D. J. Anderson & H. A. Nix), pp. 274–283. Ecological Society of Australia, Canberra.

Alam, M., Bennett, F. D. & Carl, K. P. (1971). Biological control of *Diatraea saccharalis* in Barbados by *Apanteles flavipes* and *Lixophaga diatraeae*. *Entomophaga* **16**, 151–158.

Aldrich, J. R., Numata, H., Borges, M., Bin, F., Waite, G. K. & Lusby, G. K. (1993). Artifacts and pheromone blends from *Nezara* spp. and other stink bugs (Heteroptera: Pentatomidae). *Zeitschrift für Naturforschung* **48**, 73–79.

Aldrich, J. R., Oliver, J. E., Lusby, W. R., Kochansky, J. P. & Lockwood, J. A. (1987). Pheromone strains of the cosmopolitan pest, *Nezara viridula* (Heteroptera: Pentatomidae). *Journal of Experimental Zoology* **244**, 171–175.

Allaby, M. (1995). *Facing the Future: The Case for Science*. Bloomsbury, London.

Allen, J. E., Burns, M. & Sargent, S. C. (1986). *Cataclysms on the Columbia: A Layman's Guide to the Features Produced by the Catastrophic Bretz Floods in the Pacific Northwest*. Timber Press, Portland.

Allen, W. A. & Rajotte, E. G. (1990). The changing role of extension entomology in the IPM era. *Annual Review of Entomology* 35, 379–397.

Allwood, A. J., Chinajariyawong, A., Drew, R. A. I. *et al*. (1999). Host plant records for fruit flies (Diptera: Tephritidae) in south east Asia. *Raffles Bulletin of Zoology* Supplement No. 7, 1–92.

Andrewartha, H. G. (1984). Ecology at the crossroads. *Australian Journal of Ecology* 9, 1–3.

Andrewartha, H. G. & Birch, L. C. (1954). *The Distribution and Abundance of Animals*. University of Chicago Press, Chicago.

Andrewartha, H. G. & Birch, L. C. (1984). *The Ecological Web: More on the Distribution and Abundance of Animals*. University of Chicago Press, Chicago.

Angus, R. B., Brown, R. E. & Bryant, L. J. (2000). Chromosomes and identification of the sibling species *Pterostichus nigrita* (Paykull) and *P. rhaeticus* Heer (Coleoptera: Carabidae). *Systematic Entomology* 25, 325–337.

Annecke, D. P. & Moran, V. C. (1977). Critical reviews of biological pest control in South Africa. 1. Karoo caterpillar, *Loxostege frustalis* Zeller (Lepidoptera: Pyralidae). *Journal of the Entomological Society of Southern Africa* 40, 127–145.

Annecke, D. P. & Mynhardt, M. J. (1979). On *Metaphycus stanleyi* Compere and two new species of *Metaphycus* Mercet from Africa (Hymenoptera: Encyrtidae). *Journal of the Entomological Society of Southern Africa* 42, 143–150.

Antolin, M. F. & Strong, D. R. (1987). Long-distance dispersal by a parasitoid (*Anagrus delicatus*, Mymaridae) and its host. *Oecologia* 73, 288–292.

Armstrong, K. F., Cameron, C. M. & Frampton, E. R. (1997). Fruit fly (Diptera: Tephritidae) species identification: a rapid molecular diagnostic technique for quarantine application. *Bulletin of Entomological Research* 87, 111–118.

Atanassova, P., Brookes, C. P., Loxdale, H. D. & Powell, W. (1998). Electrophoretic study of five aphid parasitoid species of the genus *Aphidius* (Hymenoptera: Braconidae), including evidence for reproductively isolated sympatric populations and a cryptic species. *Bulletin of Entomological Research* 88, 3–13.

Atkinson, W. D. (1985). Coexistence of Australian rainforest Diptera breeding in fallen fruit. *Journal of Animal Ecology* 54, 507–518.

Avise, J. C. (1994). *Molecular Markers, Natural History and Evolution*. Chapman & Hall, New York.

Ayala, F. J. (1982). *Population and Evolutionary Genetics: A Primer*. Benjamin/Cummings Publishing, Menlo Park.

Bacheler, J. S. (1995). Impact of boll weevil eradication on cotton production and insect management in Virginia and North Carolina, USA. In *Challenging the Future: Proceedings of the World Cotton Research Conference – 1, Brisbane, Australia, 1994* (eds. G. A. Constable & N. W. Forrester), pp. 405–410. CSIRO, Melbourne.

Baimai, V., Phinchongsakuldit, J., Sumrandee, C. & Tigvattananont, S. (2000). Cytological evidence for a complex of species within the taxon *Bactrocera tau* (Diptera: Tephritidae) in Thailand. *Biological Journal of the Linnean Society* **69**, 399–409.

Bajwa, W. I. & Kogan, M. (1996). *Compendium of IPM Definitions (CID): A Collection of IPM Definitions and their Citations in Worldwide IPM Literature*. Integrated Plant Protection Center (IPPC), Oregon State University, Corvallis. http://www.ippc.orst.edu/IPMdefinitions/.

Baker, P. S., Khan, A., Mohyuddin, A. I. & Waage, J. K. (1992). Overview of biological control of Lepidoptera in the Caribbean. *Florida Entomologist* **75**, 477–483.

Baker, T. C. (1993). Learning the language of insects – and how to talk back. *American Entomologist* **39**, 212–220.

Baltimore, D. (1978). Limiting science: a biologist's perspective. *Daedalus* **107**, 37–45.

Barbieri, M., Bavestrello, G. & Sara, M. (1995). Morphological and ecological differences in two electrophoretically detected species of *Cliona* (Porifera, Demospongiae). *Biological Journal of the Linnean Society* **54**, 193–200.

Barfield, C. S. & O'Neil, R. J. (1984). Is an ecological understanding a prerequisite for pest management? *Florida Entomologist* **67**, 42–47.

Barker, J. S. F. (1983). Interspecific competition. In *The Genetics and Biology of Drosophila*, vol. 3c (eds. M. Ashburner, H. L. Carson & J. N. Thompson), pp. 285–342. Academic Press, London.

Barlow, N. D., Moller, H. & Beggs, J. R. (1996). A model for the effect of *Sphecophaga vesparum vesparum* as a biological control agent of the common wasp in New Zealand. *Journal of Applied Ecology* **33**, 31–44.

Beard, J. J. & Walter, G. H. (2001). Host plant specificity in several species of generalist mite predators. *Ecological Entomology* **26**, 562–570.

Beavis, I. C. (1988). *Insects and Other Invertebrates in Classical Antiquity*. University of Exeter, Exeter.

Beeby, A. (1993). *Applying Ecology*. Chapman & Hall, London.

Beier, M. (1973). The early naturalists and anatomists during the Renaissance and seventeenth century. In *History of Entomology* (eds. R. F. Smith, T. E. Mittler & C. N. Smith), pp. 81–94. Annual Reviews, Palo Alto.

Benbrook, C. M., Groth, E., Halloran, J. M., Hansen, M. K. & Marquardt, S. (1996). *Pest Management at the Crossroads*. Consumers Union, New York.

Bennett, F. D. (1971). Current status of biological control of the small moth borers of sugar cane *Diatraea* spp. [Lep. Pyralidae]. *Entomophaga* 16, 111–124.

Berkov, A. (2002). The impact of redefined species limits in *Palame* (Coleoptera: Cerambycidae: Lamiinae: Acanthocinini) on assessments of host, seasonal, and stratum specificity. *Biological Journal of the Linnean Society* 76, 195–209.

Berlocher, S. H. & Feder, J. L. (2002). Sympatric speciation in phytophagous insects: moving beyond controversy? *Annual Review of Entomology* 47, 773–815.

Bernays, E. A. (1999). When host choice is a problem for a generalist herbivore: experiments with the whitefly, *Bemisia tabaci*. *Ecological Entomology* 24, 260–267.

Bernays, E. A. & Chapman, R. F. (1994). *Host Selection by Phytophagous Insects*. Chapman & Hall, New York.

Bernays, E. A. & Graham, M. (1988). On the evolution of host specificity in phytophagous arthropods. *Ecology* 69, 886–892.

Bernays, E. A. & Minkenberg, O. P. J. M. (1997). Insect herbivores: different reasons for being a generalist. *Ecology* 78, 1157–1169.

Bernays, E. A. & Wcislo, W. T. (1994). Sensory capabilities, information processing, and resource specialization. *Quarterly Review of Biology* 69, 187–204.

Besansky, N. J., Powell, J. R., Caccone, A., Hamm, D. M., Scott, J. A. & Collins, F. H. (1994). Molecular phylogeny of the *Anopheles gambiae* complex suggests genetic introgression between principal malaria vectors. *Proceedings of the National Academy of Sciences USA* 91, 6885–6888.

Bever, J. D., Schultz, P. A., Pringle, A. & Morton, J. B. (2001). Arbuscular mycorrhizal fungi: more diverse than meets the eye, and the ecological tale of why. *BioScience* 51, 923–931.

Bidochka, M. J., Kamp, A. M., Lavender, T. M., Dekoning, J. & De Croos, J. N. A. (2001). Habitat association in two genetic groups of the insect-pathogenic fungus *Metarhizium anisopliae*: uncovering cryptic species? *Applied and Environmental Microbiology* 67, 1335–1342.

Bilde, T. & Toft, S. (2001). The value of three cereal aphid species as food for a generalist predator. *Physiological Entomology* 26, 58–68.

Bjorkman, C. (2000). Interactive effects of host resistance and drought stress on the performance of a gall-making aphid living on Norway spruce. *Oecologia* **123**, 223–231.

Blackman, R. L. (1974). Life-cycle variation of *Myzus persicae* (Sulz.) (Hom., Aphididae) in different parts of the world, in relation to genotype and environment. *Bulletin of Entomological Research* **63**, 595–607.

Blackman, R. L. (2000). The cloning experts. *Antenna* **24**, 206–214.

Blouw, D. M. & Hagen, D. W. (1990). Breeding ecology and evidence of reproductive isolation of a widespread stickleback fish (Gasterosteidae) in Nova Scotia, Canada. *Biological Journal of the Linnean Society* **39**, 195–217.

Bocking, S. (1990). Stephen Forbes, Jacob Reighard, and the emergence of aquatic ecology in the Great Lakes Region. *Journal of the History of Biology* **23**, 461–498.

Booth, W. (1988). Revenge of the "nozzleheads". *Science* **239**, 135–137.

Borges, M., Jepson, P. C. & Howse, P. E. (1987). Long-range mate location and close-range courtship behaviour of the green stink bug, *Nezara viridula*, and its mediation by sex pheromones. *Entomologia Experimentalis et Applicata* **44**, 205–212.

Bossart, J. L. (1998). Genetic architecture of host use in a widely distributed, polyphagous butterfly (Lepidoptera: Papilionidae): adaptive inferences based on comparison of spatio-temporal populations. *Biological Journal of the Linnean Society* **165**, 279–300.

Botsford, L. W. & Jain, S. K. (1992). Population biology and its application to practical problems. In *Applied Population Biology* (eds. S. K. Jain & L. W. Botsford), pp. 1–24. Kluwer Academic Publishers, Dordrecht.

Bottrell, D. G. (1996). The research challenge for integrated pest management in developing countries: a perspective for rice in Southeast Asia. *Journal of Agricultural Entomology* **13**, 185–193.

Bouskila, A., Robertson, I. C., Robinson, M. E. *et al.* (1995). Submaximal oviposition rates in a Mymarid parasitoid: choosiness should not be ignored. *Ecology* **76**, 1990–1993.

Brady, R. H. (1982). Dogma and doubt. *Biological Journal of the Linnean Society* **17**, 79–96.

Brain, C. K. (1981). Hominid evolution and climatic change. *South African Journal of Science* **77**, 104–105.

Brézot, P., Malosse, C., Mori, K. & Renou, M. (1994). Bisabolene epoxides in sex pheromone in *Nezara viridula* (L.) (Heteroptera: Pentatomidae): role of *cis* isomer and relation to specificity of pheromone. *Journal of Chemical Ecology* **20**, 3133–3147.

Briejer, C. J. (1968). *Zilveren Sleiers en Verborgen Gevaren: Chemische Preparaten die het Leven Bedreigen*, 2nd edn. A. W. Sijthoff, Leiden.

Briese, D. T., Sheppard, A. W., Zwolfer, H. & Boldt, P. E. (1994). Structure of the phytophagous insect fauna of *Onopordum* thistles in the northern Mediterranean basin. *Biological Journal of the Linnean Society* **53**, 231–253.

Briggs, S. A. (1987). Rachel Carson: her vision and her legacy. In *Silent Spring Revisited* (eds. G. J. Marco, R. M. Hollingworth & W. Durham), pp. 3–11. American Chemical Society, Washington, DC.

Bronowski, J. (1978). *The Origins of Knowledge and Imagination.* Yale University Press, New Haven.

Brooks, H. (1978). The problem of research priorities. *Daedalus* **107**, 171–190.

Brown, A. W. A. (1961). The challenge of insecticide resistance. *Bulletin of the Entomological Society of America* **7**, 6–19.

Brown, J. K., Bird, J., Frohlich, D. R., Rosell, R. C., Bedford, I. D. & Markham, P. G. (1996). The relevance of variability within the *Bemisia tabaci* species complex to epidemics caused by subgroup III geminiviruses. In *Bemisia 1995: Taxonomy, Biology, Damage, Control and Management* (eds. D. Gerling & R. T. Mayer), pp. 77–89. Intercept, Andover.

Brown, J. K., Frohlich, D. R. & Rosell, R. C. (1995). The sweetpotato or silverleaf whiteflies: biotypes of *Bemisia tabaci* or a species complex? *Annual Review of Entomology* **40**, 511–534.

Browning, H. W. & Melton, C. W. (1987). Indigenous and exotic Trichogrammatids (Hymenoptera: Trichogrammatidae) evaluated for biological control of *Eoreuma loftini* and *Diatraea saccharalis* (Lepidoptera: Pyralidae) borers on sugarcane. *Environmental Entomology* **16**, 360–364.

Browning, J. A. (1998). One phytopathologist's growth through IPM to holistic plant health: the key to approaching genetic yield potential. *Annual Review of Phytopathology* **36**, 1–24.

Brush, S. G. (1974). Should the history of science be rated X? *Science* **183**, 1164–1172.

Burn, A. J., Coaker, T. H. & Jepson, P. C. (1987). *Integrated Pest Management.* Academic Press, London.

Bush, G. L. (1975). Sympatric speciation in phytophagous parasitic insects. In *Evolutionary Strategies of Parasitic Insects and Mites* (ed. P. W. Price), pp. 187–206. Plenum Press, New York.

Bush, G. L. (1992). Host race formation and sympatric speciation in *Rhagoletis* fruit flies (Diptera: Tephritidae). *Psyche* **99**, 335–358.

Bush, G. L. (1993). A reaffirmation of Santa Rosalia, or why are there so many kinds of *small* animals? In *Evolutionary Patterns and Processes*, vol. 14. *Linnean Society Symposium Series* (eds. D. R. Lees & D. Edwards), pp. 229–249. Academic Press, London.

Bush, G. L. (1994). Sympatric speciation in animals: new wine in old bottles. *Trends in Ecology and Evolution* **9**, 285–287.

Calkins, C. O., Klassen, W. & Liedo, P. (1994). Preface. In *Fruit Flies and the Sterile Insect Technique* (eds. C. O. Calkins, W. Klassen & P. Liedo), pp. iii–iv. CRC Press, Boca Raton.

Caltagirone, L. E. (1981). Landmark examples in classical biological control. *Annual Review of Entomology* **26**, 213–232.

Caltagirone, L. E. & Doutt, R. L. (1989). The history of the vedalia beetle importation to California and its impact on the development of biological control. *Annual Review of Entomology* **34**, 1–16.

Calvert, R. E. (1986). *Introduction to Building Management*. Butterworth, London.

Campbell, N. A. & Mahon, R. J. (1974). A multivariate study of variation in two species of rock crab of the genus *Leptograpsus*. *Australian Journal of Zoology* **22**, 417–425.

Campbell, N. A., Reece, J. B. & Mitchell, L. G. (1999). *Biology*, 5th edn. Benjamin/Cummings Publishing, Menlo Park.

Carey, J. R. (1991). Establishment of the Mediterranean fruit fly in California. *Science* **253**, 1369–1373.

Carey, J. R. (1996). The incipient Mediterranean fruit fly population in California: implications for invasion biology. *Ecology* **77**, 1690–1697.

Carr, E. H. (1987). *What is History?* Penguin, London.

Carriere, Y. (1992). Host plant exploitation within a population of a generalist herbivore, *Choristoneura rosaceana*. *Entomologia Experimentalis et Applicata* **65**, 1–10.

Carriere, Y., Dennehy, T. J., Pedersen, B. *et al.* (2001). Large-scale management of insect resistance to transgenic cotton in Arizona: can transgenic insecticidal crops be sustained? *Journal of Economic Entomology* **94**, 315–325.

Carriere, Y. & Roitberg, B. D. (1994). Trade-offs in responses to host plants within a population of a generalist herbivore, *Choristoneura rosaceana*. *Entomologia Experimentalis et Applicata* **72**, 173–180.

Carroll, D. P. & Hoyt, S. C. (1986). Hosts and habitats of parasitoids (Hymenoptera: Aphidiidae) implicated in biological control of apple aphid (Homoptera: Aphididae). *Environmental Entomology* **15**, 1171–1178.

Carson, H. L. (1989). Evolution: the pattern or the process. *Science* **245**, 872–873.

Carson, R. (1962). *Silent Spring*. Penguin, Harmondsworth.

Casagrande, R. A. (1987). The Colorado potato beetle: 125 years of mismanagement. *Bulletin of the Entomological Society of America* **33**, 142–150.

Cate, J. R. (1990). Biological control of pests and diseases: integrating a diverse heritage. In *New Directions in Biological Control: Alternatives for Suppressing Agricultural Pests and Diseases* (eds. R. R. Baker & P. E. Dunn), pp. 23–43. Alan R. Liss, New York.

Cates, R. G. (1981). Host plant predictability and the feeding patterns of monophagous, oligophagous, and polyphagous insect herbivores. *Oecologia* 48, 319–326.

Chalmers, A. F. (1999). *What is this Thing called Science? An Assessment of the Nature and Status of Science and its Methods*, 3rd edn. University of Queensland Press, Brisbane.

Chamberlin, T. C. (1897). Studies for Students: The method of multiple working hypotheses. *Journal of Geology* 5, 837–848.

Chandrasekhar, S. (1990). Science and scientific attitudes. *Nature* 344, 285–286.

Cherrett, J. M. (1977). Preface. In *Origins of Pest, Parasite, Disease and Weed Problems* (eds. J. M. Cherrett & G. R. Sagar), pp. ix–x. Blackwell Scientific Publications, Oxford.

Chitty, D. (1996). *Do Lemmings Commit Suicide? Beautiful Hypotheses and Ugly Facts*. Oxford University Press, New York.

Cittadino, E. (1990). *Nature as the Laboratory: Darwinian Plant Ecology in the German Empire, 1880–1900*. Cambridge University Press, Cambridge.

Clanchy, J. & Ballard, B. (1991). *Essay Writing for Students: A Practical Guide*, 2nd edn. Longman Cheshire, Melbourne.

Claridge, M. F. & de Vrijer, P. W. F. (1994). Reproductive behavior: the role of acoustic signals in species recognition and speciation. In *Planthoppers: Their Ecology and Management* (eds. R. F. Denno & T. J. Perfect), pp. 216–233. Chapman & Hall, New York.

Claridge, M. F., den Hollander, J. & Morgan, J. C. (1985). The status of weed-associated populations of the brown planthopper, *Nilaparvata lugens* (Stål) – host race or biological species? *Zoological Journal of the Linnean Society* 84, 77–90.

Claridge, M. F. & Nixon, G. A. (1986). *Oncopsis flavicollis* (L.) associated with tree birches (*Betula*): a complex of biological species or a host plant utilization polymorphism? *Biological Journal of the Linnean Society* 27, 381–397.

Clark, L. R., Geier, P. W., Hughes, R. D. & Morris, R. F. (1967). *The Ecology of Insect Populations in Theory and Practice*. Methuen, London.

Clarke, A. R. (1990). The control of *Nezara viridula* (L.) with introduced egg parasitoids in Australia: a review of a "landmark" example of classical biological control. *Australian Journal of Agricultural Research* 41, 1127–1146.

Clarke, A. R. (1992). Current distribution and pest status of *Nezara viridula* (L.) (Hemiptera: Pentatomidae) in Australia. *Journal of the Australian Entomological Society* 31, 289–297.

Clarke, A. R. (1993). A new *Trissolcus* Ashmead species (Hymenoptera: Scelionidae) from Pakistan: species description and its role as a biological control agent. *Bulletin of Entomological Research* 83, 523–527.

Clarke, A. R. (1995). Integrated pest management in forestry: some difficulties in pursuing the holy-grail. *Australian Forestry* **58**, 147–150.

Clarke, A. R., Allwood, A., Chinajariyawong, A. *et al.* (2001). Seasonal abundance and host use patterns of seven *Bactrocera* Marquart species (Diptera: Tephritidae) in Thailand and peninsular Malaysia. *Raffles Bulletin of Zoology* **49**, 207–220.

Clarke, A. R., Paterson, S. & Pennington, P. (1998). *Gonipterus scutellatus* Gyllenhal (Coleoptera: Curculionidae) oviposition on seven naturally co-occurring *Eucalyptus* species. *Forest Ecology and Management* **110**, 89–99.

Clarke, A. R. & Walter, G. H. (1993a). Biological control of the green vegetable bug *Nezara viridula* (L.) in eastern Australia: current status and perspectives. In *Pest Control and Sustainable Agriculture* (eds. S. A. Corey, D. J. Dall & W. M. Milne), pp. 223–225. CSIRO, Melbourne.

Clarke, A. R. & Walter, G. H. (1993b). Variegated thistle (*Silybum marianum* (L.)), a non-crop host plant of *Nezara viridula* (L.) (Heteroptera: Pentatomidae) in southeastern Queensland. *Journal of the Australian Entomological Society* **32**, 81–83.

Clarke, A. R. & Walter, G. H. (1995). "Strains" and the classical biological control of insect pests. *Canadian Journal of Zoology* **73**, 1777–1790.

Clarke, A. R., Zalucki, M. P., Madden, J. L., Patel, V. S. & Paterson, S. C. (1997). Local dispersal of the *Eucalyptus* leaf-beetle *Chrysophtharta bimaculata* (Coleoptera: Chrysomelidae), and implications for forest protection. *Journal of Applied Ecology* **34**, 807–816.

Clausen, C. P. (1936). Insect parasitism and biological control. *Annals of the Entomological Society of America* **29**, 201–223.

Clements, F. E. (1916). *Plant Succession: An Analysis of the Development of Vegetation.* Publication No. 242. Carnegie Institution of Washington, Washington, DC.

Coetzee, M. (1989). Comparative morphology and multivariate analysis for the discrimination of four members of the *Anopheles gambiae* group in southern Africa. *Mosquito Systematics* **21**, 100–116.

Coll, M. & Bottrell, D. G. (1994). Effects of nonhost plants on an insect herbivore in diverse habitats. *Ecology* **75**, 723–731.

Collier, T. R., Murdoch, W. W. & Nisbet, R. M. (1994). Egg load and the decision to host-feed in the parasitoid, *Aphytis melinus*. *Journal of Animal Ecology* **63**, 299–306.

Collins, F. H. & Paskewitz, S. M. (1996). A review of the use of ribosomal DNA (rDNA) to differentiate among cryptic *Anopheles* species. *Insect Molecular Biology* **5**, 1–9.

Common, I. F. B. (1953). The Australian species of *Heliothis* (Lepidoptera: Noctuidae) and their pest status. *Australian Journal of Zoology* **1**, 319–344.

Commoner, B. (2002). Unravelling the DNA myth. The spurious foundation of genetic engineering. *Harper's Magazine* **304**, 39–47.

Compere, H. (1961). The red scale and its insect enemies. *Hilgardia* **31**, 173–278.

Condon, M. A. & Steck, G. J. (1997). Evolution of host use in fruit flies of the genus *Blepharoneura* (Diptera: Tephritidae): cryptic species on sexually dimorphic host plants. *Biological Journal of the Linnean Society* **60**, 443–466.

Connell, J. H. (1980). Diversity and the coevolution of competitors, or the ghost of competition past. *Oikos* **35**, 131–138.

Connell, J. H. (1983). On the prevalence and relative importance of interspecific competition: evidence from field experiments. *American Naturalist* **122**, 661–696.

Conway, G. (1981). Man versus pests. In *Theoretical Ecology: Principles and Applications* (ed. R. M. May), pp. 356–386. Blackwell Scientific Publications, Oxford.

Conway, G. R. (1987). The properties of agroecosystems. *Agricultural Systems* **24**, 95–117.

Cook, J. M. (1996). A beginners' guide to molecular markers for entomologists. *Antenna* **20**, 53–62.

Coope, G. R. (1978). Constancy of insect species versus inconstancy of Quaternary environments. In *Diversity of Insect Faunas* (eds. L. A. Mound & N. Waloff), pp. 176–186. Blackwell Scientific Publications, Oxford.

Coope, G. R. (1986). The invasion and colonisation of the North Atlantic islands: a palaeoecological solution to a biogeographic problem. *Philosophical Transactions of the Royal Society of London, Series B* **314**, 619–635.

Coope, G. R. (1987). The response of late Quaternary insect communities to sudden climatic changes. In *Organization of Communities Past and Present* (eds. J. H. R. Gee & P. S. Giller), pp. 421–438. Blackwell Scientific Publications, Oxford.

Corbett, A. & Rosenheim, J. A. (1996). Impact of a natural enemy overwintering refuge and its interaction with the surrounding landscape. *Ecological Entomology* **21**, 155–164.

Corrigan, P. J. & Seneviratna, P. (1990). Occurrence of organochlorine residues in Australian meat. *Australian Veterinary Journal* **67**, 56–58.

Cottam, C. & Higgins, E. (1946). DDT and its effect on fish and wildlife. *Journal of Economic Entomology* **39**, 44–52.

Coulson, G. (1990). Habitat separation in the gray kangaroos, *Macropus giganteus* Shaw and *Macropus fuliginosus* (Desmarest) (Marsupialia: Macropodidae), in Grampians National Park, western Victoria (Australia). *Australian Mammalogy* **13**, 33–40.

Courtney, S. P. & Kibota, T. T. (1990). Mother doesn't know best: selection of hosts by ovipositing insects. In *Insect/Plant Interactions*, vol. 2 (ed. E. A. Bernays), pp. 161–188. CRC Press, Boca Raton.

Cowie, R. H. (2001). Can snails ever be effective and safe biocontrol agents? *International Journal of Pest Management* **47**, 23–40.

Cox, P. A. & Knox, R. B. (1988). Pollination postulates and two-dimensional pollination in hydrophilous monocotyledons. *Annals of the Missouri Botanical Garden* **75**, 811–818.

Coyne, J. A. (1992). Genetics and speciation. *Nature* **355**, 511–515.

Coyne, J. A. (1993). Recognizing species. *Nature* **364**, 298.

Crawley, M. J. (1986). The population biology of invaders. *Philosophical Transactions of the Royal Society of London, Series B* **314**, 711–731.

Croft, B. A. (1983). Introduction. In *Integrated Management of Insect Pests of Pome and Stone Fruits* (eds. B. A. Croft & S. C. Hoyt), pp. 1–18. Wiley-Interscience, New York.

Croft, B. A. & Hoyt, S. C. (eds.) (1983). *Integrated Management of Insect Pests of Pome and Stone Fruits*. Wiley-Interscience, New York.

Cronin, J. T. & Strong, D. R. (1993). Substantially submaximal oviposition rates by a mymarid egg parasitoid in the laboratory and field. *Ecology* **74**, 1813–1825.

Cronin, J. T. & Strong, D. R. (1994). Parasitoid interactions and their contribution to the stabilization of Auchenorrhyncha populations. In *Planthoppers: Their Ecology and Management* (eds. R. F. Denno & T. J. Perfect), pp. 400–428. Chapman & Hall, New York.

Crozier, R. H. (1975). *Animal Cytogenetics. 3. Insecta. 7. Hymenoptera*. Gebrüder Borntraeger, Berlin.

Cutkomp, L. K., Peterson, A. G. & Hunter, P. E. (1958). DDT-resistance of the Colorado potato beetle. *Journal of Economic Entomology* **51**, 828–831.

Darwin, C. (1859). *On the Origin of Species by means of Natural Selection, or the Preservation of Favoured Races in the Struggle for Life. Facsimile reprint, 1964*, Harvard University Press, Cambridge, Mass. John Murray, London.

Darwin, C. (1958). *Autobiography of Charles Darwin, 1809–1882: with original omissions restored/edited with appendix and notes by his grand-daughter, Nora Barlow*. Collins, London.

David, J. (1993). Criticism of the standard $r-K$ dichotomy, with special reference to life histories of Diplopoda. *Acta Œcologica* **14**, 129–139.

Davis, D., Pellmyr, O. & Thompson, J. N. (1992). Biology and systematics of *Greya* Busck and *Tetragma*, new genus (Lepidoptera: Prodoxidae). *Smithsonian Contributions to Zoology* **524**, 1–88.

Davis, M. A. (1981). The flight capacity of dispersing milkweed beetles, *Tetraopes tetraophthalmus*. *Annals of the Entomological Society of America* **74**, 385–386.

Davis, M. A. (1984). The flight and migration ecology of the red milkweed beetle (*Tetraopes tetraophthalmus*). *Ecology* **65**, 230–234.

Davis, M. A. (1986). Geographic patterns in the flight ability of a monophagous beetle. *Oecologia* **69**, 407–412.

Dawkins, R. (1986). *The Blind Watchmaker*. Penguin, London.

De Barro, P. J. & Hart, P. J. (2000). Mating interactions between two biotypes of the whitefly, *Bemisia tabaci* (Hemiptera: Aleyrodidae) in Australia. *Bulletin of Entomological Research* **90**, 103–112.

de Souza, K., Holt, J. & Colvin, J. (1995). Diapause, migration and pyrethroid-resistance dynamics in the cotton bollworm, *Helicoverpa armigera* (Lepidoptera: Noctuidae). *Ecological Entomology* **20**, 333–342.

De Winter, A. J. (1995). Genetic control and evolution of acoustic signals in planthoppers (Homoptera: Delphacidae). *Researches in Population Ecology* **37**, 99–104.

De Winter, A. J. & Rollenhagen, T. (1990). The importance of male and female acoustic behaviour for reproductive isolation in *Ribautodelphax* planthoppers (Homoptera: Delphacidae). *Biological Journal of the Linnean Society* **40**, 191–206.

DeBach, P. (1951). The necessity for an ecological approach to pest control on citrus in California. *Journal of Economic Entomology* **44**, 443–447.

DeBach, P. (1969). Uniparental, sibling and semi-species in relation to taxonomy and biological control. *Israel Journal of Entomology* **4**, 11–28.

DeBach, P. (1974). *Biological Control by Natural Enemies*. Cambridge University Press, London.

DeBach, P., Fisher, T. W. & Landi, J. (1955). Some effects of meteorological factors on all stages of *Aphytis lingnanensis*, a parasite of the California red scale. *Ecology* **36**, 743–751.

DeBach, P. & Rosen, D. (1991). *Biological Control by Natural Enemies*. Cambridge University Press, Cambridge.

DeLong, D. M. (1934). What shall be the objective in the training of an entomologist. *Journal of Economic Entomology* **27**, 53–58.

Delucchi, V., Rosen, D. & Schlinger, E. I. (1976). Relationship of systematics to biological control. In *Theory and Practice of Biological Control* (eds. C. B. Huffaker & P. S. Messenger), pp. 81–91. Academic Press, New York.

Dempster, J. P. (1983). The natural control of populations of butterflies and moths. *Biological Reviews* **58**, 461–481.

den Bieman, C. F. M. (1987). Host plant relations in the planthopper genus *Ribautodelphax* (Homoptera, Delphacidae). *Ecological Entomology* **12**, 163–172.

den Boer, P. J. (1970). On the significance of dispersal power for populations of Carabid-beetles (Coleoptera, Carabidae). *Oecologia* **4**, 1–28.

den Boer, P. J. (1979). Some remarks in retrospect. In *On the Evolution of Behaviour in Carabid Beetles. Miscellaneous Papers 18 (1979), Agricultural University of Wageningen, The Netherlands* (eds. P. J. den Boer, H. U. Thiele & F. Weber), pp. 213–222. H. Veenman and B. V. Zonen, Wageningen/Zoological Institute of the University of Cologne.

den Boer, P. J. (1990a). Reaction to J. Latto and C. Bernstein: regulation in natural insect populations: reality or illusion? *Acta Œcologica* **11**, 131–133.

den Boer, P. J. (1990b). The survival value of dispersal in terrestrial arthropods. *Biological Conservation* **54**, 175–192.

den Boer, P. J. & Reddingius, J. (1996). *Regulation and Stabilization Paradigms in Population Ecology*. Chapman & Hall, London.

den Hollander, J. (1995). Acoustic signals as Specific-Mate Recognition Signals in leafhoppers (Cicadellidae) and planthoppers (Delphacidae) (Homoptera: Auchenorrhyncha). In *Speciation and the Recognition Concept: Theory and Application* (eds. D. M. Lambert & H. G. Spencer), pp. 440–463. Johns Hopkins University Press, Baltimore.

Denno, R. F. & Perfect, T. J. (1994). *Planthoppers: Their Ecology and Management*. Plenum Press, New York.

Denno, R. F., Schauff, M. E., Wilson, S. W. & Olmstead, K. L. (1987). Practical diagnosis and natural history of two sibling salt marsh-inhabiting planthoppers in the genus *Prokelisia* (Homoptera: Delphacidae). *Proceedings of the Entomological Society of Washington* **89**, 687–700.

Dent, D. (1991). *Insect Pest Management*. CAB International, Wallingford.

Dent, D. R. (1997). *Methods in Ecological and Agricultural Entomology*. CAB International, Wallingford.

Dent, D. (2000). *Insect Pest Management*, 2nd edn. CAB International, Wallingford.

Dingle, H. (1996). *Migration: The Biology of Life on the Move*. Oxford University Press, New York.

Dobzhansky, T. (1951). *Genetics and the Origin of Species*. Columbia University Press, New York.

Dobzhansky, T. (1970). *Genetics of the Evolutionary Process*. Columbia University Press, New York.

Donaldson, J. S. & Walter, G. H. (1988). Effects of egg availability and egg maturity on the ovipositional activity of the parasitic wasp, *Coccophagus atratus*. *Physiological Entomology* **13**, 407–417.

Doutt, R. L. (1958). Vice, virtue and the vedalia. *Bulletin of the Entomological Society of America* **4**, 119–123.

Doutt, R. L. & Nakata, J. (1973). The *Rubus* leafhopper and its egg parasitoid: an endemic biotic system useful in grape-pest management. *Environmental Entomology* **2**, 381–386.

Downs, E. W. & Lemmer, G. F. (1965). Origins of aerial crop dusting. *Agricultural History* **39**, 123–135.

Drake, V. A., Gatehouse, A. G. & Farrow, R. A. (1995). Insect migration: a holistic conceptual model. In *Insect Migration* (eds. V. A. Drake & A. G. Gatehouse), pp. 427–457. Cambridge University Press, Cambridge.

Drès, M. & Mallet, J. (2002). Host races in plant-feeding insects and their importance in sympatric speciation. *Philosophical Transactions of the Royal Society of London, Series B* **357**, 471–492.

Drew, G. S. (1994). The scientific method revisited. *Conservation Biology* **8**, 596–597.

Drew, R. A. I. & Hancock, D. L. (1994). The *Bactrocera dorsalis* complex of fruit flies (Diptera: Tephritidae: Dacinae) in Asia. *Bulletin of Entomological Research* Supplement No. 2, 1–68.

Dubos, R. J. (1951). *Louis Pasteur: Free Lance of Science*. Victor Gollancz, London.

Dubos, R. (1988). *Pasteur and Modern Science*. Science Tech Publishers/Springer-Verlag, Berlin.

Duffy, J. E. (1996). Species boundaries, specialization, and the radiation of sponge-dwelling alpheid shrimp. *Biological Journal of the Linnean Society* **58**, 307–324.

Dunlap, T. R. (1981). *DDT: Scientists, Citizens, and Public Policy*. Princeton University Press, Princeton.

Dusenbery, D. B. (1992). *Sensory Ecology: How Organisms Acquire and Respond to Information*. W. H. Freeman, New York.

Dyer, L. A. & Floyd, T. (1993). Determinants of predation on phytophagous insects: the importance of diet breadth. *Oecologia* **96**, 575–582.

Ehler, L. E. (1990). Introduction strategies in biological control of insects. In *Critical Issues in Biological Control* (eds. M. Mackauer, L. E. Ehler & J. Roland), pp. 111–134. Intercept, Andover.

Ehler, L. E. (1991). Planned introductions in biological control. In *Assessing Ecological Risks of Biotechnology* (ed. L. R. Ginzburg), pp. 21–39. Butterworth-Heinemann, Boston.

Ehler, L. E. (1994). Parasitoid communities, parasitoid guilds, and biological control. In *Parasitoid Community Ecology* (eds. B. A. Hawkins & W. Sheehan), pp. 418–436. Oxford University Press, Oxford.

Ehler, L. E. & Bottrell, D. G. (2000). The illusion of Integrated Pest Management. *Issues in Science and Technology* **16**, 61–64.

Ehler, L. E. & Hall, R. W. (1982). Evidence for competitive exclusion of introduced natural enemies in biological control. *Environmental Entomology* **11**, 1–4.

Ehrlich, P. R. & Raven, P. H. (1964). Butterflies and plants: a study in coevolution. *Evolution* **18**, 586–608.

Eldredge, N. (1985). *Unfinished Synthesis*. Oxford University Press, New York.

Ellegård, A. (1958). *Darwin and the General Reader: The Reception of Darwin's Theory of Evolution in the British Periodical Press, 1859–1872.* Goteborg University Press, Goteborg.

Emmet, E. R. (1968). *Learning to Philosophize.* Penguin, Harmondsworth.

Farrow, R. A. (1981). Aerial dispersal of *Scelio fulgidus* (Hym.: Scelionidae), parasite of eggs of locusts and grasshoppers (Ort.: Acrididae). *Entomophaga* **26**, 349–355.

Feder, J. L., Berlocher, S. H. & Opp, S. B. (1998). Sympatric host-race formation and speciation in *Rhagoletis* (Diptera: Tephritidae): a tale of two species for Charles D. In *Genetic Structure and Local Adaptation in Natural Insect Populations: Effects of Ecology, Life History, and Behavior* (eds. S. Mopper & S. Y. Strauss), pp. 408–441. Chapman & Hall, New York.

Feder, J. L., Chilcote, C. A. & Bush, G. L. (1988). Genetic differentiation between sympatric host races of the apple maggot fly *Rhagoletis pomonella. Nature* **336**, 61–64.

Fernando, L. C. P. (1993). *Ovarian status, activity patterns and ecology of Aphytis lingnanensis Compere in Queensland citrus.* Unpublished PhD thesis. University of Queensland.

Fernando, L. C. P. & Walter, G. H. (1997). Species status of two host-associated populations of *Aphytis lingnanensis* (Hymenoptera: Aphelinidae) in citrus. *Bulletin of Entomological Research* **87**, 137–144.

Fernando, L. C. P. & Walter, G. H. (1999). Activity patterns and oviposition rates of *Aphytis lingnanensis* females, a parasitoid of California red scale *Aonidiella aurantii*: implications for successful biological control. *Ecological Entomology* **24**, 416–425.

Ferro, D. N. (1987). Insect pest outbreaks in agroecosystems. In *Insect Outbreaks* (eds. P. Barbosa & J. C. Schultz), pp. 195–215. Academic Press, London.

Feyerabend, P. (1970). Consolations for the specialist. In *Criticism and the Growth of Knowledge* (eds. I. Lakatos & A. Musgrave), pp. 197–230. Cambridge University Press, Cambridge.

Fitt, G. P. (1986a). The influence of a shortage of hosts on the specificity of oviposition behaviour in species of *Dacus* (Diptera, Tephritidae). *Physiological Entomology* **11**, 133–143.

Fitt, G. P. (1986b). The roles of adult and larval specialisations in limiting the occurrence of five species of *Dacus* (Diptera: Tephritidae) in cultivated fruits. *Oecologia* **69**, 101–109.

Fitt, G. P. (1989). The ecology of *Heliothis* species in relation to agroecosystems. *Annual Review of Entomology* **34**, 17–52.

Fitt, G. P. (1991). Host selection in the Heliothinae. In *Reproductive Behaviour of Insects* (eds. W. J. Bailey & J. Ridsdill-Smith), pp. 172–201. Chapman & Hall, London.

Flanders, S. E. (1953). Variations in susceptibility of citrus-infesting Coccids to parasitization. *Journal of Economic Entomology* **46**, 266–269.

Flint, M. L. & van den Bosch, R. (1981). *Introduction to Integrated Pest Management*. Plenum Press, New York.

Foelix, R. F. (1996). *Biology of Spiders*, 2nd edn. Oxford University Press, Oxford.

Follett, P. A. & Roderick, G. K. (1996). Genetic estimates of dispersal ability in the leucaena psyllid predator *Curinus coeruleus* (Coleoptera: Coccinellidae): implications for biological control. *Bulletin of Entomological Research* **86**, 355–361.

Foottit, R. G. (1997). Recognition of parthenogenetic insect species. In *Species: The Units of Biodiversity* (eds. M. F. Claridge, H. A. Dawah & M. R. Wilson), pp. 291–307. Chapman & Hall, London.

Forbes, S. A. (1887). The lake as a microcosm. *Bulletin of the Science Association of Peoria, Illinois* [1887], 77–87. Reprinted in Real, L. A. & Brown, J. H. (1991). *Foundations of Ecology: Classic Papers with Commentaries*, pp. 14–27. University of Chicago Press, Chicago.

Ford, J. (1974). Concepts of subspecies and hybrid zones, and their application in Australian ornithology. *Emu* **74**, 113–123.

Ford, M. J. (1982). *The Changing Climate*. George Allen and Unwin, London.

Forno, I. W. & Harley, K. L. S. (1979). The occurrence of *Salvinia molesta* in Brazil. *Aquatic Botany* **6**, 185–187.

Forrester, N. W., Cahill, M., Bird, L. J. & Layland, J. K. (1993). Management of pyrethroid and endosulfan resistance in *Helicoverpa armigera* (Lepidoptera: Noctuidae) in Australia. *Bulletin of Entomological Research – Supplement Series* **1**, i–132.

Fox, C. H. (1994). If it ain't fixed, don't break it ... *Nature* **369**, 602.

Fox, L. R. & Morrow, P. A. (1981). Specialization: species property or local phenomenon? *Science* **211**, 887–893.

Fraenkel, G. S. (1959). The raison d'etre of secondary plant substances. *Science* **129**, 1466–1470.

Fryer, G. (1996). Reflections on arthropod evolution. *Biological Journal of the Linnean Society* **58**, 1–55.

Futuyma, D. J. (1987). On the role of species in anagenesis. *American Naturalist* **130**, 465–473.

Futuyma, D. J. (1991). Evolution of host specificity in herbivorous insects: genetic, ecological, and phylogenetic aspects. In *Plant–Animal Interactions: Evolutionary Ecology in Tropical and Temperate Regions* (eds. P. W. Price, T. M. Lewinsohn, G. W. Fernandes & W. W. Benson), pp. 431–454. John Wiley, New York.

Futuyma, D. J., Keese, M. C. & Funk, D. J. (1995). Genetic constraints on macroevolution – the evolution of host affiliation in the leaf beetle genus *Ophraella*. *Evolution* **49**, 797–809.

Futuyma, D. J., Keese, M. C. & Scheffer, S. J. (1993). Genetic constraints and the phylogeny of insect–plant associations: responses of *Ophraella communa* (Coleoptera: Chrysomelidae) to host plants of its congeners. *Evolution* **47**, 888–905.

Futuyma, D. J. & Mayer, G. C. (1980). Non-allopatric speciation in animals. *Systematic Zoology* **29**, 254–271.

Futuyma, D. J. & Peterson, S. C. (1985). Genetic variation in the use of resources by insects. *Annual Review of Entomology* **30**, 217–238.

Gallo, R. C. & Montagnier, L. (1987). The chronology of AIDS research. *Nature* **326**, 435–436.

Gaston, K. J. & Lawton, J. H. (1988). Patterns in body size, population dynamics, and regional distribution of bracken herbivores. *American Naturalist* **132**, 662–680.

Gilkeson, L. A. & Hill, S. B. (1986). Diapause prevention in *Aphidoletes aphidimyza* (Diptera: Cecidomyiidae) by low-intensity light. *Environmental Entomology* **15**, 1067–1069.

Gillham, M. C. & De Vrijer, P. W. F. (1995). Patterns of variation in the acoustic calling signals of *Chloriona* planthoppers (Homoptera: Delphacidae) coexisting on the common reed *Phragmites australis*. *Biological Journal of the Linnean Society* **54**, 245–269.

Gleason, H. A. (1926). The individualistic concept of the plant association. *Bulletin of the Torrey Botanical Club* **53**, 7–26.

Gleason, H. A. (1939). The individualistic concept of the plant association. *American Midland Naturalist* **21**, 92–110.

Godfray, H. C. J. (1994). *Parasitoids: Behavioral and Evolutionary Ecology*. Princeton University Press, Princeton.

Godfray, H. C. J. & Hunter, M. S. (1994). Heteronomous hyperparasitoids, sex ratios and adaptations: a reply. *Ecological Entomology* **19**, 93–95.

Godfray, H. C. J. & Waage, J. K. (1991). Predictive modelling in biological control: the mango mealy bug (*Rastrococcus invadens*) and its parasitoids. *Journal of Applied Ecology* **28**, 434–453.

Gokhman, V. E., Timokhov, A. V. & Fedina, T. Y. (1998). First evidence for sibling species in *Anisopteromalus calandrae*. *Russian Entomological Journal* **7**, 157–162.

Gombrich, E. H. (1977). *Art and Illusion: A Study in the Psychology of Pictorial Representation*, 5th edn. Phaidon Press, London.

Gonzalez Sierra, G., Penas Merino, A. & Alonso-Herrero, E. (1996). Phenology of *Hyacinthoides non-scripta* (L.) Chouard, *Melittis melissophyllum* L. and *Symphytum tuberosum* L. in two deciduous forests in the Cantabrian mountains, Northwest Spain. *Vegetatio* **122**, 69–82.

312   References

Gordh, G. & DeBach, P. (1978). Courtship behaviour in the *Aphytis lingnanensis* group, its potential usefulness in taxonomy, and a review of sexual behavior in the parasitic Hymenoptera (Chalcidoidea: Aphelinidae). *Hilgardia* **46**, 37–75.

Gordon, D. H. & Watson, C. R. B. (1986). Identification of cryptic species of rodents (*Mastomys, Aethomys, Saccostomus*) in the Kruger National Park. *South African Journal of Science* **21**, 95–99.

Gould, F. (1998). Sustainability of transgenic insecticidal cultivars: integrating pest genetics and ecology. *Annual Review of Entomology* **43**, 701–726.

Gould, S. J. & Lewontin, R. C. (1979). The spandrels of San Marco and the Panglossian paradigm: a critique of the adaptationist programme. *Proceedings of the Royal Society of London, Series B* **205**, 581–598.

Goyer, R. A., Paine, T. D., Pashley, D. P., Lenhard, G. J., Meeker, J. R. & Hanlon, C. C. (1995). Geographic and host-associated differentiation in the fruittree leafroller (Lepidoptera: Tortricidae). *Annals of the Entomological Society of America* **88**, 391–396.

Graham, R. W., Lundelius, E. L., Jr., Graham, M. A. *et al.* (1996). Spatial response of mammals to late quaternary environmental fluctuations. *Science* **272**, 1601–1606.

Grant, J. A. (1997). IPM techniques for greenhouse crops. In *Techniques for Reducing Pesticide Use: Economic and Environmental Benefits* (ed. D. Pimentel), pp. 399–406. John Wiley, Chichester.

Grant, P. R. (1975). The classical case of character displacement. In *Evolutionary Biology*, vol. 8 (eds. T. Dobzhansky, M. K. Hecht & W. C. Steere), pp. 237–337. Plenum Press, New York.

Greathead, D. J. (1986). Parasitoids in classical biological control. In *Insect Parasitoids* (eds. J. Waage & D. J. Greathead), pp. 289–318. Academic Press, London.

Greathead, D. J. (1994). History of biological control. *Antenna* **18**, 187–199.

Green, C. A., Gordon, D. H. & Lyons, N. F. (1972). Biological species in *Praomys* (*Mastomys*) *natalensis* (Smith), a rodent carrier of Lassa virus and bubonic plague in Africa. *American Journal of Tropical Medicine and Hygiene* **27**, 627–629.

Green, C. A. & Hunt, R. H. (1980). Interpretation of variation in ovarian polytene chromosomes of *Anopheles funestus* Giles, *A. parensis* Gillies and *A. aruni. Genetica* **51**, 187–195.

Gregg, P. C. (1994). Migration of cotton pests: patterns and implications for management. In *World Cotton Research Conference No. 1*, vol. 1 (eds. G. A. Constable & N. W. Forrester), pp. 423–433. CSIRO, Brisbane.

Greig-Smith, P. (1986). Chaos or order – organization. In *Community Ecology:*

Pattern and Process (eds. J. Kikkawa & D. J. Anderson), pp. 19–29. Blackwell Scientific Publications, Oxford.

Grimm, V. (1996). A down-to-earth assessment of stability concepts in ecology: dreams, demands, and the real problems. *Seckenbergiana Maritima* **27**, 215–226.

Grimm, V. & Wissel, C. (1997). Babel, or the ecological stability discussions: an inventory and analysis of terminology and a guide for avoiding confusion. *Oecologia* **109**, 323–334.

Grmek, M. D. (1990). *History of AIDS: Emergence and Origin of a Modern Pandemic*. Princeton University Press, Princeton.

Groth, J. G. (1988). Resolution of cryptic species in Appalachian red crossbills. *Condor* **90**, 745–760.

Gu, H., Cao, A. & Walter, G. H. (2001). Host selection and utilisation of *Sonchus oleraceus* (Asteraceae) by *Helicoverpa armigera* (Lepidoptera: Noctuidae): a genetic analysis. *Annals of Applied Biology* **138**, 293–299.

Gu, H. & Walter, G. H. (1999). Is the common sowthistle (*Sonchus oleraceus*) a primary host plant of the cotton bollworm, *Helicoverpa armigera* (Lep., Noctuidae)? Oviposition and larval performance. *Journal of Applied Entomology* **123**, 99–105.

Hagen, K. S. & Franz, J. M. (1973). A history of biological control. In *History of Entomology* (eds. R. F. Smith, T. E. Mittler & C. N. Smith), pp. 433–476. Annual Reviews, Palo Alto.

Hamilton, R. M. S. & Chiswell, B. (1987). Section A: The nature and development of scientific knowledge. In *The Origins of Modern Science*. Unpublished notes for course HT155 of the School of External Studies and Continuing Education. University of Queensland, Brisbane.

Hampshire, S. (1956). *The Age of Reason: The 17th Century Philosophers – Selected, with Introduction and Interpretive Commentary by Stuart Hampshire*. New American Library, New York.

Harley, K. L. S. & Forno, I. W. (1992). *Biological Control of Weeds: A Handbook for Practitioners and Students*. Inkata Press, Melbourne.

Harpaz, I. (1973). Early entomology in the Middle East. In *History of Entomology* (eds. R. F. Smith, T. E. Mittler & C. N. Smith), pp. 21–36. Annual Reviews, Palo Alto.

Harris, C. L. (1981). *Evolution: Genesis and Revelations, with Readings from Empedocles to Wilson*. State University of New York Press, Albany.

Harris, M. (2000). Impact of integrated pest management on academia in entomology in the United States. *American Entomologist* **46**, 217–220.

Harris, M. K. (2001). IPM: What has it delivered? *Plant Disease* **85**, 112–121.

Harris, V. E. & Todd, J. W. (1980). Male-mediated aggregation of male, female and 5th-instar southern green stink bugs and concomitant attraction of a tachinid parasite, *Trichopoda pennipes*. *Entomologia Experimentalis et Applicata* **27**, 117–126.

Harrison, S. (1997). Persistent, localized outbreaks in the western tussock moth *Orgyia vetusta*: the roles of resource quality, predation and poor dispersal. *Ecological Entomology* **22**, 158–166.

Harwood, J. D., Sunderland, K. D. & Symondson, W. O. C. (2001). Living where the food is: web location by linyphiid spiders in relation to prey availability in winter wheat. *Journal of Animal Ecology* **38**, 88–99.

Hassell, M. P. (1976). *The Dynamics of Competition and Predation*. Camelot Press, Southampton.

Hassell, M. P. (1978). *The Dynamics of Arthropod Predator–Prey Systems*. Princeton University Press, Princeton.

Hassell, M. P. (1985). Insect natural enemies as regulating factors. *Journal of Animal Ecology* **54**, 323–334.

Hassell, M. P. (1986). Parasitoids and population regulation. In *Insect Parasitoids* (eds. J. Waage & D. J. Greathead), pp. 201–224. Academic Press, London.

Hassell, M. P. & Godfray, H. C. J. (1992). The population biology of insect parasitoids. In *Natural Enemies: The Population Biology of Predators, Parasites and Diseases* (ed. M. J. Crawley), pp. 265–292. Blackwell Scientific Publications, Oxford.

Hassell, M. P. & May, R. M. (1985). From individual behaviour to population dynamics. In *Behavioural Ecology: Ecological Consequences of Adaptive Behaviour. 25th Symposium of the British Ecological Society, Reading, 1984* (eds. R. M. Sibly & R. H. Smith), pp. 3–32. Blackwell Scientific Publications, Oxford.

Hassell, M. P. & Southwood, T. R. E. (1978). Foraging strategies of insects. *Annual Review of Ecology and Systematics* **9**, 75–98.

Hauser, C. L. (1987). The debate about the biological species concept – a review. *Zeitschrift für Zoologische Systematik und Evolutionsforschung* **25**, 241–257.

Hawkins, B. A. (1993). Parasitoid species richness, host mortality, and biological control. *American Naturalist* **141**, 634–641.

Hawkins, B. A., Thomas, M. B. & Hochberg, M. E. (1993). Refuge theory and biological control. *Science* **262**, 1429–1432.

Heck, K. L. (1976). Some critical considerations of the theory of species packing. *Evolutionary Theory* **1**, 247–258.

Heimpel, G. E. & Rosenheim, J. A. (1998). Egg limitation in parasitoids: a review of the evidence and a case study. *Biological Control* **11**, 160–168.

Henderson, N. R. & Lambert, D. M. (1982). No significant deviation from random mating of worldwide populations of *Drosophila melanogaster*. *Nature* **300**, 437–440.

Hengeveld, R. (1988). Mayr's ecological species criterion. *Systematic Zoology* **37**, 47–55.

Hengeveld, R. (1990). *Dynamic Biogeography*. Cambridge University Press, Cambridge.

Hengeveld, R. (1992). Potential and limitations of predicting invasion rates. *Florida Entomologist* **75**, 60–72.

Hengeveld, R. (1997). Impact of biogeography on a population-biological paradigm shift. *Journal of Biogeography* **24**, 541–547.

Hengeveld, R. (1999). Modelling the impact of biological invasions. In *Invasive Species and Biodiversity Management* (eds. O. T. Sandlund, P. J. Schei & A. Viken), pp. 127–138. Kluwer Academic Publishers, Dordrecht.

Hengeveld, R. & Walter, G. H. (1999). The two coexisting ecological paradigms. *Acta Biotheoretica* **47**, 141–170.

Hilborn, R. & Ludwig, D. (1993). The limits of applied ecological research. *Ecological Applications* **3**, 550–552.

Hilborn, R. & Stearns, S. C. (1982). On inference in ecology and evolutionary biology: the problem of multiple causes. *Acta Biotheoretica* **31**, 145–164.

Hill, S. & Turpin, T. (1994). Cultures in collision: the emergence of a new localism in academic research. In *Shifting Contexts: Transformations in Anthropological Knowledge* (ed. M. Strathern), pp. 131–152. Routledge, London.

Hobsbawm, E. (1994). *Age of Extremes: The Short Twentieth Century 1914–1991*. Abacus, London.

Hochberg, M. E. & Hawkins, B. A. (1992). Refuges as a predictor of parasitoid diversity. *Science* **255**, 973–976.

Hochberg, M. E. & Hawkins, B. A. (1993). Predicting parasitoid species richness. *American Naturalist* **142**, 671–693.

Hodek, I. (1996). Food relationships. In *Ecology of Coccinellidae*, vol. 54. Dr W. *Junk Series Entomologica* (eds. I. Hodek & A. Honek), pp. 143–238. Kluwer Academic Publishers, Dordrecht.

Hokkanen, H. M. T. (1997). Role of biological control and transgenic crops in reducing use of chemical pesticides for crop protection. In *Techniques for Reducing Pesticide Use* (ed. D. Pimentel), pp. 103–127. John Wiley, Chichester.

Hokkanen, H. & Pimentel, D. (1984). New approach for selecting biological control agents. *Canadian Entomologist* **116**, 1109–1121.

Holldobler, B. & Wilson, E. O. (1990). *The Ants*. Belknap Press of Harvard University Press, Cambridge, Mass.

Holliday, N. J. (1977). Population ecology of winter moth (*Operophtera brumata*) on apple in relation to larval dispersal and time of bud burst. *Journal of Applied Ecology* **14**, 803–813.

Holliday, N. J. (1985). Maintenance of the phenology of the winter moth. *Biological Journal of the Linnean Society* **25**, 221–234.

Holt, R. D. & Hochberg, M. E. (1997). When is biological control evolutionarily stable (or is it)? *Ecology* **78**, 1673–1683.

Holton, G. (1993). *Science and Anti-Science*. Harvard University Press, Cambridge, Mass.

Horgan, J. (1996). *The End of Science: Facing the Limits of Knowledge in the Twilight of the Scientific Age*. Little, Brown and Company, London.

Horn, D. J. & Dowell, R. V. (1979). Parasitoid ecology and biological control in ephemeral crops. In *Analysis of Ecological Systems* (eds. D. J. Horn, G. R. Stairs & R. D. Mitchell), pp. 281–306. Ohio State University Press, Columbus.

Howard, W. E. (1960). Innate and environmental dispersal of individual vertebrates. *American Midland Naturalist* **63**, 152–161.

Howarth, F. J. (1991). Environmental impacts of classical biological control. *Annual Review of Entomology* **36**, 485–509.

Hoy, M. A. (1990). Genetic improvement of arthropod natural enemies: becoming a conventional tactic? In *New Directions in Biological Control: Alternatives for Suppressing Agricultural Pests and Diseases* (eds. R. R. Baker & P. E. Dunn), pp. 405–417. Alan R. Liss, New York.

Hsiao, T. H. (1978). Host plant adaptation among geographic populations of the Colorado potato beetle. *Entomologia Experimentalis et Applicata* **24**, 237–247.

Huang, H. T. & Yang, P. (1987). The ancient cultured citrus ant. *BioScience* **37**, 665–671.

Huber, F. (1985). Approaches to insect behavior of interest to both neurobiologists and behavioral ecologists. *Florida Entomologist* **68**, 52–78.

Huffaker, C. B. & Kennett, C. E. (1966). IV. Biological control of *Parlatoria oleae* (Colvee) through the compensatory action of two introduced parasites. *Hilgardia* **37**, 283–335.

Hughes, A. J. & Lambert, D. M. (1984). Functionalism, structuralism and "ways of seeing". *Journal of Theoretical Biology* **111**, 787–800.

Hughes, R. D. (1981). The Australian bushfly: a climate-dominated nuisance pest of man. In *The Ecology of Pests: Some Australian Case Histories* (eds. R. L. Kitching & R. E. Jones), pp. 177–191. CSIRO, Melbourne.

Hull, D. L. (1988). *Science as a Process*. University of Chicago Press, Chicago.

Hunt, R. H., Coetzee, M. & Fettene, M. (1998). The *Anopheles gambiae* complex: a new species from Ethiopia. *Transactions of the Royal Society of Tropical Medicine and Hygiene* **92**, 231–235.

Huntley, B. (1991). How plants respond to climate change: migration rates, individualism and the consequences for plant communities. *Annals of Botany* **67**, 15–22.

Hurst, L. D., Hamilton, W. D. & Ladle, R. J. (1992). Covert sex. *Trends in Ecology and Evolution* **7**, 144–145.

Hutchinson, G. E. (1948). Circular causal systems in ecology. *Annals of the New York Academy of Sciences* **50**, 221–246.

Itami, J. K., Craig, T. P. & Horner, J. D. (1998). Factors affecting gene flow between the host races of *Eurosta solidaginis*. In *Genetic Structure and Local Adaptation in Natural Insect Populations: Effects of Ecology, Life History, and Behavior* (eds. S. Mopper & S. Y. Strauss), pp. 375–407. Chapman & Hall, New York.

Jaenike, J. (1981). Criteria for ascertaining the existence of host races. *American Naturalist* **117**, 830–834.

Janz, N., Nyblom, K. & Nylin, S. (2001). Evolutionary dynamics of host-plant specialization: a case study of the tribe Nymphalini. *Evolution* **55**, 783–796.

Jermy, T. (1984). Evolution of insect/host plant relationships. *American Naturalist* **124**, 609–630.

Jervis, M. A. & Copland, M. J. W. (1996). The life cycle. In *Insect Natural Enemies: Practical Approaches to their Study and Evaluation* (eds. M. A. Jervis & N. Kidd), pp. 63–161. Chapman & Hall, London.

Jones, D. P. (1973). Agricultural entomology. In *History of Entomology* (eds. R. F. Smith, T. E. Mittler & C. N. Smith), pp. 307–332. Annual Reviews, Palo Alto.

Jones, O. T. (1994). The academia–Industry continuum. *Antenna* **18**, 58–62.

Jones, R. E. (1981). The cabbage butterfly, *Pieris rapae* L.: 'a just sense of how not to fly'. In *The Ecology of Pests: Some Australian Case Histories* (eds. R. L. Kitching & R. E. Jones), pp. 217–228. CSIRO, Melbourne.

Jones, R. E. & Kitching, R. L. (1981). Why an ecology of pests. In *The Ecology of Pests: Some Australian Case Histories* (eds. R. E. Jones & R. L. Kitching), pp. 1–5. CSIRO, Melbourne.

Jones, T. H., Godfray, H. C. J. & Hassell, M. P. (1996). Relative movement patterns of a tephritid fly and its parasitoid wasps. *Oecologia* **106**, 317–324.

Jones, V. P. (1995). Reassessment of the role of predators and *Trissolcus basalis* in biological control of southern green stink bug (Hemiptera: Pentatomidae) in Hawaii. *Biological Control* **5**, 566–572.

Judson, O. P. & Normark, B. B. (1996). Ancient asexual scandals. *Trends in Ecology and Evolution* **11**, 41–45.

Kambhampati, S., Black, W. C., IV & Rai, K. S. (1992). Random amplified polymorphic DNA of mosquito species and populations (Diptera: Culicidae): techniques, statistical analysis and applications. *Journal of Medical Entomology* **29**, 939–945.

Kaminer, B. (1988). On Albert Szent-Gyorgyi. *Biological Bulletin* **174**, 192–213.

Kareiva, P. (1990). Establishing a foothold for theory in biocontrol practice: using models to guide experimental design and release protocols. In *New Directions in Biological Control: Alternatives for Suppressing Agricultural Pests and Diseases* (eds. R. R. Baker & P. E. Dunn), pp. 65–81. Alan R. Liss, New York.

Kareiva, P. (1994). Ecological theory and endangered species. *Ecology* 757, 583.

Kassen, R. (2002). The experimental evolution of specialists, generalists and the maintenance of diversity. *Journal of Evolutionary Biology* 15, 173–190.

Keller, A. (1985). Has science created technology? *Minerva* 22, 160–182.

Keller, M. A. (1984). Reassessing evidence for competitive exclusion of introduced natural enemies. *Environmental Entomology* 13, 192–195.

Kennedy, G. G. & Storer, N. P. (2000). Life systems of polyphagous arthropod pests in temporally unstable cropping systems. *Annual Review of Entomology* 45, 467–493.

Kennedy, G. G. & Sutton, T. B. (2000). *Emerging Technologies for Integrated Pest Management: Concepts, Research, and Implementation.* American Phytopathological Society, St Paul, Minn.

Kennedy, J. S. (1967). Behaviour as physiology. In *Insects and Physiology* (eds. J. W. L. Beament & J. E. Treherne), pp. 249–265. Oliver and Boyd, Edinburgh.

Kennedy, J. S. (1992). *The New Anthropomorphism.* Cambridge University Press, New York.

Kfir, R. (1994). Attempts at biological control of the stem borer *Chilo partellus* (Swinhoe) (Lepidoptera: Pyralidae) in South Africa. *African Entomology* 2, 67–68.

Kimura, M. T., Beppu, K., Ichijô, N. & Toda, M. J. (1978). Bionomics of Drosophilidae (Diptera) in Hokkaido. II. *Drosophila testacea. Kontyû* 46, 585–595.

King, W. V. & Gahan, J. B. (1949). Failure of DDT to control house flies. *Journal of Economic Entomology* 42, 405–409.

Kirsch, J. A. W. & Poole, W. E. (1967). Serological evidence for speciation in the grey kangaroo, *Macropus giganteus* Shaw, 1790 (Marsupialia: Macropodidae). *Nature* 215, 1097.

Kisimoto, R. (1976). Synoptic weather conditions inducing long-distance immigration of planthoppers, *Sogatella furcifera* Horvath and *Nilaparvata lugens* (Stål). *Ecological Entomology* 1, 95–109.

Kisimoto, R. & Rosenberg, L. J. (1994). Long-distance migration in delphacid planthoppers. In *Planthoppers: Their Ecology and Management* (eds. R. F. Denno & T. J. Perfect), pp. 302–322. Chapman & Hall, New York.

Kisimoto, R. & Sogawa, K. (1995). Migration of the brown planthopper *Nilaparvata lugens* and the white-backed planthopper *Sogatella furcifera* in East Asia: the role of weather and climate. In *Insect Migration* (eds. V. A. Drake & A. G. Gatehouse), pp. 67–91. Cambridge University Press, Cambridge.

Kitaysky, A. S. & Golubova, E. G. (2000). Climate change causes contrasting trends in reproductive performance of planktivorous and piscivorous alcids. *Journal of Animal Ecology* **69**, 248–262.

Kitching, R. L. (1981). The geography of the Australian Papilionoidea. In *Ecological Biogeography of Australia*, vol. 2, part 3, *Terrestrial Invertebrates* (ed. A. Keast), pp. 977–1005. W. Junk, The Hague.

Klemm, M. (1959). Entomologie und Pflanzenschutz in China. *Nachrichtenblatt des Deutschen Pflanzenschutzdienstes der Braunschweig* **11**, 121–124.

Kmiec, E. (1999). Gene therapy. *American Scientist* **87**, 240–248.

Knowles, L. L., Futuyma, D. J., Eanes, W. F. & Rannala, B. (1999). Insight into speciation from historical demography in the phytophagous beetle genus *Ophraella*. *Evolution* **53**, 1846–1856.

Knowlton, N. & Jackson, J. B. C. (1994). New taxonomy and niche partitioning on coral reefs: jack of all trades or master of some. *Trends in Ecology and Evolution* **9**, 7–9.

Knox, A. G. (1992). Species and pseudospecies: the structure of Crossbill populations. *Biological Journal of the Linnean Society* **47**, 325–335.

Knox, B., Ladiges, P., Evans, B. & Saint, R. (2001). *Biology*, 2nd edn. McGraw-Hill Australia, Roseville.

Kogan, M. (1988). Integrated pest management theory and practice. *Entomologia Experimentalis et Applicata* **49**, 59–70.

Kogan, M. (1996). Areawide management of major pests: is the concept applicable to the *Bemisia* complex? In *Bemisia 1995: Taxonomy, Biology, Damage, Control and Management* (eds. D. Gerling & R. T. Mayer), pp. 643–657. Intercept, Andover.

Kogan, M. (1998). Integrated pest management: historical perspectives and contemporary developments. *Annual Review of Entomology* **43**, 243–270.

Kogan, M. (1999). Integrated Pest Management: constructive criticism or revisionism. *Phytoparasitica* **27**, 93–96.

Kokkinn, M. J. & Davis, A. R. (1986). Secondary production: shooting a halcyon for its feathers. In *Limnology in Australia* (eds. P. de Dekker & W. D. Williams), pp. 251–261. CSIRO & Junk, East Melbourne.

Konishi, M. & Ito, Y. (1973). Early entomology in East Asia. In *History of Entomology* (eds. R. F. Smith, T. E. Mittler & C. N. Smith), pp. 1–20. Annual Reviews, Palo Alto.

Krebs, C. J. (1995). Two paradigms of population regulation. *Wildlife Research* **22**, 1–10.

Kuhn, T. S. (1962). Historical structure of scientific discovery. *Science* **136**, 760–764.

Kutchan, T. M. (2001). Ecological arsenal and developmental dispatcher: the paradigm of secondary metabolism. *Plant Physiology* **125**, 58–60.

Labeyrie, V. (1988). Scientific and artistic research in applied entomology. *Entomologia Experimentalis et Applicata* **49**, 17–24.

Lambert, D. M., Centner, M. R. & Paterson, H. E. H. (1984). Simulation of the conditions necessary for the evolution of species by reinforcement. *South African Journal of Science* **80**, 308–311.

Lambert, D. M., Michaux, B. & White, C. S. (1987). Are species self-defining? *Systematic Zoology* **36**, 196–205.

Lambert, D. M. & Millar, C. D. (1995). DNA science and conservation. *Pacific Conservation Biology* **2**, 21–38.

Lambert, D. M. & Paterson, H. E. H. (1982). Morphological resemblance and its relationship to genetic distance measures. *Evolutionary Theory* **5**, 291–300.

Lambert, D. M. & Spencer, H. G. (1995). *Speciation and the Recognition Concept: Theory and Application.* Johns Hopkins University Press, Baltimore.

Lane, S. D., Mills, N. J. & Getz, W. M. (1999). The effects of parasitoid fecundity and host taxon on the biological control of insect pests: the relationship between theory and data. *Ecological Entomology* **24**, 181–190.

Larsen, L. M., Nielsen, J. K. & Sørensen, H. (1992). Host plant recognition in monophagous weevils: specialization of *Ceutorhynchus inaffectatus* to glucosinolates from its host plant *Hesperis matronalis*. *Entomologia Experimentalis et Applicata* **64**, 49–55.

Larsson, S. & Ekbom, B. (1995). Oviposition mistakes in herbivorous insects – confusion or a step towards a new host plant? *Oikos* **72**, 155–160.

Lawton, J. H. (1986). The effect of parasitoids on phytophagous insect communities. In *Insect Parasitoids* (eds. J. Waage & D. Greathead), pp. 265–287. Academic Press, London.

Lawton, J. H. (1989). Book Review: 'Disentangling the bank'. *Nature* **339**, 517.

Lawton, J. H. (1991). Ecology as she is done, and could be done. *Oikos* **16**, 289–290.

Lawton, J. H. & Strong, D. R. (1981). Community patterns and competition in folivorous insects. *American Naturalist* **118**, 317–338.

Leather, S. R. (1991). Feeding specialisation and host distribution of British and Finnish *Prunus* feeding macrolepidoptera. *Oikos* **60**, 40–48.

Lemon, R. W. (1994). Insecticide resistance. *Journal of Agricultural Science* **122**, 329–333.

Lensink, R. (1997). Range expansion of raptors in Britain and the Netherlands since the 1960s: testing an individual-based diffusion model. *Journal of Animal Ecology* **66**, 811–826.

Lewin, R. (1986). Supply-side ecology. *Science* **234**, 25–27.

Lewis, W. J., Jones, R. L., Gross, H. R. J. & Norlund, D. A. (1976). The role of kairomones and other behavioral chemicals in host finding by parasitic insects. *Behavioural Biology* **16**, 267–289.

Lewis, W. J. & Norlund, D. A. (1984). Behaviour-modifying chemicals to enhance natural enemy effectiveness. In *Biological Control in Agricultural IPM Systems* (eds. M. A. Hoy & D. C. Herzog), pp. 89–100. Academic Press, London.

Lewis, W. J., Vet, L. E. M., Tumlinson, J. H., van Lenteren, J. C. & Papaj, D. R. (1990). Variations in parasitoid foraging behavior: essential element of a sound biological control theory. *Environmental Entomology* **19**, 1183–1193.

Lewontin, R. C. (1978). Adaptation. *Scientific American* **239**, 212–230.

Lewontin, R. C. (1991). *Biology as Ideology: The Doctrine of DNA*. HarperPerennial, New York.

Lima, S. L. & Zollner, P. A. (1996). Towards a behavioral ecology of ecological landscapes. *Trends in Ecology and Evolution* **11**, 131–135.

Lindeman, R. L. (1942). The trophic–dynamic aspect of ecology. *Ecology* **23**, 399–418.

Lindquist, D. A., Butt, B., Feldmann, U., Gingrich, R. E. & Economopoulos, A. (1990). Current status and future prospects for genetic methods of insect control or eradication. In *Pesticides and Alternatives: Innovative Chemical and Biological Approaches to Pest Control* (ed. J. E. Casida), pp. 69–88. Elsevier, New York.

Lindsay, D. R. (1986). The training of agricultural scientists. In *Science for Agriculture: The Way Ahead* (ed. L. W. Martinelli), pp. 145–151. Australian Institute of Agricultural Science, Melbourne.

Linn, C. E., Young, M. S., Gendle, M., Glover, T. J. & Roelofs, W. L. (1997). Sex pheromone blend discrimination in two races and hybrids of the European corn borer moth, *Ostrinia nubilalis*. *Physiological Entomology* **22**, 212–223.

Liss, W. J., Gut, L. J., Westigard, P. H. & Warren, C. E. (1986). Perspectives on arthropod community structure, organization, and development in agricultural crops. *Annual Review of Entomology* **31**, 455–478.

Lively, C. M. (1993). Rapid evolution by biological enemies. *Trends in Ecology and Evolution* **8**, 345–346.

Lodge, D. M. (1993). Biological invasions – lessons for ecology. *Trends in Ecology and Evolution* **8**, 133–137.

Loehle, C. (1987). Hypothesis testing in ecology: psychological aspects and the importance of theory maturation. *Quarterly Review of Biology* **62**, 397–409.

Logan, M. S., Iverson, S. J., Ruzzante, D. E. *et al.* (2000). Long term diet differences between morphs in tritrophically polymorphic *Percichthys trucha* (Pisces: Percichthyidae) populations from the southern Andes. *Biological Journal of the Linnean Society* **69**, 599–616.

Lomnicki, A. (1978). Individual differences between animals and the natural regulation of their numbers. *Journal of Animal Ecology* **47**, 461–475.

Lomnicki, A. (1988). *Population Ecology of Individuals*. Princeton University Press, Princeton.

Louda, S. M. (1998). Ecology of interactions needed in biological control practice and policy. *Bulletin of the British Ecological Society* **29**, 8–11.

Louis, C., Pintureau, B. & Chapelle, L. (1993). Recherches sur l'origine de l'unisexualité: la thermothérapie élimine á la fois rickettsieset parthénogenése thélytoque chez un Trichogramme (Hym., Trichogrammatidae). *Comptes Rendu de l'Academie des Sciences* **316**, 27–33.

Lovtrup, S. (1984). The eternal battle against empiricism. *Revista di Biologia* **77**, 183–209.

Loxdale, H. D. & Lushai, G. (1998). Molecular markers in entomology. *Bulletin of Entomological Research* **88**, 577–600.

Lu, W., Kennedy, G. G. & Gould, F. (2001). Genetic analysis of larval survival and larval growth of two populations of *Leptinotarsa decemlineata* on tomato. *Entomologia Experimentalis et Applicata* **99**, 143–155.

Luck, R. F. (1990). Evaluation of natural enemies for biological control: a behavioural approach. *Trends in Ecology and Evolution* **5**, 196–199.

Ludwig, D. (1994). Bad ecology leads to bad public policy. *Trends in Ecology and Evolution* **9**, 411.

Luttrell, R. G. (1994). Cotton pest management: part 2. A US perspective. *Annual Review of Entomology* **39**, 527–542.

Lyle, C. (1947). Achievements and possibilities in pest eradication. *Journal of Economic Entomology* **40**, 1–8.

Lynch, J. F. (1989). Community ecology of salamanders. *Evolution* **43**, 1127–1128.

Macedo, N., de Araújo, J. R. & Botelho, S. M. P. (1993). Sixteen years of biological control of *Diatraea saccaralis* (Fabr.) (Lepidoptera: Pyralidae) by *Cotesis flavipes* (Cam.) (Hymenoptera: Braconidae), in the state of Sao Paulo, Brazil. *Anais da Sociedade Entomologica do Brasil* **22**, 441–448.

Magee, B. (1973). *Popper*. Fontana/Collins, London.

Mahon, R. J., Miethke, P. M. & Mahon, J. A. (1982). The evolutionary relationships of three forms of the jarrah leaf miner, *Perthida glyphopa* (Common) (Lepidoptera: Incurvariidae). *Australian Journal of Zoology* **30**, 243–249.

Makni, H., Marrakchi, M. & Pasteur, N. (2000). Biochemical characterization of sibling species in Tunisian *Mayetiola* (Diptera: Cecidomyiidae). *Biochemical Systematics and Ecology* **28**, 101–109.

Mallet, J. (1995). A species definition for the modern synthesis. *Trends in Ecology and Evolution* **10**, 294–299.

Margolis, H. (1993). *Paradigms and Barriers: How Habits of Mind Govern Scientific Beliefs*. University of Chicago Press, Chicago.

Marohasy, J. (1996). Host shifts in biological weed control: real problems, semantic difficulties or poor science? *International Journal of Pest Management* **42**, 71–75.

Marohasy, J. (1998). The design and interpretation of host-specificity tests for weed biological control with particular reference to insect behaviour. *Biocontrol* **19**, 13–20.

Martinat, P. J. & Barbosa, P. (1987). Relationship between host-plant acceptability and suitability in newly eclosed first-instar gypsy moths, *Lymantria dispar* (L.) (Lepidoptera: Lymantriidae). *Annals of the Entomological Society of America* **80**, 141–147.

Maslow, A. H. (1954). Problem centering vs means centering in science. In *Motivation and Personality* (eds. W. G. Holtzman & G. Murphy), pp. 12–18. Harper & Row, London.

Masters, J. C. (1991). Loud calls of *Galago cassicaudatus* and *G. garnettii* and their relation to habitat structure. *Primates* **32**, 153–167.

Masters, J. C., Lambert, D. M. & Paterson, H. E. H. (1984). Scientific prejudice, reproductive isolation, and apartheid. *Perspectives in Biology and Medicine* **28**, 107–116.

Masters, J. C. & Rayner, R. J. (1993). Competition and macroevolution: the ghost of competition yet to come. *Biological Journal of the Linnean Society* **49**, 87–98.

Masters, J. C. & Reyner, R. J. (1996). The recognition concept and the fossil record: putting the genetics back into phylogenetic species. *South African Journal of Science* **92**, 225–231.

Matteson, P. C. (2000). Insect pest management in tropical Asian irrigated rice. *Annual Review of Entomology* **45**, 549–574.

Matteson, P. C., Gallagher, K. D. & Kenmore, P. E. (1994). Extension of integrated pest management for planthoppers in Asian irrigated rice: empowering the user. In *Planthoppers: Their Ecology and Management* (eds. R. F. Denno & T. J. Perfect), pp. 656–685. Chapman & Hall, New York.

Matthews, G. A. (2000). *Pesticide Application Methods*, 3rd edn. Blackwell Science, Oxford.

Matthews, G. A. & Hislop, E. C. (1993). *Application Technology for Crop Protection*. CAB International, Wallingford.

Matthews, M. (1999). *Heliothine Moths of Australia: A Guide to Pest Bollworms and Related Noctuid Groups*. CSIRO, Collingwood.

May, A. W. S. (1953). Queensland host records for the Dacinae. *Queensland Journal of Agricultural Science* **10**, 36–79.

Mayden, R. L. (1997). A hierarchy of species concepts: the denouement in the saga of the species problem. In *Species: The Units of Biodiversity* (eds. M. F. Claridge, H. A. Dawah & M. R. Wilson), pp. 381–424. Chapman & Hall, London.

Maynard Smith, J. (1966). Sympatric speciation. *American Naturalist* **100**, 637–650.

Mayr, E. (1961). Cause and effect in biology. *Science* **134**, 1501–1506.

Mayr, E. (1963). *Animal Species and Evolution*. Harvard University Press, Cambridge, Mass.

Mayr, E. (1976). Species concepts and definitions. In *Topics in the Philosophy of Biology*, vol. 27. *Boston Studies in the Philosophy of Science* (eds. M. Grene & E. Mendelsohn), pp. 353–371. D. Reidel, Dordrecht.

Mayr, E. (1982). *The Growth of Biological Thought*. Harvard University Press, Cambridge, Mass.

Mayr, E. & Provine, W. B. (1980). *The Evolutionary Synthesis*. Harvard University Press, Cambridge, Mass.

McGeoch, M. A. & Wossler, T. C. (2000). Range expansion and success of the weed biocontrol agent *Trichilogaster acaciaelongifoliae* (Froggatt) (Hymenoptera: Pteromalidae) in South Africa. *African Entomology* **8**, 273–280.

McIntosh, R. P. (1975). H.A. Gleason – "Individualistic Ecologist" 1882–1975: his contributions to ecological theory. *Bulletin of the Torrey Botanical Club* **102**, 253–273.

McIntosh, R. P. (1985). *The Background of Ecology: Concept and Theory*. Cambridge University Press, Cambridge.

McIntosh, R. P. (1987). Pluralism in ecology. *Annual Review of Ecology and Systematics* **18**, 321–341.

McIntosh, R. P. (1995). H.A. Gleason's 'Individualistic concept' and theory of animal communities: a continuing controversy. *Biological Reviews* **70**, 317–357.

McKenzie, H. L. (1937). Morphological differences distinguishing California red scale, yellow scale, and related species. *University of California Publications in Entomology* **6**, 323–326.

Medawar, P. B. (1984). *Pluto's Republic, incorporating the Art of the Soluble and Induction and Intuition in Scientific Thought*. Oxford University Press, Oxford.

Meijer, J. (1974). A comparative study of the immigration of carabids (Coleoptera, Carabidae) into a new polder. *Oecologia* **16**, 185–208.

Merryweather, J. & Fitter, A. (1995). Arbuscular mycorrhiza and phosphorus as controlling factors in the life history of *Hyacinthoides non-scripta* (L.) Chouard ex Rothm. *New Phytologist* **129**, 629–636.

Metcalf, R. L. & Luckman, W. H. (eds.) (1994). *Introduction to Insect Pest Management*. John Wiley, New York.

Meyer, A. (1990). Ecological and evolutionary consequences of the trophic polymorphism in *Cichlasoma citrinellum* (Pisces: Cichlidae). *Biological Journal of the Linnean Society* **39**, 279–299.

Michelbacher, A. E. (1945). The importance of economic entomology. *Journal of Economic Entomology* **38**, 129–130.

Michelbacher, A. E. & Smith, R. F. (1943). Some natural factors limiting the abundance of the alfalfa butterfly. *Hilgardia* **15**, 369–397.

Miklas, N., Renou, M., Malosse, I. & Malosse, C. (2000). Repeatability of pheromone blend composition in individual males of the southern green stink bug, *Nezara viridula*. *Journal of Chemical Ecology* **26**, 2473–2485.

Miller, J. C. & Ehler, L. E. (1990). The concept of parasitoid guild and its relevance to biological control. In *Critical Issues in Biological Control* (eds. M. MacKauer, L. E. Ehler & J. Roland), pp. 159–169. Intercept, Andover.

Mills, N. J. (1990). Are parasitoids of significance in endemic populations of forest defoliators? Some experimental observations from Gypsy Moth, *Lymantria dispar* (Lepidoptera: Lymantriidae). In *Population Dynamics of Forest Insects* (eds. A. D. Watt, S. R. Leather, M. D. Hunter & N. A. C. Kidd), pp. 265–274. Intercept, Andover.

Mills, N. J. (1992). Parasitoid guilds, life-styles, and host ranges in the parasitoid complexes of Tortricoid Hosts (Lepidoptera: Tortricoidea). *Environmental Entomology* **21**, 230–239.

Mills, N. J. (1994a). Biological control: some emerging trends. In *Individuals, Populations and Patterns in Ecology* (eds. S. R. Leather, A. D. Watt, N. J. Mills & K. F. A. Walters), pp. 213–222. Intercept, Andover.

Mills, N. J. (1994b). The structure and complexity of parasitoid communities in relation to biological control. In *Parasitoid Community Ecology* (eds. B. A. Hawkins & W. Sheehan), pp. 397–417. Oxford University Press, Oxford.

Milne, M. & Walter, G. H. (1997). The significance of prey in the diet of the phytophagous thrips, *Frankliniella schultzei*. *Ecological Entomology* **22**, 74–81.

Milne, M. & Walter, G. H. (1998a). Host species and plant part specificity of the polyphagous onion thrips, *Thrips tabaci* (Thysanoptera: Thripidae), in an Australian cotton-growing area. *Australian Journal of Entomology* **37**, 115–119.

Milne, M. & Walter, G. H. (1998b). Significance of mite prey in the diet of the onion thrips *Thrips tabaci* Lindeman (Thysanoptera: Thripidae). *Australian Journal of Entomology* **37**, 120–124.

Milne, M. & Walter, G. H. (2000). Feeding and breeding across host plants within a locality by the widespread thrips *Frankliniella schultzei*, and the invasive potential of polyphagous herbivores. *Diversity and Distributions* **6**, 243–257.

Mitchell, D. S. (1972). The kariba weed: *Salvinia molesta*. *British Fern Gazette* **10**, 251–252.

Mitchell, P. B. (1991). Historical perspectives on some vegetation and soil changes in semi-arid New South Wales. *Vegetatio* **91**, 169–182.

Mitter, C., Poole, R. W. & Matthews, M. (1993). Biosystematics of the Heliothinae (Lepidoptera: Noctuidae). *Annual Review of Entomology* **38**, 207–225.

Mohyuddin, A. I. (1991). Utilization of natural enemies for the control of insect pests of sugarcane. *Insect Science and Application* **12**, 19–26.

Mohyuddin, A. I., Inayatullah, C. & King, E. C. (1981). Host selection and strain occurrence in *Apanteles flavipes* (Cameron) (Hymenoptera; Braconidae) and its bearing on biological control of graminaceous stem-borers (Lepidoptera: Pyralidae). *Bulletin of Entomological Research* **71**, 575–581.

Moore, J. A. (1987). The not so silent spring. In *Silent Spring Revisited* (eds. G. J. Marco, R. M. Hollingworth & W. Durham), pp. 15–24. American Chemical Society, Washington, DC.

Mopper, S. & Strauss, S. Y. (1998). *Genetic Structure and Local Adaptation in Natural Insect Populations: Effects of Ecology, Life History, and Behavior*. Chapman & Hall, New York.

Moran, C. & Shaw, D. D. (1977). Population cytogenetics of the genus *Caledia* (Orthoptera: Acrididae). *Chromosoma* **63**, 181–204.

Moran, N. A. (1992). The evolution of aphid life cycles. *Annual Review of Entomology* **37**, 321–348.

Morton, E. S. (1975). Ecological sources of selection on avian sounds. *American Naturalist* **109**, 17–34.

Mowry, B. (1985). From Galen's theory to William Harvey's theory: a case study in the rationality of scientific theory change. *Studies in the History and Philosophy of Science* **16**, 49–82.

Muggleton, J. (1983). Relative fitness of Malathion-resistant phenotypes of *Oryzaephilus surinamensis* L. (Coleoptera: Silvanidae). *Journal of Applied Ecology* **20**, 245–254.

Murdoch, W. W. (1992). Ecological theory and biological control. In *Applied Population Biology* (eds. S. K. Jain & L. W. Botsford), pp. 197–221. Kluwer Academic Publishers, Dordrecht.

Murdoch, W. W. (1994). Population regulation in theory and practice. *Ecology* **75**, 271–287.

Murdoch, W. W. & Briggs, C. J. (1996). Theory for biological control: recent developments. *Ecology* **77**, 2001–2013.

Murdoch, W. W., Briggs, C. J. & Nisbet, R. M. (1996a). Competitive displacement and biological control in parasitoids: a model. *American Naturalist* **148**, 807–826.

Murdoch, W. W., Briggs, C. J. & Nisbet, R. M. (1997). Dynamical effects of host size- and parasitoid state-dependent attacks by parasitoids. *Journal of Animal Ecology* **66**, 542–556.

Murdoch, W. W., Chesson, J. & Chesson, P. L. (1985). Biological control in theory and practice. *American Naturalist* **125**, 344–366.

Murdoch, W. W., Swarbrick, S. L., Luck, R. F., Walde, S. & Yu, D. S. (1996b). Refuge dynamics and metapopulation dynamics: an experimental test. *American Naturalist* **147**, 424–444.

Murray, B. G. (1999). Can the population regulation controversy be buried and forgotten? *Oikos* **84**, 148–152.

Myers, J. H., Higgins, C. & Kovacs, E. (1989). How many insect species are necessary for the biological control of insects? *Environmental Entomology* **18**, 541–547.

Myers, J. H., Savoie, A. & van Randen, E. (1998). Eradication and pest management. *Annual Review of Entomology* **43**, 471–491.

Myers, J. H., Smith, J. N. M. & Elkinton, J. S. (1994). Biological control and refuge theory. *Science* **265**, 811.

Narang, S. K., Tabachnick, W. J. & Faust, R. M. (1993). Complexities of population genetic structure and implications for biological control programs. In *Applications of Genetics to Arthropods of Biological Control Significance* (eds. S. K. Narang, A. C. Bartlett & R. M. Faust), pp. 10–52. CRC Press, Boca Raton.

National Health and Medical Research Council (1993). *Cyclodiene Insecticide Use in Australia*. Commonwealth Government Printer for National Health and Medical Research Council, Canberra.

Navajas, M., Tsagkarakov, A., Lagnel, J. & Perrot-Minnot, M. J. (2000). Genetic differentiation in *Tetranychus urticae* (Acari: Tetranychidae): polymorphism, host races or sibling species? *Experimental and Applied Acarology* **24**, 365–376.

Neuffer, G. (1990). Zur Abundanz und Gradation der San-José-Schildaus *Quadraspidiotus pernicicosus* Comst. und deren Gegenspieler *Prospaltella perniciosi* Tow. *Gesunde Pflanzen* **42**, 89–96.

Newman, E. I. (1993). *Applied Ecology*. Blackwell Scientific Publications, Oxford.

Newton, P. J. (1988). Movement and impact of *Trichogrammatoidea cryptophlebiae* Nagaraja (Hymenoptera: Trichogrammatoidea) in citrus orchards after inundative releases against the false codling moth, *Cryptophlebia leucotreta* (Meyrick) (Lepidoptera: Tortricidae). *Bulletin of Entomological Research* **78**, 85–99.

Norton, G. A. & Mumford, J. D. (1993). *Decision Tools for Pest Management*. CAB International, Wallingford.

Norton, R. A. & Palmer, S. C. (1991). The distribution, mechanisms and evolutionary significance of parthenogenesis in oribatid mites. In *The Acari: Reproduction, Development and Life-history Strategies* (eds. R. Schuster & P. W. Murphy), pp. 107–136. Chapman & Hall, London.

Novotny, V. (1994). Association of polyphagy in leafhoppers (Auchenorrhyncha, Hemiptera) with unpredictable environments. *Oikos* 70, 223–232.

Nylin, S. (2001). Life history perspectives on pest insects: what's the use? *Austral Ecology* 26, 507–517.

Obata, S. (1986). Mechanisms of prey finding in the aphidophagous ladybird beetle, *Harmonia axyridis* (Coleoptera: Coccinellidae). *Entomophaga* 31, 303–311.

Oliver, J. H. (1988). Crisis in biosystematics of arthropods. *Science* 240, 967.

O'Neil, R. J. (1989). Comparison of laboratory and field measurements of the functional response of *Podisus maculiventris* (Heteroptera: Pentatomidae). *Journal of the Kansas Entomological Society* 62, 148–155.

O'Neil, R. J. (1990). Functional response of arthropod predators and its role in the biological control of insect pests in agricultural systems. In *New Directions in Biological Control: Alternatives for Suppressing Agricultural Pests and Diseases* (eds. R. R. Baker & P. E. Dunn), pp. 83–96. Alan R. Liss, New York.

O'Neill, R. V. (2001). Is it time to bury the ecosystem concept? (With full military honors of course!). *Ecology* 82, 3275–3284.

O'Neill, W. M. (1969). The motion of the blood. I. Case study. In *Fact and Theory: An Aspect of the Philosophy of Science*, pp. 5–14. Sydney University Press, Sydney.

Ordish, G. (1976). *The Constant Pest: A Short History of Pests and their Control*. Peter Davies, London.

Orr, D. W. (1994). Technological fundamentalism. *Conservation Biology* 8, 335–337.

Ospovat, D. (1981). *The Development of Darwin's Theory: Natural History, Natural Theology, and Natural Selection, 1838–1859*. Cambridge University Press, Cambridge.

Osteen, C. D. & Szmedra, P. I. (1989). Agricultural pesticide use trends and policy issues. *USDA Agricultural Economic Report* 622, 1–87.

Ota, D. & Čokl, L. (1991). Mate location in the southern green stink bug *Nezara viridula* (Heteroptera: Pentatomidae) mediated through substrate-borne signals on ivy. *Journal of Insect Behavior* 4, 441–447.

Oyeyele, S. & Zalucki, M. P. (1990). Cardiac glycosides and oviposition by *Danaus plexippus* on *Asclepias fruticosa* in south-east Queensland (Australia), with notes on the effects of plant nitrogen content. *Ecological Entomology* 15, 177–185.

Panagiotakopulu, E., Buckland, P. C., Day, P. M., Sarpaki, A. A. & Doumas, C. (1995). Natural insecticides and insect repellents in antiquity: a review of the evidence. *Journal of Archaeological Science* 22, 705–710.

Panizzi, A. R. (1997). Wild hosts of pentatomids: ecological significance and role in their pest status on crops. *Annual Review of Entomology* 42, 99–122.

Papacek, D. & Smith, D. (1998). Sustaining IPM as a commercial service to citrus growers. In *Pest Management – Future Challenges. Proceedings of the Sixth Australasian Applied Entomological Research Conference*, vol. 1 (eds. M. P. Zalucki, R. A. I. Drew & G. G. White), pp. 106–112. University of Queensland Printery, Brisbane.

Parmesan, C., Ryrholm, N., Stefanescu, C. *et al*. (1999). Poleward shifts in geographical ranges of butterfly species associated with regional warming. *Nature* 399, 579–583.

Pashley, D. P., Hammond, A. M. & Hardy, T. N. (1992). Reproductive isolating mechanisms in fall armyworm host strains (Lepidoptera, Noctuidae). *Annals of the Entomological Society of America* 85, 400–405.

Paskewitz, S. M. & Collins, F. H. (1990). Use of the polymerase chain reaction to identify mosquito species of the *Anopheles gambiae* complex. *Medical and Veterinary Entomology* 4, 367–373.

Paterson, H. E. H. (1964). Direct evidence for the specific distinctness of forms A, B, and C of the *Anopheles gambiae* complex. *Revista di Malariologia* 43, 192–196.

Paterson, H. E. (1973). Animal species studies. *Journal of the Royal Society of Western Australia* 56, 31–36.

Paterson, H. E. H. (1978). More evidence against speciation by reinforcement. *South African Journal of Science* 74, 369–371.

Paterson, H. E. (1980). A comment on "mate recognition systems". *Evolution* 34, 330–331.

Paterson, H. E. H. (1981). The continuing search for the unknown and the unknowable: a critique of contemporary ideas on speciation. *South African Journal of Science* 77, 113–119.

Paterson, H. E. H. (1985). The recognition concept of species. In *Species and Speciation* (ed. E. S. Vrba), pp. 21–29. Transvaal Museum, Pretoria.

Paterson, H. E. H. (1986). Environment and species. *South African Journal of Science* 82, 62–65.

Paterson, H. E. H. (1988). On defining species in terms of sterility: problems and alternatives. *Pacific Science* 42, 65–71.

Paterson, H. E. H. (1989). A view of species. In *Dynamic Structures in Biology* (eds. B. Goodwin, A. Sibatani & G. Webster), pp. 77–88. Edinburgh University Press, Edinburgh.

Paterson, H. E. H. (1991). The recognition of cryptic species among economically important insects. In *Heliothis: Research Methods and Prospects* (ed. M. P. Zalucki), pp. 1–10. Springer-Verlag, New York.

Paterson, H. E. (1993a). Botha de Meillon and the *Anopheles gambiae* complex. In *Entomologist Extraordinary: A Festschrift in honour of Botha de Meillon* (ed. M. Coetzee), pp. 39–46. South African Institute for Medical Research, Johannesburg.

Paterson, H. E. H. (1993b). Animal species and sexual selection. In *Evolutionary Patterns and Processes* (eds. D. R. Lees & D. Edwards), pp. 209–228. Academic Press for the Linnean Society of London, London.

Paterson, H. E. H. (1993c). *Evolution and the Recognition Concept of Species: Collected Writings*. Johns Hopkins University Press, Baltimore.

Paterson, H. E. H. & Macnamara, M. (1984). The recognition concept of species. *South African Journal of Science* **80**, 312–318.

Paterson, H. E., Paterson, J. S. & van Eeden, G. J. (1963). A new member of the *Anopheles gambiae* complex. *Medical Proceedings* **9**, 414–418.

Pedigo, L. P. (1995). Closing the gap between IPM theory and practice. *Journal of Agricultural Entomology* **12**, 171–181.

Pedigo, L. P. (1999). *Entomology and Pest Management*, 3rd edn. Prentice Hall, Upper Saddle River, N.J.

Perkins, J. H. (1982). *Insects, Experts and the Insecticide Crisis: The Quest for New Pest Management Strategies*. Plenum Press, New York.

Perring, T. M., Cooper, A. D., Rodriguez, R. J., Farrar, C. A. & Bellows, T. S. (1993). Identification of a whitefly species by genomic and behavioural studies. *Science* **259**, 74–77.

Perutz, M. (1991). *Is Science Necessary? Essays on Science and Scientists*. Oxford University Press, Oxford.

Peters, R. H. (1991). *A Critique for Ecology*. Cambridge University Press, Cambridge.

Petrie, H. G. (1976). Do you see what I see? The epistemology of interdisciplinary inquiry. *Journal of Aesthetic Education* **10**, 29–43.

Pielou, E. C. (1991). *After the Ice Age: The Return of Life to Glaciated North America*. University of Chicago Press, Chicago.

Pierce, G. J. & Ollason, J. G. (1987). Eight reasons why optimal foraging is a complete waste of time. *Oikos* **49**, 111–117.

Pigott, C. D. (1990). The influence of evergreen coniferous nurse-crops on the field layer in two woodland communities. *Journal of Applied Ecology* **27**, 448–459.

Pimentel, D. (1997). *Techniques for Reducing Pesticide Use*. John Wiley, Chichester.

Pimentel, D. (1997). Pest management in agriculture. In *Techniques for Reducing Pesticide Use – Economic and Environmental Benefits* (ed. D. Pimentel), pp. 1–10. John Wiley, New York.

Pimentel, D., Andow, D., Dyson-Hudson, R. *et al.* (1980). Environmental and social costs of pesticides: a preliminary assessment. *Oikos* **34**, 126–140.

Pimentel, D. & Lehman, H. (1993). *The Pesticide Question*. Chapman & Hall, New York.

Pimm, S. L. (1991). *The Balance of Nature? Ecological Issues in the Conservation of Species and Communities*. University of Chicago Press, Chicago.

Pinto, J. D. & Stouthamer, R. (1994). Systematics of the Trichogrammatidae with emphasis on *Trichogramma*. In *Biological Control with Egg Parasitoids* (eds. E. Wajnberg & S. A. Hassan), pp. 1–36. CAB International, Wallingford.

Pinto, J. D., Stouthamer, R., Platner, G. R. & Oatman, E. R. (1991). Variation in reproductive compatibility in *Trichogramma* and its taxonomic significance (Hymenoptera: Trichogrammatidae). *Annals of the Entomological Society of America* **84**, 37–46.

Pitelka, L. F. (1994). Ecosystem response to elevated CO_2. *Trends in Ecology and Evolution* **9**, 204–207.

Platt, J. R. (1964). Strong inference. *Science* **146**, 347–353.

Polaszek, A. & Walker, A. K. (1991). The *Cotesia flavipes* species complex: parasitoids of cereal stem borers in the tropics. *Redia* **74**, 335–341.

Popper, K. R. (1983). *Realism and the Aim of Science: From the Postscript to the Logic of Scientific Discovery (1956)*. Hutchinson, London.

Price, P. W. (1972). Methods of sampling and analysis for predictive results in the introduction of entomophagous insects. *Entomophaga* **17**, 211–222.

Price, P. W. (1984). The concept of the ecosystem. In *Ecological Entomology* (eds. C. B. Huffaker & R. L. Rabb), pp. 19–50. John Wiley, New York.

Price, P. W. (1991). Evolutionary theory of host and parasitoid interactions. *Biological Control* **1**, 83–93.

Price, P. W. (1996). Empirical research and factually based theory: what are their roles in entomology? *American Entomologist* **42**, 209–214.

Price, P. W., Gaud, W. S. & Slobodchikoff, C. N. (1984). Introduction: is there a new ecology? In *A New Ecology: Novel Approaches to Interactive Systems* (eds. P. W. Price, C. N. Slobodchikoff & W. S. Gaud), pp. 1–11. John Wiley, New York.

Price, P. W. & Waldbauer, G. P. (1994). Ecological aspects of pest management. In *Introduction to Insect Pest Management* (eds. R. L. Metcalf & W. H. Luckman), pp. 35–72. John Wiley, New York.

Purves, W. K., Orians, G. H., Heller, H. C. & Sadava, D. (1998). *Life: The Science of Biology*, 5th edn. Sinauer Associates, Sunderland, Mass.

Quicke, D. L. J. (1997). *Parasitic Wasps*. Chapman & Hall, London.

Quine, W. V. & Ullian, J. S. (1978). *The Web of Belief*, 2nd edn. Random House, New York.

Radcliffe, E. B. & Hutchison, W. D. (2002). *Radcliffe's IPM World Textbook.* University of Minnesota, St Paul. http://ipmworld.umn.edu/.

Rainey, R. C. (1989). *Migration and Meteorology: Flight Behaviour and the Atmospheric Environment of Locusts and Other Migrant Pests.* Clarendon Press, Oxford.

Ramirez, O. A. & Mumford, J. D. (1995). The role of public policy in implementing IPM. *Crop Protection* 14, 565–572.

Rao, S. V. & DeBach, P. (1969). Experimental studies on hybridization and sexual isolation between some *Aphytis* species (Hymenoptera: Aphelinidae). I. Experimental hybridization and an interpretation of evolutionary relationships among the species. *Hilgardia* 39, 515–553.

Rapport, D. J. (1991). Myths in the foundations of economics and ecology. *Biological Journal of the Linnean Society* 44, 185–202.

Ratcliffe, L. M. & Grant, P. R. (1985). Species recognition in Darwin's finches (*Geospiza*, Gould). III. Male responses to playback of different song types, dialects and heterospecific songs. *Animal Behaviour* 33, 290–307.

Readshaw, J. L. (1989). The influence of seasonal temperatures on the natural regulation of the screwworm, *Cochliomyia hominivorax*, in the southern U.S.A. *Medical and Veterinary Entomology* 3, 159–168.

Rees, D. A. (1993). Time for scientists to pay their dues. *Nature* 363, 203–204.

Reeve, J. D. (1987). Foraging behavior of *Aphytis melinus*: effects of patch density and host size. *Ecology* 68, 530–538.

Reeve, J. D. (1988). Environmental variability, migration, and persistence in host–parasitoid systems. *American Naturalist* 132, 810–836.

Rehbock, P. F. (1983). *The Philosophical Naturalists: Themes in Early Nineteenth-century British Biology.* University of Wisconsin Press, Madison.

Remaudière, G. & Naumann-Etienne, K. (1991). Découverte au Pakistan de l'hôte primaire de *Rhopalosiphum maidis* (Fitch) (Hom. Aphididae). *Compte Rendus de l'Academie d'Agriculture Française* 77, 61–62.

Rey, J. R. & McCoy, E. D. (1979). Application of island biogeographic theory to pests of cultivated crops. *Environmental Entomology* 4, 577–582.

Richards, K. T. (1968). *A study of the insect pest complex of the Ord River irrigation area.* Unpublished MSc thesis. University of Western Australia, Perth.

Richardson, B. J., Baverstock, P. R. & Adams, M. (1986). *Allozyme Electrophoresis: A Handbook for Animal Systematics and Population Studies.* Academic Press, London.

Ricklefs, R. E. (1987). Community diversity: relative roles of local and regional processes. *Science* 235, 167–171.

Ricklefs, R. E. (1989). Speciation and diversity: the integration of local and regional processes. In *Speciation and its Consequences* (eds. D. Otte & J. A. Endler), pp. 599–622. Sinauer Associates, Sunderland, Mass.

Rochat, J. & Gutierrez, A. P. (2001). Weather-mediated regulation of olive scale by two parasitoids. *Journal of Animal Ecology* **70**, 476–490.

Roderick, G. K. (1992). Postcolonization evolution of natural enemies. In *Selection Criteria and Ecological Consequences of Importing Natural Enemies* (eds. W. C. Kauffman & J. R. Nechols), pp. 71–86. Entomological Society of America, Lanham, Maryland.

Roelofs, W. L., Du, J.-W., Tang, X.-H., Robbins, P. S. & Eckenrode, C. J. (1985). Three European corn borer populations in New York based on sex pheromones and voltinism. *Journal of Chemical Ecology* **11**, 829–836.

Roland, J. (1994). After the decline: what maintains low winter moth density after successful biological control? *Journal of Animal Ecology* **63**, 392–398.

Roman, S. (1993). Problem-solving. *Nature* **363**, 576.

Romesburg, H. C. (1981). Wildlife science: gaining reliable knowledge. *Journal of Wildlife Management* **45**, 293–313.

Room, P. M. (1986). *Salvinia molesta* – a floating weed and its biological control. In *The Ecology of Exotic Animals and Plants* (ed. R. L. Kitching), pp. 164–186. John Wiley, Brisbane.

Room, P. P., Harley, K. L. S., Forno, I. W. & Sands, D. P. A. (1981). Successful biological control of the floating weed salvinia. *Nature* **294**, 78–80.

Rose, S. (1997). *Lifelines: Biology, Freedom, Determinism*. Penguin Books, London.

Rosen, D. (1978). The importance of cryptic species and specific identifications as related to biological control. In *Biosystematics in Agriculture*, vol. 2 (ed. J. A. Romberger), pp. 23–35. John Wiley, New York.

Rosen, D. (1985). Biological control. In *Comprehensive Insect Physiology, Biochemistry and Pharmacology*, vol. 12, *Insect Control* (eds. G. A. Kerkut & L. I. Gilbert), pp. 413–464. Pergamon Press, Oxford.

Rosen, D. (1986). The role of taxonomy in effective biological control programs. *Agriculture, Ecosystems and Environment* **15**, 121–129.

Rosenbaum, H. C. & Deinard, A. S. (1998). Caution before claim: an overview of microsatellite analysis in ecology and evolutionary biology. In *Molecular Approaches to Ecology and Evolution* (eds. R. DeSalle & B. Schierwater), pp. 87–106. Birkhäuser Verlag, Basel.

Rosenberg, A. (1985). *The Structure of Biological Science*. Cambridge University Press, Cambridge.

Rosenheim, J. A. & Rosen, D. (1991). Foraging and oviposition decisions in the parasitoid *Aphytis lingnanensis*: distinguishing the influences of egg load and experience. *Journal of Animal Ecology* **60**, 873–893.

Rousset, F. & Raymond, M. (1991). Cytoplasmic incompatibility in insects: why sterilize females? *Trends in Ecology and Evolution* **6**, 54–57.

Rouvray, D. H. (1997). The treatment of uncertainty in the sciences. *Endeavour* **21**, 154–158.

Ruberson, J. R. (1999). *Handbook of Pest Management*. Marcel Dekker, New York.

Ruberson, J. R., Tauber, M. J. & Tauber, C. A. (1989). Intraspecific variability in Hymenopteran parasitoids: comparative studies of two biotypes of the egg parasitoid *Edovum puttleri* (Hymenoptera: Eulophidae). *Journal of the Kansas Entomological Society* **62**, 189–202.

Rungrojwanich, K. & Walter, G. H. (2000). The Australian fruit fly parasitoid *Diachasmimorpha kraussii* (Fullaway): mating behavior, modes of sexual communication and crossing tests with *D. longicaudata* (Ashmead) (Hymenoptera: Braconidae: Opiinae). *Pan-Pacific Entomologist* **76**, 12–23.

Rural Industries Research and Development Corporation (1990). *The Market for Australian Produced Organic Food*. Rural Industries Research and Development Corporation, Kingston, ACT, Australia.

Ruse, M. (1976). The scientific methodology of William Whewell. *Centaurus* **20**, 227–257.

Rushton, S. P., Lurz, P. W. W., Gurnell, J. & Fuller, R. (2000). Modelling the spatial dynamics of parapoxvirus disease in red and grey squirrels: a possible cause of the decline in the red squirrel in the UK? *Journal of Applied Ecology* **37**, 997–1012.

Russell, B. (1961). *History of Western Philosophy*, 2nd edn. George Allen and Unwin, London.

Ryan, M. A. (1994). Damage to pawpaw trees by the banana-spotting bug, *Amblypelta lutescens lutescens* (Distant) (Hemiptera: Coreidae), in North Queensland. *International Journal of Pest Management* **40**, 280–282.

Ryan, M. A., Cokl, A. & Walter, G. H. (1996). Differences in vibratory sound communication between a Slovenian and an Australian population of *Nezara viridula* (L.) (Heteroptera: Pentatomidae). *Behavioural Processes* **36**, 183–193.

Ryan, M. A., Moore, C. J. & Walter, G. H. (1995). Individual variation in pheromone composition in *Nezara viridula* (Heteroptera: Pentatomidae): how valid is the basis for designating "pheromone strains"? *Comparative Biochemistry and Physiology B* **111**, 189–193.

Ryan, M. A. & Walter, G. H. (1992). Sound communication in *Nezara viridula* (L.) (Heteroptera: Pentatomidae): further evidence that signal transmission is substrate-borne. *Experientia* **48**, 1112–1115.

Sabrosky, C. W. (1955). The interrelations of biological control and taxonomy. *Journal of Economic Entomology* **48**, 710–714.

Sadeghi, H. & Gilbert, F. (2000a). Aphid suitability and its relationship to oviposition preference in predatory hoverflies. *Journal of Animal Ecology* **69**, 771–784.

Sadeghi, H. & Gilbert, F. (2000b). Oviposition preferences of aphidophagous hoverflies. *Ecological Entomology* **25**, 91–100.

Saetre, G. P., Moum, T., Bures, S., Kral, M., Adamjan, M. & Moreno, J. (1997).

A sexually selected character displacement in flycatchers reinforces premating isolation. *Nature* **387**, 589–592.

Sale, P. F. (1988). Perception, pattern, chance and the structure of reef fish communities. *Environmental Biology of Fishes* **21**, 3–15.

Saloniemi, I. (1993). An environmental explanation for the character displacement pattern in *Hydrobia* snails. *Oikos* **67**, 75–80.

Salt, G. (1935). Experimental studies in insect parasitism. III – Host selection. *Proceedings of the Royal Society of London, Series B* **117**, 413–435.

Salt, G. (1968). The resistance of insect parasitoids to the defense reactions of their hosts. *Biological Reviews* **43**, 200–232.

Samways, M. J. (1989). Climate diagrams and biological control: an example from the areography of the ladybird *Chilocorus nigritus* (Fabricius, 1798) (Insecta, Coleoptera, Coccinellidae). *Journal of Biogeography* **16**, 345–351.

Sasal, P., Trouve, S., Muller-Graf, C. & Morand, S. (1999). Specificity and host predictability: a comparative analysis among monogean parasites of fish. *Journal of Animal Ecology* **68**, 437–444.

Sayaboc, P. D. (1994). *Investigation of the behaviour and host relationships of* Anisopteromalus calandrae *(Howard) (Hymenoptera: Pteromalidae), a parasitoid of stored grain beetles*. Unpublished MAgrSci thesis. University of Queensland.

Sayyed, A. H. (2001). Fitness costs and stability of resistance to *Bacillus thuringiensis* in a field population of the diamondback moth *Plutella xylostella* L. *Ecological Entomology* **26**, 502–508.

Schaefer, C. W. & Ahmad, I. (2000). Cotton stainers and their relatives (Pyrrhocoroidae: Pyrrhocoridae and Largidae). In *Heteroptera of Economic Importance* (eds. C. W. Schaefer & A. R. Panizzi), pp. 271–307. CRC Press, Boca Raton.

Schaefer, C. W. & Panizzi, A. R. (2000). *Heteroptera of Economic Importance*. CRC Press, Boca Raton.

Schamroth, L. (1978). The cardiovascular system: Harvey and before. *South African Journal of Science* **74**, 423–426.

Schilthuizen, M. (2000). Dualism and conflicts in understanding speciation. *BioEssays* **22**, 1134–1141.

Schlötterer, C. & Pemberton, J. (1998). The use of microsatellites for genetic analysis of natural populations – a critical review. In *Molecular Approaches to Ecology and Evolution* (eds. R. DeSalle & B. Schierwater), pp. 71–86. Birkhäuser Verlag, Basel.

Schneider, D. (1987). Plant recognition by insects: a challenge for neuro-ethological research. In *Insects–Plants* (eds. V. Labeyrie, G. Fabres & D. Lachaise), pp. 117–123. W. Junk, Dordrecht.

Schoonhoven, L. M., Jermy, T. & van Loon, J. J. A. (1998). *Insect–Plant Biology*. Chapman & Hall, London.

Schoub, B. D. (1994). *AIDS and HIV in Perspective: A Guide to Understanding the Virus and its Consequences*. Cambridge University Press, Cambridge.

Schröter, C. & Kirchner, O. (1896). Die Vegetation des Bodensees. *Schriften: Vereins für Geschichte des Bodensees und seiner Umgebung* 25, 1–119.

Schröter, C. & Kirchner, O. (1902). Die Vegetation des Bodensees. *Schriften: Vereins für Geschichte des Bodensees und seiner Umgebung* 35, 1–86.

Schumacher, P., Weyeneth, A., Weber, D. C. & Dorn, S. (1997). Long flights in *Cydia pomonella* L. (Lepidoptera: Tortricidae) measured by a flight mill: influence of sex, mated status and age. *Physiological Entomology* 22, 149–160.

Scriber, J. M. (1986). Origins of the regional feeding abilities in the tiger swallowtail butterfly: ecological monophagy and the *Papilio glaucus australis* subspecies in Florida. *Oecologia* 71, 94–103.

Seigler, D. S. (1998). *Plant Secondary Metabolism*. Kluwer Academic Publishers, Boston.

Seymour, J. E. & Sands, D. P. A. (1993). Green vegetable bug (*Nezara viridula* [L.]) (Hemiptera: Pentatomidae) in Australian pecans. In *Pest Control and Sustainable Agriculture* (eds. S. A. Corey, D. J. Dall & W. M. Milne), pp. 226–228. CSIRO, Melbourne.

Shield, J. W. (1959). Population studies in the littoral at Rottnest Island. *Journal of the Royal Society of Western Australia* 42, 89.

Shiva, V. (1991). *The Violence of the Green Revolution: Third World Agriculture, Ecology and Politics*. Third World Network, Penang.

Shrader-Frechette, K. (1996). Methodological rules for four classes of scientific uncertainty. In *Scientific Uncertainty and Environmental Problem Solving* (ed. J. Lemons), pp. 12–39. Blackwell Science, Oxford.

Simberloff, D. (1976). Trophic structure determination and equilibrium in an arthropod community. *Ecology* 57, 395–398.

Simberloff, D. (1980). A succession of paradigms in ecology: essentialism to materialism and probabilism. *Synthese* 43, 3–39.

Simberloff, D. (1982). The status of competition theory in ecology. *Annales Zoologici Fennici* 19, 241–253.

Simberloff, D. (1983). Competition theory, hypothesis testing, and other community ecological buzzwords. *American Naturalist* 122, 626–635.

Simberloff, D. (1984). The great god of competition. *The Sciences* 24, 17–22.

Simberloff, D. (1990). Hypothesis, errors, and statistical assumptions. *Herpetologica* 46, 351–357.

Simberloff, D. (1992). Conservation of pristine habitats and unintended effects of biological control. In *Selection Criteria and Ecological Consequences of Importing Natural Enemies* (eds. W. C. Kauffman & J. R. Nechols), pp. 103–117. Entomological Society of America, Lanham, Maryland.

Simberloff, D. & Dayan, T. (1991). The guild concept and the structure of ecological communities. *Annual Review of Ecology and Systematics* **22**, 115–143.

Simberloff, D. & Stiling, P. (1996). How risky is biological control? *Ecology* **77**, 1965–1974.

Simmonds, F. J. & Greathead, D. J. (1977). Introductions and pest and weed problems. In *Origins of Pest, Parasite, Disease and Weed Problems. 18th Symposium of the British Ecological Society, Bangor* (eds. J. M. Cherrett & G. R. Sagar), pp. 109–124. Blackwell Scientific Publications, Oxford.

Sinclair, M. (1988). *Marine Populations: An Essay on Population Regulation and Speciation*. Washington Sea Grant Program, University of Washington Press, Seattle.

Sinclair, M. & Solemdal, P. (1988). The development of "population thinking" in fisheries biology between 1878 and 1930. *Aquatic Living Resources* **1**, 189–213.

Slobodkin, L. B. (1988). Intellectual problems of applied biology. *BioScience* **38**, 337–342.

Smiley, J. T. (1985). Are chemical barriers necessary for evolution of butterfly–plant associations? *Oecologia* **65**, 580–583.

Smith, A. D. M. & Maelzer, D. A. (1986). Aggregation of parasitoids and density-independence of parasitism in field populations of the wasp *Aphytis melinus* and its host, the red scale *Aonidiella aurantii*. *Ecological Entomology* **11**, 425–434.

Smith, H. S. (1941). Racial segregation in insect populations and its significance in applied entomology. *Journal of Economic Entomology* **34**, 1–13.

Smith, H. S. & DeBach, P. (1942). The measurement of the effect of entomophagous insects on population densities of their hosts. *Journal of Economic Entomology* **35**, 845–849.

Smith, R. F. & Allen, W. W. (1954). Insect control and the balance of nature. *Scientific American* **190**, 38–42.

Smith, R. F., Apple, J. L. & Bottrell, D. G. (1976). The origins of integrated pest management concepts for agricultural crops. In *Integrated Pest Management* (eds. J. L. Apple & R. F. Smith), pp. 1–16. Plenum Press, New York.

Smith, R. F., Mittler, T. E. & Smith, C. N. (1973). *History of Entomology*. Annual Reviews, Palo Alto.

Smith, S. M. (1996). Biological control with *Trichogramma*: advances, successes, and potential of their use. *Annual Review of Entomology* **41**, 375–406.

Smocovitis, V. B. (1996). *Unifying Biology: The Evolutionary Synthesis and Evolutionary Biology*. Princeton University Press, Princeton.

Soehardjan, M. (1989). Transfer of technology in the utilization of *Curinus coeruleus* Mulsant. In *International Workshop: Leucaena psyllid – Problems and Management*, pp. 181–183. Bogor, Indonesia.

Southwood, T. R. E. (1978). *Ecological Methods with Particular Reference to the Study of Insect Populations*. Chapman & Hall, London.

Southwood, T. R. E. (1988). Tactics, strategies and templets. *Oikos* 52, 3–18.

Southwood, T. R. E., Hassell, M. P., Reader, P. M. & Rogers, D. J. (1989). Population dynamics of the viburnum whitefly (*Aleurotrachelus jelinekii*). *Journal of Animal Ecology* 58, 921–942.

Speight, M. R., Hunter, M. D. & Watt, A. D. (1999). *Ecology of Insects: Concepts and Applications*. Blackwell Science, Oxford.

Spencer, H. G., Lambert, D. M. & McArdle, B. H. (1987). Reinforcement, species, and speciation: a reply to Butlin. *American Naturalist* 130, 958–962.

Spencer, H. G., McArdle, B. H. & Lambert, D. M. (1986). A theoretical investigation of speciation by reinforcement. *American Naturalist* 128, 241–262.

Spradbery, J. P. (1994). Screw worm fly: a tale of two species. *Agricultural Zoology Reviews* 6, 1–62.

Stearns, S. C. (1977). The evolution of life history traits: a critique of the theory and a review of the data. *Annual Review of Ecology and Systematics* 8, 145–171.

Stehr, F. W. (1975). Parasitoids and predators in pest management. In *Introduction to Insect Pest Management* (eds. R. L. Metcalf & W. H. Luckman), pp. 147–188. John Wiley, New York.

Steiner, W. M. (1993). Genetics and insect biotypes. In *Applications of Genetics to Arthropods of Biological Control Significance* (eds. S. K. Narang, A. C. Bartlett & R. M. Faust), pp. 1–17. CRC Press, Boca Raton.

Stern, V. M., Smith, R. F., van den Bosch, R. & Hagen, K. S. (1959). The integrated control concept. *Hilgardia* 29, 81–101.

Stevens, J. E. (1994). Science and religion at work. *BioScience* 44, 60–64.

Stich, H. F. (1963). An experimental analysis of the courtship pattern of *Tipula oleracea* (Diptera). *Canadian Journal of Zoology* 41, 99–109.

Stiling, P. D. (1987). The frequency of density dependence in insect host–parasitoid systems. *Ecology* 68, 844–856.

Stiling, P. (1988). Density-dependent processes and key factors in insect populations. *Journal of Animal Ecology* 57, 581–593.

Stoner, K. A., Sawyer, A. J. & Shelton, A. M. (1986). Constraints to the implementation of IPM programs in the USA: a course outline. *Agriculture, Ecosystems and Environment* 17, 253–268.

Stouthamer, R., Luck, R. F. & Hamilton, W. D. (1990). Antibiotics cause parthenogenetic *Trichogramma* (Hymenoptera/Trichogrammatidae) to revert to sex. *Proceedings of the National Academy of Sciences USA* **87**, 2424–2427.

Strickland, E. H. (1945). Could the widespread use of DDT be a disaster? *Entomological News* **56**, 85–88.

Strong, D. R. (1983). Natural variability and the manifold mechanisms of ecological communities. *American Naturalist* **122**, 636–660.

Strong, D. R. (1988). Parasitoid theory: from aggregation to dispersal. *Trends in Ecology and Evolution* **3**, 277–280.

Strong, D. R. (1990). Interface of the natural enemy and environment. In *New Directions in Biological Control: Alternatives for Suppressing Agricultural Pests and Diseases* (eds. R. R. Baker & P. E. Dunn), pp. 57–64. Alan R. Liss, New York.

Strong, D. R., Lawton, J. H. & Southwood, T. R. E. (1984). *Insects on Plants*. Blackwell Scientific Publications, Oxford.

Surtees, G. (1977). The prediction of pest and disease problems in man. In *Origins of Pest, Parasite, Disease and Weed Problems. 18th Symposium of the British Ecological Society, Bangor* (eds. J. M. Cherrett & G. R. Sagar), pp. 347–363. Blackwell Scientific Publications, Oxford.

Sutherland, W. J. (1996). *From Individual Behaviour to Population Ecology*. Oxford University Press, Oxford.

Sutherst, R. W., Maywald, G. F. & Skarratt, D. B. (1995). Predicting insect distributions in a changed climate. In *Insects in a Changing Environment* (eds. R. Harrington & N. E. Stork), pp. 59–91. Academic Press, London.

Sweet, M. H. (2000). Seed and chinch bugs (Lygaeoidea). In *Heteroptera of Economic Importance* (eds. C. W. Schaefer & A. R. Panizzi), pp. 143–264. CRC Press, Boca Raton.

Symondson, W. O. C. & Hemingway, J. (1997). Biochemical and molecular techniques. In *Methods in Ecological and Agricultural Entomology* (eds. D. R. Dent & M. P. Walton), pp. 293–350. CAB International, Wallingford.

Symondson, W. O. C., Sunderland, K. D. & Greenstone, M. H. (2002). Can generalist predators be effective biocontrol agents? *Annual Review of Entomology* **47**, 561–594.

Szmedra, P. I. (1991). Pesticide use in agriculture. In *CRC Handbook of Pest Management in Agriculture*, vol. 1 (ed. D. Pimentel), pp. 649–677. CRC Press, Boca Raton.

Tait, E. J. (1987). Planning an integrated pest management system. In *Integrated Pest Management* (eds. A. J. Burn, T. H. Coaker & P. C. Jepson), pp. 190–207. Academic Press, New York.

Tammaru, T., Kaiteniemi, P. & Ruohomaki, K. (1995). Oviposition choices of *Epirrita autumnata* (Lepidoptera: Geometridae) in relation to its eruptive population dynamics. *Oikos* **74**, 296–304.

Tansley, A. G. (1935). The use and abuse of vegetational concepts and terms. *Ecology* **16**, 284–307.

Tatarenkov, A. & Johannesson, K. (1998). Evidence of a reproductive barrier between two forms of the marine periwinkle *Littorina fabalis* (Gastropoda). *Biological Journal of the Linnean Society* **63**, 349–365.

Tatarenkov, A. & Johannesson, K. (1999). Micro- and macrogeographic allozyme variation in *Littorina fabalis*: do sheltered and exposed forms hybridize? *Biological Journal of the Linnean Society* **67**, 199–212.

Tauber, M. J., Hoy, M. A. & Herzog, D. C. (1984). Biological control in agricultural IPM systems: a brief overview of the current status and future prospects. In *Biological Control in Agricultural IPM Systems* (eds. M. A. Hoy & D. C. Herzog), pp. 3–9. Academic Press, London.

Taylor, L. R. (1986). Synoptic dynamics, migration and the Rothamsted insect survey. *Journal of Animal Ecology* **55**, 1–38.

Templeton, A. R. (1987). Species and speciation. *Evolution* **41**, 233–235.

Templeton, A. R. (1989). The meaning of species and speciation: a genetic perspective. In *Speciation and its Consequences* (eds. D. Otte & J. A. Endler), pp. 3–27. Sinauer Associates, Sunderland, Mass.

Thompson, J. N. (1988). Evolutionary ecology of the relationship between oviposition preference and performance of offspring in phytophagous insects. *Entomologia Experimentalis et Applicata* **47**, 3–14.

Thompson, J. N. (1993). Preference hierarchies and the origin of geographic specialization in host use in swallowtail butterflies. *Evolution* **47**, 1585–1594.

Thompson, J. N. (1994). *The Coevolutionary Process.* University of Chicago Press, Chicago.

Thompson, P. A. & Cox, S. A. (1978). Germination of the bluebell (*Hyacinthoides non-scripta* (L.) Chouard) in relation to distribution and habitat. *Annals of Botany* **42**, 51–62.

Thorne, G. N. (1986). Confessions of a narrow-minded applied biologist, or why do interdisciplinary research? *Annals of Applied Biology* **108**, 205–217.

Thorpe, K. W. (1985). Effects of height and habitat type on egg parasitism by *Trichogramma minutum* and *T. pretiosum* (Hymenoptera: Trichogrammatidae). *Agriculture, Ecosystems and Environment* **12**, 117–126.

Toda, M., Okada, S., Ota, H. & Hikida, T. (2001). Biochemical assessment of evolution and taxonomy of the morphologically poorly diverged geckos, *Gekko yakuensis* and *G. hokuensis* (Reptilia: Squamata) in Japan, with special reference to their occasional hybridization. *Biological Journal of the Linnean Society* **73**, 153–165.

Travis, J. (1989). The role of optimizing selection in natural populations. *Annual Review of Ecology and Systematics* **20**, 279–296.

Tsagkarakou, A., Navajas, M., Papaioannou-Souliotis, P. & Pasteur, N. (1998). Gene flow among *Tetranychus urticae* (Acari: Tetranychidae) populations in Greece. *Molecular Ecology* 7, 71–79.

Tuchman, B. W. (1978). *A Distant Mirror: The Calamitous 14th Century*. Macmillan, London.

Tumlinson, J. H., Turlings, T. C. J. & Lewis, W. J. (1992). The semiochemical complexes that mediate insect parasitoid foraging. *Agricultural Zoology Reviews* 5, 221–252.

Turchin, P. (1999). Population regulation: a synthetic view. *Oikos* 84, 153–159.

Turchin, P. (2001). Does population ecology have general laws? *Oikos* 94, 17–26.

Vadas, R. L. (1994). The anatomy of an ecological controversy: honey-bee searching behavior. *Oikos* 69, 158–166.

van Alphen, J. J. M. & Vet, L. E. M. (1986). An evolutionary approach to host finding and selection. In *Insect Parasitoids* (eds. J. Waage & D. Greathead), pp. 23–61. Academic Press, London.

Van den Bosch, F., Hengeveld, R. & Metz, J. A. J. (1992). Analysing the velocity of animal range expansion. *Journal of Biogeography* 19, 135–150.

van Driesche, R. G. & Bellows, T. S. (1996). *Biological Control*. Chapman & Hall, London.

van Emden, H. F. & Peakall, D. B. (1996). *Beyond Silent Spring: Integrated Pest Management and Chemical Safety*. Chapman & Hall, London.

van Huizen, T. H. P. (1977). The significance of flight activity in the life cycle of *Amara plebeja* Gyll. (Coleoptera, Carabidae). *Oecologia* 29, 27–41.

van Huizen, T. H. P. (1979). Individual and environmental factors determining flight in Carabid beetles. In *On the Evolution of Behaviour in Carabid Beetles. Miscellaneous Papers 18 (1979), Agricultural University of Wageningen, The Netherlands* (eds. P. J. den Boer, H. U. Thiele & F. Weber), pp. 199–212. H. Veenman and B. V. Zonen, Wageningen/Zoological Institute of the University of Cologne.

van Klinken, R. D. & Walter, G. H. (1996). The ecology of organisms that breed in a divided and ephemeral habitat: insects of fallen rainforest fruit. *Acta Œcologica* 17, 405–420.

van Lenteren, J. C. (1980). Evaluation of control capabilities of natural enemies: does art have to become science? *Netherlands Journal of Zoology* 30, 369–381.

van Lenteren, J. C. & Woets, J. (1988). Biological and integrated pest control in greenhouses. *Annual Review of Entomology* 33, 239–270.

Van Valen, L. (1965). Morphological variation and width of ecological niche. *American Naturalist* 99, 377–390.

Varley, G. C., Gradwell, G. R. & Hassell, M. P. (1973). *Insect Population Ecology: An Analytical Approach*. Blackwell Scientific Publications, Oxford.

Velasco, L. R. I. (1990). *The influence of host plants in the population dynamics of the green vegetable bug* (Nezara viridula *Linn.*) *in S.E. Queensland, Australia.* Unpublished PhD thesis. University of Queensland.

Velasco, L. R. I. & Walter, G. H. (1992). Availability of different host plant species and changing abundance of the polyphagous bug *Nezara viridula* (Hemiptera: Pentatomidae). *Environmental Entomology* 21, 751–759.

Velasco, L. R. I. & Walter, G. H. (1993). Influence of temperature on survival and reproduction of *Nezara viridula* (L.) (Hemiptera: Pentatomidae). *Journal of the Australian Entomological Society* 32, 225–228.

Vet, L. E. M. & Dicke, M. (1992). Ecology of infochemical use by natural enemies in a tritrophic context. *Annual Review of Entomology* 37, 141–172.

Vet, L. E. M., Lewis, W. J., Papaj, D. R. & van Lenteren, J. C. (1990). A variable-response model for parasitoid foraging behavior. *Journal of Insect Behavior* 3, 471–490.

Vinson, S. B. (1977). Behavioral chemicals in the augmentation of natural enemies. In *Biological Control by Augmentation of Natural Enemies* (eds. R. L. Ridgway & S. B. Vinson), pp. 237–279. Plenum Publishing Corporation, Texas.

Vinson, S. B. (1991). Chemical signals used by parasitoids. *Redia* 74, 15–42.

Vinson, S. B. (1998). The general host selection behavior of parasitoid Hymenoptera and a comparison of initial strategies utilized by larvaphagous and oophagous species. *Biological Control* 11, 79–96.

Vorley, V. T. & Wratten, S. D. (1987). Migration of parasitoids (Hymenoptera: Braconidae) of cereal aphids (Hemiptera: Aphididae) between grassland, early-sown cereals and late-sown cereals in southern England. *Bulletin of Entomological Research* 77, 555–568.

Vrba, E. S. (1980). Evolution, species and fossils: how does life evolve? *South African Journal of Science* 76, 61–84.

Vrba, E. S. (1995). Species as habitat-specific, complex systems. In *Speciation and the Recognition Concept: Theory and Application* (eds. D. M. Lambert & H. G. Spencer), pp. 3–44. Johns Hopkins University Press, Baltimore.

Waage, J. K. (1979). Foraging for patchily-distributed hosts by the parasitoid, *Nemeritis canescens*. *Journal of Animal Ecology* 48, 353–371.

Waage, J. (1990). Ecological theory and the selection of biological control agents. In *Critical Issues in Biological Control* (eds. M. Mackauer, L. E. Ehler & J. Roland), pp. 135–157. Intercept, Andover.

Waage, J. K. & Greathead, D. J. (1988). Biological control: challenges and opportunities. *Philosophical Transactions of the Royal Society of London, Series B* 318, 111–128.

Waage, J. K. & Hassell, M. P. (1982). Parasitoids as biological control agents – a fundamental approach. In *Parasites as Biological Control Agents* (eds. R. M.

Anderson & E. U. Canning), pp. 5–144. Cambridge University Press, Cambridge.

Waage, J. K. & Mills, N. J. (1992). Biological control. In *Natural Enemies: The Population Biology of Predators, Parasites and Diseases* (ed. M. J. Crawley), pp. 412–430. Blackwell Scientific Publications, Oxford.

Waite, G. K. & Huwer, R. K. (1998). Host plants and their role in the ecology of the fruitspotting bugs *Amblypelta nitida* Stål and *Amblypelta lutescens lutescens* (Distant) (Hemiptera: Coreidae). *Australian Journal of Entomology* 37, 340–349.

Wallace, B. (1985). Reflections on some insect pest control procedures. *Bulletin of the Entomological Society of America* 31, 8–13.

Walter, G. H. (1988a). Activity patterns and egg production in *Coccophagus bartletti*, an aphelinid parasitoid of scale insects. *Ecological Entomology* 13, 95–105.

Walter, G. H. (1988b). Competitive exclusion, coexistence and community structure. *Acta Biotheoretica* 37, 281–313.

Walter, G. H. (1988c). Heteronomous host relationships in aphelinids – evolutionary pathways and adaptive significance (Hymenoptera: Chalcidoidea). In *Parasitic Hymenoptera Research* (ed. U. K. Gupta), pp. 313–326. Brill, Leiden.

Walter, G. H. (1991). What is resource partitioning? *Journal of Theoretical Biology* 150, 137–143.

Walter, G. H. (1993a). The concept of interaction in ecological theory. In *Lectures in Theoretical Biology: The Second Stage* (eds. K. Kull & T. Tiivel), pp. 131–148. Estonian Academy of Sciences, Tallinn.

Walter, G. H. (1993b). Mating behaviour of two closely related *ochraceus*-group *Coccophagus* species (Hymenoptera: Aphelinidae). *African Entomology* 1, 15–24.

Walter, G. H. (1993c). Oviposition behaviour of diphagous parasitoids (Hymenoptera, Aphelinidae): a case of intersexual resource partitioning? *Behaviour* 124, 73–87.

Walter, G. H. (1995a). Book Review: 'Parasitoids – Behavioral and Evolutionary Ecology'. *Journal of the Australian Entomological Society* 34, 6, 12.

Walter, G. H. (1995b). Species concepts and the nature of ecological generalizations about diversity. In *Speciation and the Recognition Concept: Theory and Application* (eds. D. M. Lambert & H. G. Spencer), pp. 191–224. Johns Hopkins University Press, Baltimore.

Walter, G. H. & Benfield, M. D. (1994). Temporal host plant use in three polyphagous Heliothinae, with special reference to *Helicoverpa punctigera* (Wallengren) (Noctuidae: Lepidoptera). *Australian Journal of Ecology* 19, 458–465.

Walter, G. H. & Donaldson, J. S. (1994). Heteronomous hyperparasitoids, sex ratios and adaptations. *Ecological Entomology* **19**, 89–92.

Walter, G. H. & Hengeveld, R. (2000). The structure of the two ecological paradigms. *Acta Biotheoretica* **48**, 15–46.

Walter, G. H., Hulley, P. E. & Craig, A. J. F. K. (1984). Speciation, adaptation and interspecific competition. *Oikos* **43**, 246–248.

Walter, G. H. & Paterson, H. E. H. (1994). The implications of palaeontological evidence for theories of ecological communities and species richness. *Australian Journal of Ecology* **19**, 241–250.

Walter, G. H. & Paterson, H. E. H. (1995). Levels of understanding in ecology: interspecific competition and community ecology. *Australian Journal of Ecology* **20**, 463–466.

Walter, G. H. & Zalucki, M. P. (1999). Rare butterflies and theories of evolution and ecology. In *Biology of Australian Butterflies. Monographs on Australian Lepidoptera*, vol. 6 (eds. R. L. Kitching, E. Scheermeyer, R. E. Jones & N. E. Pierce), pp. 349–368. CSIRO, Melbourne.

Wangersky, P. J. (1978). Lotka–Volterra population models. *Annual Review of Ecology and Systematics* **9**, 189–218.

Wapshere, A. J. (1989). A testing sequence for reducing rejection of potential biological control agents for weeds. *Annals of Applied Biology* **114**, 515–526.

Ward, L. K. & Spalding, D. F. (1993). Phytophagous British insects and mites and their food-plant families: total numbers and polyphagy. *Biological Journal of the Linnean Society* **49**, 257–276.

Waterhouse, D. F. & Norris, K. R. (1987a). *Biological Control: Pacific Prospects*. Inkata Press, Melbourne.

Waterhouse, D. F. & Norris, K. R. (1987b). *Nezara viridula* (Linnaeus). In *Biological Control: Pacific Prospects* (eds. D. F. Waterhouse & K. R. Norris), pp. 81–89. Inkata Press, Melbourne.

Waterhouse, D. F. & Sands, D. P. A. (2001). *Classical Biological Control of Arthropods in Australia*. Australian Centre for International Agricultural Research, Canberra.

Waters, J. M., Esa, Y. B. & Wallis, G. P. (2001). Genetic and morphological evidence for reproductive isolation between sympatric populations of *Galaxias* (Teleostei: Galaxiidae) in South Island, New Zealand. *Biological Journal of the Linnean Society* **73**, 287–298.

Way, M. J. (1977). Pest and disease status in mixed stands vs. monocultures: the relevance of ecosystem stability. In *Origins of Pest, Parasite, Disease and Weed Problems. 18th Symposium of the British Ecological Society, Bangor* (eds. J. M. Cherrett & G. R. Sagar), pp. 127–138. Blackwell Scientific Publications, Oxford.

Webb, J. W. & Moran, V. C. (1978). The influence of the host plant on the population dynamics of *Acizzia russellae* (Homoptera: Psyllidae). *Ecological Entomology* **3**, 313–321.

Weiner, J. (1995). On the practice of ecology. *Journal of Ecology* **83**, 153–158.

Wellington, W. G. (1977). Returning the insect to insect ecology: some consequences for pest management. *Environmental Entomology* **6**, 1–70.

White, C. S. & Lambert, D. M. (1994). Genetic differences among pheromonally distinct New Zealand leafroller moths. *Biochemical Systematics and Ecology* **22**, 329–339.

White, I. M. & Elson-Harris, M. M. (1994). *Fruit Flies of Economic Significance: Their Identification and Bionomics*. CAB International, Wallingford.

White, M. J. D. (1970). Heterozygosity and genetic polymorphism in parthenogenetic animals. In *Essays in Evolution and Genetics in Honor of Theodosius Dobzhansky* (eds. M. C. Hecht & W. C. Steere), pp. 237–262. Appleton-Century-Crofts, New York.

White, M. J. D. (1973). *Animal Cytology and Evolution*. Cambridge University Press, Cambridge.

White, T. C. R. (1969). An index to measure weather-induced stress of trees associated with outbreaks of psyllids in Australia. *Ecology* **50**, 905–909.

White, T. C. R. (1970a). Some aspects of the life history, host selection, dispersal, and oviposition of adult *Cardiaspina densitexta* (Homoptera: Psyllidae). *Australian Journal of Zoology* **18**, 105–117.

White, T. C. R. (1970b). The nymphal stage of *Cardiaspina densitexta* (Homoptera: Psyllidae) on leaves of *Eucalyptus fasciculosa*. *Australian Journal of Zoology* **18**, 273–293.

White, T. C. R. (2001). Opposing paradigms: regulation or limitation of populations? *Oikos* **93**, 148–152.

Whitten, M. J. (1992). Pest management in 2000: what we might learn from the twentieth century. In *Pest Management and the Environment in 2000* (eds. A. A. S. A. Kadir & H. S. Barlow), pp. 9–44. CAB International, Wallingford.

Whittle, C. P., Bellas, T. E. & Bishop, A. L. (1991). Sex pheromone of lucerne leafroller, *Merophyas divulsana* (Walker) (Lepidoptera: Tortricidae): evidence for two distinct populations. *Journal of Chemical Ecology* **17**, 1883–1895.

Whorton, J. C. (1974). *Before Silent Spring: Pesticides and Public Health in Pre-DDT America*. Princeton University Press, Princeton.

Wieland, B. (1985). Towards an economic theory of scientific revolutions – a cynical view. *Erkenntnis* **23**, 79–95.

Wiens, J. H. (1977). On competition and variable environments. *American Scientist* **65**, 590–597.

Wigglesworth, V. B. (1955). The contribution of pure science to applied biology. *Annals of Applied Biology* **42**, 34–44.

Wiklund, C. (1981). Generalist vs. specialist oviposition behaviour in *Papilio machaon* (Lepidoptera) and functional aspects on the hierarchy of oviposition preferences. *Oikos* **36**, 163–170.

Wilkerson, R. C., Parsons, T. J., Albright, D. G., Klein, T. A. & Braun, M. J. (1993). Random amplified polymorphic DNA (RAPD) markers readily distinguish cryptic mosquito species (Diptera: Culicidae: *Anopheles*). *Insect Molecular Biology* **1**, 205–211.

Williams, G. C. (1966). *Adaptation and Natural Selection: A Critique of Some Current Evolutionary Thought.* Princeton University Press, Princeton.

Williams, T. & Hails, R. S. (1994). Biological control and refuge theory. *Science* **265**, 811–812.

Williamson, M. (1996). *Biological Invasions.* Chapman & Hall, London.

Wilson, E. O. (1971). *The Insect Societies.* Belknap Press of Harvard University Press, Cambridge, Mass.

Wilson, F. & Woolcock, L. T. (1960). Temperature determination of sex in a parthenogenetic parasite, *Ooencyrtus submetallicus* (Howard) (Hymenoptera: Encyrtidae). *Australian Journal of Zoology* **8**, 153–169.

Wilson, L. J. & Bauer, L. R. (1993). Species composition and seasonal abundance of thrips (Thysanoptera) on cotton in the Namoi Valley. *Journal of the Australian Entomological Society* **32**, 187–192.

Wint, W. (1983). The role of alternative host-plant species in the life of a polyphagous moth, *Operophtera brumata* (Lepidoptera: Geometridae). *Journal of Animal Ecology* **52**, 439–450.

Witmer, M. C. & Cheke, A. S. (1991). The dodo and the tambalacoque tree: an obligate mutualism reconsidered. *Oikos* **61**, 133–137.

Wolda, H. (1983). "Long-term" stability of tropical insect populations. *Researches in Population Ecology* Supplement No. 3, 112–126.

Wolda, H. (1995). The demise of the population regulation controversy. *Researches on Population Ecology* **37**, 91–93.

Wolf, D. E., Takebayashi, N. & Rieseberg, L. H. (2001). Predicting the risk of extinction through hybridization. *Conservation Biology* **15**, 1039–1053.

Wood, T. K. (1993). Diversity in the New World Membracidae. *Annual Review of Entomology* **38**, 409–435.

Woolhouse, M. E. J. & Harmsen, R. (1987). Just how unstable are agroecosystems? *Canadian Journal of Zoology* **65**, 1577–1580.

Worner, S. P. (1988). Ecoclimatic assessment of potential establishment of exotic pests. *Journal of Economic Entomology* **81**, 973–983.

Worster, D. (1977). *Nature's Economy: A History of Ecological Ideas.* Cambridge University Press, Cambridge.

Wratten, S. D. (1987). The effectiveness of native natural enemies. In *Integrated Pest Management* (eds. A. J. Burn, T. H. Coaker & P. C. Jepson), pp. 89–112. Academic Press, London.

Young, C. M. (1986). Novelty of "supply-side ecology". *Science* **235**, 415–416.

Youngson, R. (1998). *Scientific Blunders: A Brief History of How Wrong Scientists Can Sometimes Be...* Robinson, London.

Zalom, F. G. (1993). Reorganizing to facilitate the development and use of integrated pest management. *Agriculture, Ecosystems and Environment* **46**, 245–256.

Zalucki, M. P., Brower, L. P. & Malcolm, S. B. (1990). Oviposition by *Danaus plexippus* in relation to cardenolide content of three *Asclepias* species in the southeastern USA. *Ecological Entomology* **15**, 231–240.

Zalucki, M. P., Daglish, G., Firempong, S. & Twine, P. (1986). The biology and ecology of *Heliothis armigera* (Hubner) and *H. punctigera* Wallengren (Lepidoptera: Noctuidae) in Australia: what do we know? *Australian Journal of Zoology* **34**, 779–844.

Zalucki, M. P. & Kitching, R. L. (1982). Temporal and spatial variation of mortality in field populations of *Danaus plexippus* L. and *D. chrysippus* L. larvae (Lepidoptera: Nymphalidae). *Oecologia* **53**, 201–207.

Zalucki, M. P., Murray, D. A. H., Gregg, P. C., Fitt, G. P., Twine, P. H. & Jones, C. (1994). Ecology of *Helicoverpa armigera* (Hübner) and *H. punctigera* (Wallengren) in the inland of Australia: larval sampling and host plant relationships during winter and spring. *Australian Journal of Zoology* **42**, 329–346.

Zwolfer, H. (1971). The structure and effect of parasite complexes attacking phytophagous host insects. In *Dynamics of Populations – Proceedings of the Advanced Study Institute on 'Dynamics of Numbers in Populations', Oosterbeek, The Netherlands, 7–18 September 1970* (eds. P. J. den Boer & G. R. Gradwell), pp. 405–418. Centre for Agricultural Publishing and Documentation, Wageningen.

Index